The Names of Science

The Names of Science

Terminology and Language in the History of the Natural Sciences

HELGE KRAGH

OXFORD
UNIVERSITY PRESS

OXFORD
UNIVERSITY PRESS

Great Clarendon Street, Oxford, OX2 6DP,
United Kingdom

Oxford University Press is a department of the University of Oxford.
It furthers the University's objective of excellence in research, scholarship,
and education by publishing worldwide. Oxford is a registered trade mark of
Oxford University Press in the UK and in certain other countries

Published in the United States of America by Oxford University Press
198 Madison Avenue, New York, NY 10016, United States of America

British Library Cataloguing in Publication Data
Data available

Library of Congress Control Number: 2024932240

ISBN 9780198917458
ISBN 9780198917441 (pbk.)

DOI: 10.1093/9780198917472.001.0001

Printed and bound by
CPI Group (UK) Ltd, Croydon, CR0 4YY

To Lucas and Philip, my beloved grandchildren

Preface

This book is about the relationship between science and language through history with an emphasis on the technical terms used in science and in many cases originating in specific scientific and historical contexts. Like other communicative cultural forms, science depends crucially on the chosen language, which in the case of science is typically specialized and in some respects differs markedly from everyday speech and writing. Scientists have always been concerned with how to communicate their results and to propose new terms that suitably describe their discoveries and ideas. As pointed out by an American chemist in a study of Louis Pasteur's contributions to the language of stereochemistry: 'Examining the history of the language can provide insights into the origins and development of the science itself, since the history of a science is echoed in the development of its language' (Gal, 2019).

The inevitable linguistic components of the natural sciences have been examined by a large number of authors predominantly with backgrounds in philosophy, linguistics, and educational and communication studies. However, only a small part of this literature pays serious attention to the historical dimension and to the concrete research works in which scientists indirectly and more or less consciously have acted as linguists and neologists. This is what Lavoisier did in the eighteenth century and Faraday, Lyell, Maxwell, and others did in the following century.

The part of the science-language literature written by scientists, science teachers, and historians of science is on the whole insufficient and scattered, insofar that it mostly consists of separate studies of particular names and words in this or that branch of science. I have in a couple of previous works contributed to the genre with studies on etymological and terminological issues related to twentieth-century physics and cosmology (see Kragh 2014a, 2014b, 2023 in the Bibliography). This book is broader and more ambitious as it aims at a partial integration of elements of linguistics and elements of the history of science. However, the focus is throughout on the sciences and how they have evolved historically and not on linguistics. Concepts from the science of language merely serve as an additional instrument that provides the historical development with a new and, I believe, fruitful perspective. Of course, the kind of terminologically oriented history of science I have in mind is merely a supplement to the traditional history of science, not an alternative to it.

The present work is more comprehensive than earlier studies both chronologically and as far as the spectrum of the sciences is concerned. There are chapters and sections covering important contributions in diverse sciences from the Renaissance to the twenty-first century. They range from the chemical phlogiston theory to the dark energy of modern cosmology. On the other hand, although I discuss examples from both the life sciences and the inorganic sciences, far most are from the latter category including not only physics, chemistry, and astronomy, but also geology and allied earth sciences. When I only deal cursorily with the vast field of the biomedical sciences it is simply because I know too little about it and its complex history. The same is the case with mathematics and its language, of which I have little to say. As a result of the focus on the inorganic sciences, inevitably there are some important terminological aspects of the history of science that are left out, such as the development of botanical and zoological nomenclature.

While English is today the undisputed international language of science, this was not the case in the past when Latin, French, and German were sometimes as widely or more widely used in scientific communications. Nonetheless, for reasons of simplicity and consistency, not to mention my limited knowledge of foreign languages, I have chosen to largely limit the book to the English language. This is not to deny that in a longer historical perspective science is a multilingual enterprise with massive contributions from the non-English speaking world. Each of these other languages—French, German, Arabic, and more—has its own history in the context of the development of science, but they cannot possibly be covered in a book like the present one.

I have of course benefited from and made use of the existing literature as written by authors with a background in either linguistics or science and its history. As mentioned, much of this literature is scattered and not easily identified. The comprehensive bibliography at the end of the book will hopefully serve as a resource for further studies on the historical relationship between language and science. This relationship can sometimes be illustrated quantitatively by using electronic resources such as Google Scholar and Web of Science, which I do routinely. In some cases, I have used information from the Google Ngram Viewer based on a very large corpus of scanned books mostly of a nonscientific kind. Yet another source of considerable value is the electronic edition of the *Oxford English Dictionary* (OED) which I likewise refer to throughout the book. In addition, I have made frequent use of early scientific dictionaries such as John Harris' *Lexicon Technicum* from 1704 and Samuel Johnson's *Dictionary of the English Language* from 1755, both of which sources are available online.

Apart from a brief introductory list of useful linguistic terms, the book is structured in six chapters which by and large, but only by and large, are devoted to separate sciences, namely physics, chemistry, and astronomy. The words associated with the historical development of geology, or what is today often called the earth sciences, are mostly dealt with in a section placed at the end of Chapter 1. When I have found it natural and relevant, I have included several and often lengthy digressions dealing with other sciences or with linguistic concepts of a more general nature such as eponymy, metaphors, oxymorons, and acronomy. Thus, in Chapter 1 there is one section focusing on the words 'atom' and 'molecule', and another section with a survey of how eponyms, metaphors, and pseudonyms have been used through the history of science. Section 2.2 dealing with words for electrical particles includes the term 'gene' belonging to a very different science, and Section 5.2, mainly on chemistry, refers to some of the terms in computer language. The many and perhaps confusingly many digressions and sidelights are used to include topics I could not find place for elsewhere.

Chapter 2 is primarily about words and names in the electrical and electromagnetic sciences from about 1700 to the late twentieth century, including key words such as 'ion' and 'electron' and with digressions on some of the popular prefixes and suffixes that appear in scientific names, *super-* and *meta-* among them. The origin and fate of the terms 'chain reaction', 'plasma', and 'transistor' are considered in more detail. Elementary particles whether real or imagined are dealt with in Chapter 3, which among other things looks at the drastic change in the physics vocabulary that occurred in the post-World War II era. While new words for particles, phenomena, and instruments had traditionally been based on Greek or Latin roots, now they were increasingly unconventional and often whimsical (quark, gluon, flavour, and anyon are but a few examples).

Terminological novelties belonging to twentieth-century physics are further considered in Chapter 4, which, however, starts with a section focusing on terms related to nineteenth-century thermodynamics, not only well-known words such as energy, entropy, and enthalphy but also those used in low-temperature studies, the so-called cryogenic sciences. Section 4.2 is mostly devoted to Röntgen's X-rays alias Röntgen rays and the startling phenomenon of radioactivity. The chapter also includes a fairly detailed Section 4.3 on 'quantum languages', that is, words and phrases coined as a result of the new quantum mechanics and its predecessor known as the old quantum theory.

As suggested in Chapter 5, the history of chemistry from its humble beginning in alchemical arts to its later status as perhaps the largest and most useful of all natural sciences is a rich resource for historical-linguistic studies. The chemical revolution in the late eighteenth century, which is dealt with in sections 5.2 and 5.3, relied to a considerable extent on the reformed chemical language introduced by Lavoisier and a group of other French chemists. It is from this reform we have the terms oxygen and hydrogen and many more. Each of the 118 chemical elements in the periodic table has its own naming history, and some of those histories are told in sections 5.4 and 5.5. The history of chemistry also provides instructive cases of so-called semantic shifts where, for example, a word originating in a scientific context migrates to everyday language with a different meaning. To 'pass a litmus test' or to 'have a catalysing effect' are well-known phrases with roots in chemical science.

The subject of the last chapter is the terminology associated with the world outside the Earth as studied by astronomers, astrophysicists, and cosmologists. Some of these words, once neologisms, refer to branches of science (e.g. astrochemistry, asteroseismology) and others to celestial objects (e.g. satellite, quasar, interstellar object). The latter category is examined in Section 6.2 which includes not only the planets and their moons, but also and in some detail William Herschel's coining of 'asteroid' for the new objects discovered in the early nineteenth century. The hypothesis of extraterrestrial life has resulted in several new words and more recently in a *lingua cosmica*, an artificial language supposedly understandable to all higher beings in the universe. Sections 6.4 and 6.5 focus on cosmology in both its older and modern meaning. To explain to a lay public what the expanding universe, the Big Bang, and the inflation scenario were all about, metaphors proved essential. The book ends with some of the words associated with modern cosmology such as 'multiverse' and 'dark energy', both of which entered the lexicon of science in the early twenty-first century.

Although the book refers to many technical terms and difficult things, it does not presuppose much scientific knowledge. It is primarily concerned with the words of science and their meanings at different times, not with science per se. In the few cases where equations and mathematical symbols enter, these are carefully explained. Given that the book is fairly elementary from a scientific point of view, it will be accessible to, and presumably of interest to, a broad readership with a basic insight in the history of science. Indeed, I have intentionally written it with the purpose of reaching both a humanist audience (historians, philosophers) and a scientific audience. I want to acknowledge the helpful assistance of Kader Ahmad, librarian at the Niels Bohr Institute in Copenhagen, who has provided me with not easily attainable sources from foreign libraries. I also want to thank three anonymous reviewers for critical comments to the manuscript and suggestions of how to improve it.

Helge Kragh

Contents

List of Figures

Introduction

Some linguistic terms

The following is a brief list of linguistic terms appearing in the text and exemplified with references to their use in science language. Further examples follow in the separate sections. The term linguistics and related terms (linguist etc.) are derived from the Latin *lingua* meaning tongue.

- **Acronym**. Greek, from *acr* (height, summit) and *onym* (name), as in *Acropolis*. A word usually formed by the initial letters of a longer word or phrase, such as IUPAC (*International Union of Pure and Applied Chemistry*) and SUSY (*supersymmetry*). The verb 'acronymize' is used when a phrase is turned into an acronym. Many but not all acronyms can be pronounced as words. Examples of the latter category are TV (*television*), DNA (*deoxyribonucleic acid*), GPS (*Global Positioning System*), and the eponym EPR (*Einstein–Podolsky–Rosen*). Abbreviations of this kind are sometimes called *initialisms* (or *alphabetisms*) and not acronyms, but at other times initialisms are included in the acronymic umbrella term. Some acronyms which cannot be pronounced as a whole contain a part which can and they are spoken of accordingly. PFAS (*per- and polyfluoroalkyl substances*) is a plural word for a family of environmentally harmful fluorinated organic compounds. In spoken language it is P-FAS.
- **Affix**. Many compound words used in science and elsewhere are made up from a root and an *affix* with the latter normally being attached to either the beginning of the root word (*prefix*) or to its end (*suffix*). The affixes in science words are typically derived from Greek and Latin. Common prefixes are *bio-* (life) and *hydro-* (water), and among the suffixes are *-graph* (write) and *-ology* (study of). Some affixes can appear as both prefix and suffix, e.g. *exothermic*, *thermonuclear*; *graphite*, *telegraph*.
- **Antonym**. Greek, from *anti* (opposite, against) and *onym* (name). A word that means the opposite of another word. An antonym is thus the opposite of a synonym. Examples are hot–cold, big–small, and charged–neutral. 'Unscientific' and 'pseudoscientific' are antonyms of 'scientific', and the antonym of the *exo-* prefix is the *endo-* prefix.
- **Connotation**. A word typically suggests more meanings and images than the one to which it explicitly or literally refers (its *denotation*). In particular situations the word implies or suggests different connotations. When we say that a person is independent it may mean, depending on the context, that he or she is unfriendly. The term 'physiology' once had the connotation of what we now call physics (Section 1.6).

- **Demonym**. Greek, *demos* (people) and *onym* (name). The word refers to a group of people in relation to a particular place or limited territory. The people can be real or imagined and the place anything from a street to a galaxy. Californian is a demonym and so are European and Martian (an inhabitant of Mars).
- **Eponym**. Greek, *epi* (upon, over) and *onym* (name). A person after whom something is named. Thus, Marxism, Newtonian, and America are eponyms because the words refer to Karl Marx, Isaac Newton, and Amerigo Vespucci, respectively. The person needs not be real, but can be mythical or imagined: planet Jupiter is named after the Roman god. Eponyms are frequent in all branches of science, where they are found in numerous laws, instruments, objects, concepts, units, and equations. They have often been formed to honour the person in question, typically a scientist, and for this reason they may be controversial, either for political reasons or because the eponymous label is associated with a priority claim. Several eponyms refer not to a single person, but to two or more, such as in 'Boyle–Mariotte law' and 'Einstein–Podolsky–Rosen paradox'.
- **Etymology**. Greek word meaning 'analysis of a word to find its true origin'. The etymologist traces the development of a word in several languages hoping to find its common ancestral form, its *etymon*. Since etymology is also about the evolution of words through history, the study belongs to historical linguistics.
- **Euphony**. Greek, from *eu* (good, sweet) and *phone* (sound, speech). A euphonious word or phrase is pleasing to the ear. If it sounds decidedly unpleasant, it is a *cacophony* (Greek *kakos* = bad, evil). A word which is euphonious in one language may be cacophoneous in another language. An example is 'oryctognosy' used in nineteenth century geology (Section 1.5). When scientists coin a new word, they often take into consideration whether it is euphonious.
- **Glossary**. A list, usually alphabetical, of words or terms in a special field and their definitions or other explanations of their meaning. A glossary of meteorology lists terms and concepts relevant to this science and its allied sciences.
- **Homonym**. From Greek 'having the same name' (*homos* = same). If two or more words are spelled or pronounced in the same way, but are different with respect to meanings and origin, they are homonyms. The word 'current' may refer to an electric current but also mean that something is recent or up to date. There is no relation either historically or conceptually between the two meanings. If the homonyms are spelled differently, but have the same sound, they are called *homophones*, e.g. Sun, son and flower, flour. The biological word 'gene' in plural and the trousers 'jeans' are homophonous.
- **Hybrid word**. A word that etymologically derives from two or more languages, in English typically by combining Greek and Latin parts. Many words in science and technology are of this kind, such as television (Greek–Latin), chloroform (Greek–Latin), and positron (Latin–Greek). The German–English (or German–Latin) word 'eigenvalue' used in quantum physics is a nonclassical hybrid (Section 4.3). The formation of hybrid words is sometimes called hybridization.
- **Metaphor**. Greek *metaphora* composed of *meta* (over, across) and *pherein* (to carry), when a word is transferred from its original meaning to another one.

Metaphoric language is figurative as it compares two things of which only one is true or factual. To say about a person that she has a 'heart of gold' is a metaphor since the heart is not really made of gold. Likewise, to say about a nuclear particle that it 'tunnel' through an energy barrier is a metaphor. It is widely recognized that metaphors are necessary when it comes to translating specialized scientific language to the public domain, but also that they can be misleading if taken too seriously. The universe did not literally begin in a 'big bang'. Many of the specialized terms used in science are themselves metaphors, as when the particles or forces keeping the atomic nucleus together are called 'gluons'.

- **Metonymy**. From Greek 'change of name'. A metonym is a kind of figure of speech in which a concept or thing is referred to by a quite different name associated with the concept or thing in question, as when 'Pentagon' is used for the U.S. Department of Defence or a person says 'Keep your nose out of my affairs.' Metonyms and metaphors are related but different.

- **Misnomer**. A word that is incorrect or wrongly applied, in the sense that it refers to things that are not (or are no longer) the case. Misnomers are very frequent in both ordinary and scientific language. The name 'Greenland' is a misnomer—'Whiteland' would have been better. In many cases they are harmless as they are known to be factually incorrect. A common source is that the name became adopted before the true nature of the object or concept was known. The name oxygen implies that the element is present in all acids, which it is not; the Hubble constant is not a constant, nor is the light year a unit of time; there are no planets in a planetary nebula.

- **Neologism**. From Greek *neo* (new) and *logos* (speech). A word or phrase which has recently been coined and is in the process of entering common use eventually to become part of accepted language. Neologisms (or sometimes 'neologies') are typically introduced, in science and elsewhere, when there are no existing words that adequately cover a new object, phenomenon, or concept. They can be freely invented but are usually based on already known words. Scientists have throughout history proposed neologisms when faced with new discoveries; witness words such as ion, electron, and superconductivity. The term 'scientist' is itself a neologism (Section 1.3). An old word used in a wholly new sense is sometimes counted as a neologism.

- **Nomenclature**. From Latin *nomen* (name) and *calare* (to call). A systematized body of names and terms including rules and conventions to form them. Nomenclature systems have been and still are particularly important in branches of natural history (botany, zoology, mineralogy) and also in chemistry and astronomy. Today they are parts of all empirical sciences, if more important in some branches than in others.

- **Nonce word**. A word coined for a particular use and unlikely ever to become an accepted part of the language. A nonce, also known as an *occasionalism*, only turns up at a single occasion such as did 'oreston' for the positive electron in 1934 (Section 3.1). Some nonce words are essentially meaningless and easily disposable. They appear in the history of science as proposals for the name of new things. What was originally a nonce may later be commonly adopted and then lose its status as a nonce.

- **Onomastics**. From Greek word meaning 'belonging to naming'. The formal study of the history and origin of proper and personal names. In 1871, Maxwell regretted his lack of 'onomastic power' (Section 2.3).
- **Oxymoron**. From Greek, *oksus* (sharp, keen) and *moros* (dull, stupid). A word or a phrase which is self-contradictory (a *contradictio in terminis*), that is, where one part has a meaning opposite to that of the other part. The phrase may consist of two antonyms. To say that a figure is a 'round square' or someone a 'married bachelor' is an oxymoron. With increasing knowledge, a term may lose its oxymoronic status, such as has been the case with 'bioinorganic' and 'dark star'. A bachelor is by definition unmarried and 'married bachelor' therefore forever an oxymoron.
- **Philology**. From Greek meaning 'love of word(s)'. The academic study of language and literary texts traditionally with a focus on the historical development. As a major discipline of the humanities, philology consists of several subdisciplines with different approaches to the study of language.
- **Polysemy**. From Greek, 'many signs'. Refers to a word or phrase with multiple coexisting meanings. In contrast to homonymy, these meanings are historically and conceptually related if often in an obscure manner. A 'plane' can mean a flat surface but also an aircraft. Apart from the familiar meaning of 'neighbourhood', it is also a mathematical term referring to the points close to a given point. 'Earth' means one thing in astronomy and another in chemistry. The opposite of polysemy is *monosemy*, the relatively rare case of words with only a single meaning. Scientific words such as sulphuric acid and plate tectonics are ideally monosemic.
- **Protologism**. When a new word is coined for something, but the word is not yet known outside a small group of people and has perhaps not yet appeared in print, it is a protologism. As soon as it becomes better known and adopted by a wider community, it changes into a neologism appearing in glossaries and dictionaries. The term 'proton' existed before Rutherford coined it, but only in the form of a protologism (Section 3.3).
- **Pseudonym**. From Greek meaning 'false name'. A fictitious name that a person or a group of persons take on instead of their true name or names. Pseudonyms are names used for certain occasions and with a certain purpose, and they do not include ordinary name changes or minor changes in a name such as including a middle name or not. They are common among authors of fiction literature, where a pseudonym is sometimes known as a *nom de plum* or pen name. Strangely, pseudonyms can also be found in earlier scientific literature (Section 1.1).
- **Rhetoric**. Greek–Latin for the art of an orator. An ancient art of discourse focusing on persuasive arguments in verbal or written communications. It is about the pragmatic uses of language, not its meaning. Although rhetoric is mostly associated with political, commercial, and cultural language, it plays a significant role in other domains as well, science included. The term rhetoric is frequently used derogatively, as in the phrase 'it's nothing but empty rhetoric.'
- **Semantics**. From Greek *semantikos* (significant, important). The study of the meaning of language, its words, phrases, and sentences, and the relationship

between form and meaning. Semantics is a major field not only in linguistics but also, in a different version, in theories of meaning belonging to logic and philosophy. One of the problems in semantics is concerned with the origin of meaning in language. Is it to be found in language itself or is it defined by the user?

- **Synonym**. From Greek meaning 'name(s) that are similar'. The prefix *syn-* means 'together'. A word or phrase that means the same as another word or phrase. Two words can be synonyms, but an isolated word cannot be either a synonym or an antonym. The terms are relational. Chemistry and alchemy are today considered very different, if not as antonyms, but for a period of time they were synonyms or nearly so. Whereas 'antielectron' is a synonym of 'positron', in the past this was not always the case (Section 3.1).

- **Tautology**. From Greek word meaning 'repeating what is said'. *Tautos* = the same. A statement that is necessarily true, such as 'either the electron is charged, or it is neutral'. In rhetoric a tautology often means an unnecessary repetition of words ('a pedestrian travelling on foot' or 'an electrically charged ion'), just two different ways of saying the same thing. In the second meaning a tautology is also called a *pleonasm*.

- **Terminology**. Hybrid word (Latin–Greek) coined in 1770. The more or less systematic study of terms in a particular subject field whether scientific or not. While terminology is a branch of linguistics, in the domains of science it is the involved scientists and their organizations who take care of the terminology and how to use the relevant terms.

- **Toponym**. From Greek, *topos* (place) and *onym* (name). As eponyms refer to people, toponyms refer to terms named after a city, country, or some other location. Usually the place or location is geographical (*geonym*), but it can be anything, a star or a fantasy world. Berkelium, element number 97, named after Berkeley, California, is a toponym and so are 'Vienna Circle' and 'Copenhagen Interpretation'. The original version of quantum mechanics was often called 'Göttingen mechanics' (Section 4.3).

- **Vocabulary**. From medieval Latin, 'a list of words'. A vocabulary is a body of words employed by a language or a group of people. It can be highly specialized, as it is in all branches of science. The 'vocabulary of chemistry' refers to the glossary of chemistry and other words commonly used by professional chemists.

1
Issues of science, history, and language

The words used in scientific communication make up a subject of confusing complexity, not only because there are so many of them, but also because they and their meanings have, as a rule, changed considerably—sometimes completely—over time. The *atom* discussed by John Dalton in the early part of the nineteenth century was not the same as the one envisaged by Democritus in Greek antiquity and it differed even more from the particle as understood by modern scientists. The class of names are associated with a broad variety of topics, relating to objects, phenomena, instruments, processes, theories, laws, equations, concepts, units, and much more. Moreover, they appear in the way in which science is and has been organized in various disciplines and subdisciplines such as neurology, biochemistry, mineralogy, and astrophysics. Some of the names are eponymous labels honouring great (and sometimes not-so-great) scientists, while others have classical Greek or Latin roots and others again are proper neologisms freely invented. Most names just happen to be adopted by a majority of the scientific community, but there are also those—such as the names of exoplanets, chemical compounds, zoological species, and the constants of nature—that are officially decided by nomenclature committees and the like. However, this is a practice of relatively recent origin.

This chapter starts with some considerations concerning the names of the various sciences and the languages, from Latin in the Renaissance to English in modern times, in which scientists have predominantly communicated. It also refers to literary practices of the past such as pseudonymous or anonymous authors and the use of anagrams. Section 1.2 is a general survey of eponyms and metaphors illustrated by a variety of examples from the history of science. This is a subject that will reappear in later chapters. Names and words associated with the concepts of atoms and molecules are discussed in Section 1.3, to be followed by a more detailed account of how key terms such as 'science', 'scientist', and 'physicist' entered the vocabulary. As many of the present scientific disciplines did not exist in the past, so there were in the past sciences—or candidates for recognized sciences—that do not exist any longer. The terms related to some of these 'lost sciences' are briefly considered in Section 1.5. Terminological issues of what is today called the earth sciences are the subject of the final section in which the development of geological words and concepts is fragmentarily followed over a period of some four hundred years.

1.1 Linguistic aspects of science words

In a paper of 1935, the leading American linguist Leonard Bloomfield reflected on the nature of scientific language and how it differed from the language of the man in the street. 'The forms of the scientist's speech are so peculiar in vocabulary and syntax that most members of his speech-community do not understand them', he observed. Bloomfield further commented:

> In a brief utterance the scientist manages to say things which in ordinary language would require a vast amount of talk. ... Moreover, the scientist manages to say very complex things: if his statements were put into ordinary language, the phrases would become so involved ... that the hearers would not 'understand'. ... It is evident that the speech-forms of the scientist constitute a highly specialized linguistic phenomenon. To describe and evaluate this phenomenon is first and foremost a problem of linguistics.[1]

Literary and cultural writers are rarely impressed by the prose of scientific texts or the words that scientists invent and adopt. Here is one complaint from an author of a book on the language of science:

> Euphony has never seemed to the scientist to be a feature worthy of much consideration. Beauty of any sort cannot be measured and scientists are devoted to measuring ... It almost seems as if scientists preferred ugly words, for in large numbers they coin and use such cacophonous terms as *pycnogonid*, *antibody*, or *betatron*; hideous, all three of them, yet typical of science ... There are so many words in the vocabulary of science which are undeniably ugly.[2]

Although there is some truth in this allegation, and especially when it comes to modern science, it is not true in general that scientists do not or did not care about language and words. As we shall see in the chapters that follow, scientists in the past often endeavoured to construct words that were not only precise and etymologically correct but also pleasing to the ear. According to the same author, humour, emotions, and figures of speech are wholly absent from scientific prose:

> The scientist does not write in metaphors; metonymy or satire might mar the clearness he prizes so highly ... Surely the scientist, besides being tone-deaf, lacks something that would testify to common humanity. For again, the language of science makes no provision for the slightest gleam of humour.

Again, this stereotype may be approximately true of most modern science as communicated in *Nature*, *Journal of Biological Chemistry*, or *Physical Review*, but it is definitely not a fair description of scientific language in the pre-1850 era. As evidenced by many examples, the claim that scientists systematically avoid metaphors is plain wrong (Section 1.2). On the other hand, it is also a gross distortion to claim, and especially for modern scientific forms of communication, that 'for the most part, scientists and technologists are ... creating a periphrastic, suggestive vocabulary more

---

[1] Bloomfield (1935).

[2] Savory (1967), p. 68 and p. 118. 'Pycnogonids' is a zoological term for a class of marine spider-like animals with long legs and small bodies. 'Betatron' was a name for a class of high-energy particle accelerators developed in the 1940s (Section 3.3). For 'antibody', see Section 3.1.

typical of poetry than of formal notation.'[3] Readers with a thirst for poetry should look elsewhere than in the pages of *Physical Review* or similar high-ranking scientific journals.

The American poet alison Deming reflected in an essay from 1998 on the relations between science and poetry. On the one hand, she was fascinated by the 'beautiful particularity and musicality' of the language of science as well as by its many neologisms. 'I am wooed by words such as "hemolymph", "zeolite", "cryptogram", "sclera", "xenotransplant", and "endolithic", and I long to save them from the tedious syntax in which most science writing imprisons them.' On the other hand, Deming realized that science and poetry use language in a fundamentally different manner:

> [Science] uses language for verification and counts on words to have a meaning so specific that they will not be colored by feelings and biases. Science uses language as if it were another form of measurement—exact, definitive, and logical. The unknown, for science, is in nature. Poetry uses language itself as the object . . . and counts on the imprecision of words to create accidental meanings and resonances. The unknown, for poetry, is in language.[4]

It is problematic to speak of *a* language of science and to disregard, as Deming implicitly does, that the language of scientific communications has changed materially over time. The historical dimension obviously needs to be taken into account. With regard to invented words or neologisms, these are coined all the time, in science as elsewhere. However, most of them pass unnoticed or, if noticed at all, they are only briefly taken up and soon dropped. As nicely expressed by an American author: 'When it comes to neologisms, supply far outstrips demand. Coined words are like swarms of salmon eggs: few hatch, fewer mature, and only a handful make it upstream. Even those who do survive seldom endure.'[5]

It will be useful first to introduce some general aspects relating to the names of the different sciences. Why do important territories in the scientific landscape such as mechanics, arithmetic, geology, and astronomy have names characterized by different suffixes? To underline the unity of the various sciences, wouldn't it be preferable with a common suffix? Perhaps, but this is not how the words have evolved historically. In his important work *Philosophy of the Inductive Sciences* from 1840, William Whewell included a lengthy chapter on 'Aphorisms Concerning the Language of Science' in which he said:

> The name of many sciences ends on *ics* after the analogy of *Mathematics, Metaphysics*; or *Optics, Mechanics*. But these in most other languages, as in our own formerly, have the singular form *Optice, l'Optique, Optik, Optick*: and though we now write *Optics*, we make such words of the single number: 'Newton's Opticks is

---

[3] Van Dyke (1992), p. 394. A periphrastic writing or speech is wordy and unnecessarily convoluted. From Greek, 'to express in a roundabout manner'. For a detailed comparative investigation of the language, style, and structure in scientific articles from the seventeenth to the twentieth century, see Gross, Harmon, and Reidy (2002).

[4] Deming (1998).

[5] Keyes (2021), p. 10.

an example'. As, however, this connexion in new words is startling, as when we say, 'Thermo-electrics is now much cultivated', it appears better to employ the singular form, after the analogy of Logic and *Rhetoric*, when we have words to construct. Hence we may call the science of languages *Linguistic*, as it is called by the best German writers, for instance, William von Humboldt.[6]

In due course Whewell's noun 'linguistic' became used as an adjective, whereas the scientific study of languages became known as 'linguistics'. None of the English words were much used during the nineteenth century. In 1861 the German-British philologist Friedrich Max Müller gave a series of lectures to the Royal Institution in which he argued that the science of language belonged to the physical sciences. Should the new science be honoured with a name of its own? Not according to Müller: 'We hear it spoken of as Comparative Philology, Scientific Etymology, Phonology, and Glossology. In France it has received the convenient, but somewhat barbarous, name of *Linguistique*. ... I myself prefer the simple designation of the Science of Language, though in these days of high-sounding titles, this plain name will hardly meet with general acceptance.'[7]

Mecha*nics* is a noun, whereas mecha*nic* or more often mechan*ical* is generally an adjective. The same holds for mathematics with mathematical as the most common adjectival form, whereas we speak of arithme*tic* and not arithme*tics* as the scientific study of the numbers and the operations connecting them. As explained in an older glossary of scientific terms:

> As in Greek, adjectives of this kind are sometimes taken as nouns, e.g. a *critic*, a *mechanic*, a *soporific*, an *emetic*, a *hypnotic*. Before the fifteenth century such nouns were used as the names of certain arts and sciences, e.g. *music, logic, arithmetic*. Later the plural forms -*ics* became more usual, e.g. *mechanics, optics, dynamics*. Adjectives are often formed from such nouns by the use of -*al*, e.g. *mechanical, optical*.[8]

Of course, far from all sciences end with -*ic* or -*ics*. In many cases the suffix is based on the Greek word for knowledge *logos*, giving rise to -*logy* and -*ology* names, where the -*o* is a stem vowel. Examples are zoology, physiology, geology, histology, and cosmology. The term anthropology, in the meaning of 'speaking or discoursing of men' first appeared in English in 1656 (*anthrōpos* = human being).[9] The same glossary says:

> Following the pattern of such words as astrology, genealogy, theology, -logy is now widely used in forming the names of departments of knowledge (e.g. various sciences) and may be freely translated as 'study of'. ... Some of the modern inventions

---

[6] Whewell (1840), vol. 1, p. cxiv. The reference is to the German philosopher and linguist Wilhelm von Humboldt, the elder brother of the naturalist and geographer Alexander von Humboldt.

[7] Müller (1862), p. 14. In 1868, Müller became Oxford's first professor in 'comparative philology'.

[8] Flood (1960), p. 84. Whereas a soporific induces sleep, an emetic induces vomiting (*sopor* = sleep in Latin; *emein* = to vomit in Greek).

[9] Blount (1707), first edition of 1656, *OED*.

(e.g. weatherology, sexology) are horrible words and it is to be hoped that they will not be permanently accepted into the language.[10]

Horrible or not, the term 'weatherology' has never been seriously considered as a scientific term or a substitute for meteorological forecasting. It does appear in a few scientific papers, though. In a 1943 review paper on meteorology and climatology, an American meteorologist referred a couple of times to 'weatherology', but always in quotation marks to indicate that he did not endorse the term, which has the character of a protologism and not a neologism.[11] Contrariwise, the term sexology has long been accepted as an interdisciplinary science belonging to medicine and psychology in particular. In this sense it dates from the early twentieth century.

Astronomy could as well have been called astrology if the latter name had not been occupied by another field no longer counted as a science. But astronomy it is and neither astrology nor 'astronomics' or 'astronics', the latter two being nonexistent words. The names of sciences ending on -nomy or -nomics are based on the Greek word nomos in the sense of something systematically arranged or subject to law. Example of the first class are, apart from astronomy, agronomy and economy, whereas ergonomics and genomics (the study of genomes) are examples of the second class.

There are, in addition, some sciences carrying the suffix -graphy derived from the Greek graphein (γραφειν) meaning 'to write'. Etymologically these branches of science are about drawing, representing, or recording parts of nature. The two-dimensional crystalline form of carbon known as graphite is used in pencils. Well-known examples of -graphy sciences are crystallography and geography, and older examples include 'cometography' and 'selenography', the descriptive studies of comets and the Moon, respectively. For 'cosmography', see Section 6.4. Apart from the mentioned sciences there are some with well-established names which do not end on a recognizable suffix of which chemistry is a prominent example (Section 5.1).

In a longer historical perspective there is no single language of science, but several languages. While currently English is unquestionably the scientific lingua franca, much like Latin (or Neo-Latin) was during the Renaissance and early modern period, the dominance of English is a relatively modern phenomenon. During the scientific revolution and well into the nineteenth century, English was not an international language with which the majority of foreign scientists were acquainted. By 1700, the population of the United Kingdom was 8.6 million, less than that of Spain, Italy, and Germany, and much less than that of France (21.5 million). Most of the text in Henry Oldenburg's important *Philosophical Transactions* unofficially associated with the Royal Society was in English, which proved to be a problem for many foreigners. In 1672, the Danish natural philosopher Erasmus Bartholin expressed in a letter to Oldenburg his 'great longing for the Latin edition of your Philosophical Transactions, in order that everyone may read them'.[12] Under the title *Acta Philosophica* a series of

---

[10] Flood (1960), p. 101. Savory (1967), p. 90, agrees that sexology and also musicology are strange words on line with weatherology.
[11] Thornthwaite (1943).
[12] Hall and Hall (1965–86), vol. 9, p. 403.

volumes was actually published in the course of the 1670s with Latin translations of the English journal.

During the early part of the seventeenth century a few of the key figures of the scientific revolution began writing in the vernacular rather than in Latin, their reasons being idealistic and sometimes patriotic. Bacon's *De Augmentis Scientiarum* from 1632 was an expanded Latinized version of his *The Advancement of Learning* originally published in English in 1605. It is well known that Galileo wrote most of his important books, including the famous *Dialogo* (1632), in Italian to make them accessible to non-academics. Descartes wrote his first published work, the *Discours de la Méthode* (1637), in French and not Latin. An even earlier example is the Flemish engineer, mathematician, and natural philosopher Simon Stevin, who wrote all his works in Dutch and left it to others to translate them into Latin. Not only did Stevin aim to communicate with uneducated sailors and merchants, he also suggested that the Dutch language, because of its structure and rules for compounding words from monosyllables, was the best of all to express verbally the discoveries and methods of the new science.[13] Moreover, he believed that Dutch was the original language of mankind spoken in the Garden of Eden—the mythical age long before classical antiquity when all philosophy and science was known to people.

Stevin, Bacon, Galileo, Descartes, and other early natural philosophers writing in their own languages were exceptional, however. The general trend towards the vernacular came later and even then many prominent scientists preferred writing their works in Latin. Consider the case of Denmark–Norway, where scholars found it difficult to imagine the Danish language serving as a means of scientific communication. When the chemist and philologist Ole Borch (Olaus Borrichius) in 1686 published a book that aimed to instruct the general public on how to use domestic plants for medicinal purposes, he wrote the text in Latin. As a result, *De Usu Plantarum Indigenarum* had to be translated into Danish. Borch's contemporary Erasmus Bartholin argued eloquently in *De Studio Linguae Danicae* (1674) that the native tongue of the Danes might 'shed light, and not insignificantly so, upon our knowledge of nature'.[14] Ironically as well as typically, he expressed his linguistic patriotism in Latin instead of Danish. Despite his rhetoric of linking science with the common folk, Bartholin published every one of his academic works in Latin.

In Britain, France, and Italy, most scientific and other learned texts were written in the vernacular by the late seventeenth century. The process of emancipation from Latin was slower in the German-speaking countries where it took until the 1750s before academic works in German clearly outnumbered those published in Latin. While in 1754 a quarter of all books printed in Germany was in Latin, by 1787 the fraction had dwindled to a tenth. Generally, during the eighteenth century scientists increasingly began writing in the vernacular rather than in Latin.[15] To mention but one example, in 1745 the newly established Royal Danish Academy of Sciences and Letters decided to use the country's own language in its proceedings series entitled

---

[13] Martin (2015). Van Berkel, Van Helden, and Palm (1999), pp. 18–20. See Brink (1989) for Stevin's views of a reformed Dutch language.

[14] Kragh et al. (2008), pp. 57–8.

[15] Gunnarsson (2011). See also Moon and Spencer (1948b).

Skrifter (Writings) and that despite the fact that all its members were fluent in Latin and used to writing in that language. Initially the proceedings appeared with a parallel edition in Latin (called *Scriptorum*), but after a few years the practice was discontinued.[16] As a result of the language policy, most of the contributions to *Skrifter* failed to attract attention outside Scandinavia.

The situation was largely the same for other smaller European languages such as Polish, Portuguese, Swedish, and Hungarian. Latin was on decline, French and the vernacular on the rise. When Hans Christian Ørsted announced his discovery of electromagnetism nearly seventy years later (Section 2.3), it was in the form of a leaflet written in Latin, a language that at the time had all but disappeared from the physics and chemistry literature. It may have been the last time that a major scientific discovery was communicated in Latin.

Latest by the mid-eighteenth century French had replaced Latin as the preferred language of science and culture in large parts of continental Europe. Frederick II of Prussia was a Francophile. When he reconstituted the Berlin Academy of Sciences in 1744, by royal decree the official language was French, not German. According to Voltaire, in an *éloge* of 1752, 'French, which is the common language of Europe, and which has been enriched with all these new and necessary expressions, is much more appropriate than Latin for spreading all this new knowledge throughout the world.'[17] Voltaire's compatriot and friend, the mathematician Jean le Rond d'Alembert, never wrote in Latin. In a preface to the great *Encyclopédie* (of which he was a co-editor together with Denis Diderot), he said about the switch from Latin to French:

> Thus England has imitated us; Germany, where Latin seemed to have taken refuge, is beginning gradually to lose the use of it. I have no doubt that Germany will soon be followed by the Swedes, the Danes, and the Russians. So before the end of the eighteenth century, a philosopher who will wish to truly educate himself about the discoveries of his predecessors will have to burden his memory with seven or eight different languages.[18]

D'Alembert realized that there was no single substitute for Latin as a universal language, not even French, and that the trend toward the vernacular carried with it disadvantages as well as advantages.

Significant scientific contributions through history are limited to at most a dozen living languages of which French, German, English, Dutch, and Russian have been the most important. According to historian Michael Gordin, 'There is no other sphere of human cultural activity—trade, poetry, politics, what have you—that takes place in such a limited set of tongues.'[19] In the fin-de-siècle period almost 90% of all science publications were in either English, German, or French, with the first language in the top. And yet, at the time it was far from evident that English would eventually become the universal language of science. From about 1910 to 1925 German took

---

[16] Kragh et al. (2008), p. 146.
[17] Quoted in Roelli (2011), p. 348.
[18] Translation in Terrall (2017), p. 639.
[19] Gordin (2017), p. 4.

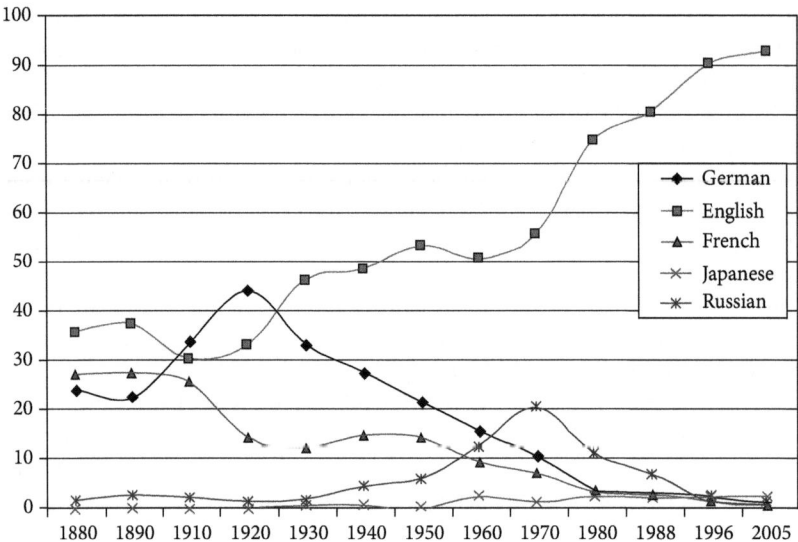

Figure 1.1 The percentage of the global scientific literature published from 1880 to 2005 in different languages.

Reproduced from Ammon (2012), *Applied Linguistics Review,* 3: 333–355.

over the top position after which its share of the world production steadily declined. Today the hegemony of English is no less complete than it was for Latin at the time of Tycho Brahe (Figure 1.1). The dominance is considerably less marked in the social sciences and the humanities, though. While in the period 1980–96 Russian-language publications in the natural sciences dropped from 10.8 to 2.1%, and those in German from 2.5 to 1.2%, publications in English rose from 74.6 to 90.7%.

Obviously, the languages of science do not reflect and never have reflected the distribution of spoken languages in the world. In the early twentieth century, Chinese Mandarin was spoken by more people than English, French, and German together, and yet the percentage of scientific literature written in Mandarin was close to nil. While German with 58 million native speakers was one of the chief languages of science (about 30%), Spanish with about 330 million speakers was not. Today, there are approximately 1.45 billion people who can speak English, more or less, but of these only 373 million have it as their first language. There are 475 million people who speak Spanish as their first language and 929 million who speak Mandarin.[20]

Of course, the historical trend in global rank order depends on the specific field or discipline. For example, from about 1850 to 1920 French was very popular in astronomy and mathematics but rather unpopular in biology. In the earlier period from approximately 1810 to 1850, Swedish was a surprisingly attractive language to many chemists, which was mainly a result of Jöns Jacob Berzelius' influential textbook *Lärbok i Kemien* and his later and even more influential *Årsberättelse*

---

[20] Crystal (1987), pp. 286–8.

(Annual Survey), a series of annual reviews of the latest developments in chemistry and physics that started in 1821. Berzelius deliberately used his status and many publications to maintain Swedish as a viable scientific language in the face of French dominance in particular.[21] Although his reviews were translated into German, and the later volumes also into French, many chemists were eager to read them as soon as they were published in Swedish.

Still around 1960 English was only moderately dominant in the large field of zoology. In the decade 1955–64, 38.6% of all zoological publications were written in this language while German, French, and Russian followed with 14.7, 11.6, and 10.8%, respectively. Thus, these three languages contributed to the world literature with approximately the same weight as English. Today the percentage of English-language publications is undoubtedly much higher.[22] The figures may be compared with those of 1800–29 when German (35%) and French (29%) were the most popular languages in the zoological sciences followed by English (14%) and Latin (7%).

As a result of Germany's defeat in World War I, Germany and its allied nations (Austria, Hungary, Bulgaria) were kept out of most international scientific conferences until about 1926. Not only was German or 'teutonic' science subject to an international boycott, German language was also not allowed at many conferences and in journals published by the victorious nations. As stated in an article of 1919 in the journal *Science* published by the American Association for the Advancement of Science: 'German is, without doubt, a barbarous language only just emerging from the stage of the primitive Gothic character, and I venture to suggest that it would be to the advantage of science to treat it as such from the date August 1, 1914.' The proposal was in part inspired by an earlier one by Lord Walsingham, a British entomologist:

None but a German would use the German language by preference for scientific descriptions of species or genera. ... To those Germans, if any there be, who are honestly well disposed, and who put the interests of science before the greed for world-domination, it can be no hardship to publish their descriptions in the English or French language, with which the great majority of their scientific workers are more or less intimately acquainted.[23]

Nonetheless, many of the important scientific advances in the period were reported in German-language periodicals which French, American, and British scientists needed to be able to read if they wanted to keep abreast with the development.

This was definitely the case in theoretical physics, where the leading journal was *Zeitschrift für Physik* founded in 1920. The language of this very successful journal was German and its success was a strong argument for the importance of German as an international language of physics. By the mid-1920s, it was evident that French and British efforts to limit German as a scientific language had failed. When a paper in

---

[21] Beckman (2016). Abbri (2019). As noted in Gordin (2017), p. 45, the interest of chemists and mineralogists in Swedish as a scientific language goes back to Torbern Bergman some forty years before Berzelius.

[22] Savory (1967), p. 152.

[23] Hampson (1919). Walsingham (1918). On the boycott of German language, see Gordin (2017), pp. 173–80.

English appeared in the *Zeitschrift* in 1925 without even translating it into German, it caused vehement protests from many German physicists with the result that its editor, the able physicist Karl Scheel, had to promise that all papers had to be in the German language. With the emergence of quantum mechanics in 1925–6, reading knowledge of German simply became a *conditio sine qua non* for those who wanted to join the competitive race of harvesting the fruits of the foundation created by Heisenberg, Schrödinger, and other German-speaking physicists (Section 4.3).

One thing is the language in which works in science are written; another is the names of the authors as they appear in those works. One might think that *pseudonyms* (Greek for 'false names') have no place in scientific literature but history shows otherwise. In fact, we do not need to go far back in time to find prominent examples of authors who chose to publish their works under a pseudonym. Starting in the 1930s there appeared a long series of books and articles on pure mathematics written by the remarkably productive but nonexistent Nicolas Bourbaki. The name was a collective pseudonym for an influential and largely anonymous group of mostly French mathematicians including luminaries such as Henri Cartan, Jean Dieudonné, and André Weil. Bourbaki has been described as the greatest mathematician who never was. Inspired by Bourbaki, two Princeton mathematicians published papers under the pseudonyms 'H. Pétard' and 'E. S. Pondiczery'.[24]

The Bourbaki collective was in many ways exceptional, given that most pseudonymous scientists have been individuals who for some reason or other have decided to write under another name. For example, in the nineteenth century there are several examples of women scientists who felt forced to write under a male pseudonym to be recognized and have their works published. The French mathematician Sophie Germain kept a correspondence with the famous Carl Friedrich Gauss during the Napoleonic Wars but signed her letters 'Monsieur LeBon'. It took a couple of years before Gauss realized that Monsieur LeBon belonged to the female sex.

In this and similar cases, the pseudonym could hide the identity of the author or correspondent, but in other cases the use of a pseudonym did not imply anonymity nor was it meant to do so. Some pseudonyms referred to the author by an invented description which in the early period was typically in Greek or Latin and later could be anything. The Scottish mathematician William Wallace once wrote a piece in the *Ladies' Diary* under the name 'Peter Puzzle'. The contemporary polymath Thomas Young, known for the wave theory of light and much more, used no less than fourteen pseudonyms and initials including 'XXXX', 'Hydrophilus', and 'Apsophus'.[25]

One of the first books ever on so-called recreational mathematics, and possibly the most successful in this genre, was published in 1624 with the abbreviated title *Récréations Mathématiques*. The book appeared in an English translation in 1633 and during the seventeenth century more than seventy editions and translations were published. It is in this work that the term 'thermometer' first appears in print. Who was the author of this early bestseller? No author was given on the title page of the original French version, but the dedication of the book stated that it was written by a certain H. van Etten, a Flemish student of the French Jesuit astronomer and

---

[24] See Barany (2020).
[25] Craik (1999). Lagemann (1956).

mathematician Jean Leurechon. Until recently it has been assumed that Leurechon was the true author of *Récréations* and that for some reason he published it under the pseudonym H. van Etten. However, it is now believed that Hendrik or Heinrich van Etten was a real person and the author of *Récréations*, or that he possibly wrote it in collaboration with his teacher Leurechon.[26]

During the seventeenth century, pseudonyms were particularly common in chemistry and alchemy where they might be used as a way of guarding the author's discoveries, recipes, and trade secrets. A noteworthy example is the American-born George Starkey who wrote a series of texts under his own name but even more texts, and much better known, under the pseudonym Eirenaeus Philalethes.[27] He never revealed that the two authors were one and the same. Newton was much influenced by Philalethes' works which were also read by other eminent natural philosophers during the scientific revolution such as Boyle, Locke, and Leibniz. They all believed that Philalethes was or had been a real person. Confusingly, at about the same time another British 'chymist', Thomas Vaughan, used an almost identical pseudonym, namely Eugenius Philalethes, and was consequently sometimes thought to be author of the works of Starkey alias Philalethes, the 'lover of truth' as the name means.

In contrast to the word pseudonym, the related word *anonym* is rarely used except as the root of other words such as anonymous, anonymity, and anonymously. In practice a pseudonym is often considered a synonym of anonym although the latter word means 'without a name' and thus would exclude also pseudonymous names. If a written text appears with no author name at all, whether the true name or a pen name, it is truly anonymous. In the past it was not unusual for scientists and other learned people to publish works anonymously and only later (if ever) reveal their identity.

The French scientist and philosopher Pierre Louis Maupertuis, a towering figure in the Enlightenment, published several of his books anonymously, among them *Vénus Physique* (1745) and *Système de la Nature* (1751).[28] The reason was not so much to protect him from the authorities as it was part of a marketing strategy. At least in the salons of Paris, speculations about the author's identity often wetted the appetite of potential readers and made anonymously published books more attractive. Well after many readers had identified Maupertuis as the author, he refrained from openly admitting his authorship. Another example is young Immanuel Kant, who in Königsberg in 1755 published a remarkable book in which he proposed a new and later very influential picture of the universe, the so-called 'island universe theory' (Section 6.4). The author's name appeared nowhere in the *Universal Natural History and Theory of the Heavens*, as the English title reads.

Although the term 'natural history' appearing in the title of Kant's work was the same as used by Pliny the Elder and other ancient authors, it had a substantially different meaning as it implied how nature (in casu the universe) had changed over time. In the older *historia naturalis* tradition, history as given by the Greek ἱστορία meant essentially knowledge or investigation of some subject area, which could be nature or something else. The word did not involve as a necessary element a temporal

---

[26] Eamon (1994), pp. 307–8.
[27] Newman (1994).
[28] Terrall (2006).

or dynamical dimension, such as it did according to Kant and later Enlightenment authors (e.g. Buffon, see Section 1.2).

In the seventeenth and eighteenth centuries, anonymous contributions to scientific journals were frequent and not considered very problematic. The early issues of *Philosophical Transactions* were full of letters and similar contributions from unnamed authors, a policy that the editor Henry Oldenburg found acceptable and even welcomed. In other cases, the author was known but not formally appearing as such. The format of an article or report was typically that its content was given through the voice of the editor, as when Oldenburg reported in the first issue of *Philosophical Transactions*: 'The Ingenious Mr. Hook [Hooke] did, some moneths since, intimate to a friend of his, that he had, with an excellent twelve foot Telescope, observed ... a small Spot in the biggest of the 3 obscurer *Belts of Jupiter*.'[29]

The reasons why scientific authors wanted to publish anonymously varied from one author to another. In some cases, the identification of the author might have harmed his reputation and in other cases he might just have wanted to avoid anticipated controversy.[30] During the first half of the 1770s there appeared at least three anonymous attacks in French scientific journals on the then-popular phlogiston theory, one of which is believed to have been written by young Lavoisier (Section 5.3). With the emergence of more professional and specialized journals in the nineteenth century the days of anonymous and pseudonymous articles came to an end. Today no recognized scientific journal would dream of accepting a paper without the correct name of the author.

Nor would a modern science journal accept a discovery claim in the form of an *anagram*, a cryptic method of communication popular during the scientific revolution. The Greek *ana* (ἀνά) means up again, back, or throughout. An anagram is a word or phrase formed by rearranging the letters of an original word or phrase. 'No crelet', 'loncetre', and 'nortcele' are all anagrams of 'electron', which in the latter case is simply the real name spelled backwards. Constructed words of this kind are sometimes called *anadromes* or, if used as a pseudonym, *ananyms* (notwen is an ananym for Newton). In the seventeenth century such games of letters were often used to suggest a discovery without revealing its nature in plain language.[31] The chosen anagram would seem to be sheer nonsense, but when decoded it made sense to the correspondent or prepared reader.

Galileo communicated his important discovery of Venus' phases in an anagram to Kepler and thus stated his priority without giving any details. The unscrambled version of the anagram was *Cynthiae figuras aemulatur mater amorum* meaning 'shapes of Cynthia [the Moon] are imitated by the mother of love [Venus]'.[32] When Christiaan Huygens in 1656 discovered the ring of Saturn, he likewise announced it in anagrammatic form:

aaaaaaaccccccdeeeeeghiiiiiiiilllllmmnnnnnnnnnnnooooooppqrrstttttuuuuu

[29] *Philosophical Transactions*, March 1665, p. 3. Bazerman (1988), p. 75.
[30] Kronick (1988).
[31] Bagioli (2012).
[32] Drake (1984).

Or, when translated: *Annulo cingitur, tenui, plano, nusquam cohaerente, ad eclipti-cam inclinato*, which in English reads 'It is surrounded by a thin flat ring, nowhere touching, and inclined to the ecliptic'.[33] To mention yet another example, when Robert Hooke in 1676 wanted his learned colleagues to know that he had found a simple relation between the extension of a spring x and the force necessary to produce it F, he first did it by means of an alphabetically ordered anagram of the same kind that Huygens had used. According to Hooke's law ($F = kx$), the force is proportional to the extension. In the original anagrammatic form, he stated it as a meaningless series of letters:

ceiiinossssttuv

Two years later he revealed the correct decoding in a book, namely the Latin phrase *ut tensio sic vis* meaning 'as the extension, so the force'.

As a curious later example of an anagram combined with a pseudonym, consider the important and controversial book *The Unseen Universe* published anonymously in 1875 by the Manchester physicists Balfour Stewart and Peter Guthrie Tait. The key message of the book—a contribution to natural philosophy and metaphysics rather than physics—was that there coexisted with the material universe a spiritual universe and that the two were connected by means of energy transfers. By 1888 *The Unseen Universe* had appeared in fourteen editions of which the first three were anonymous. However, the names of the authors were no deep secret. On 20 May 1875 *Nature* included a review of the book 'which rumours attribute to a co-partnery of two distinguished physicists'.

Without revealing their names, Stewart and Tait advertised their forthcoming work in a letter titled 'An Anagram' in *Nature* of 15 October 1874:

> The practice of enclosing discoveries in sealed packets and sending them to Academies, seems so inferior to the old one of Huygens, that the following is sent you for publication in the old conserved form: $A^8C^3DE^{12}F^4GH^6I^6L^3M^3$ $N^5O^6PR^4S^5T^{14}U^6V^2WXY^2$

The letter was signed 'West', which presumably stood for 'We S[tewart] T[ait]'. When the book was published, the anagram was decoded as follows: 'Thought conceived to affect the matter of another universe simultaneously with this may explain a future state'.[34]

1.2 Eponyms and metaphors in the history of science

An *eponym* (Greek ἐπώνυμος, 'given as a name'), is a relationship between two named things of which one is usually taken to be a real or imagined person. It can be a name 'derived from the name of a real or mythical person' and also refer to 'the person

[33] Van Helden (2006).
[34] Stewart and Tait (1875), p. 159.

for which such a name is derived'.[35] To illustrate the double meaning, Darwinism is an eponym named after Charles Robert Darwin and at the same time Darwin is an eponym of Darwinism. Names of this kind are abundant in all branches of science and have been so for more than half a millennium. The scientific 'things' named after a person or persons cover a broad spectrum including theories, constants, laws, equations, objects, instruments, phenomena, units, and more. Although usually named after scientists and inventors, in some cases science eponyms refer to people with no connection whatsoever to the world of science. A comet is named after Caesar, an asteroid after Bismarck, a trilobite after Mike Jagger, and a beetle after Arnold Schwarzenegger. Some eponymous diseases are named after patients and not after the physician who first identified the disease. The skin disease called Mortimer's disease was described in 1898 by the English pathologist Jonathan Hutchinson, who named it after Mrs. Mortimer, one of his patients.

Eponyms vary more or less arbitrarily between the apostrophized or possessive form (Planck's constant) and the non-possessive form (Planck constant).[36] Sometimes both forms appear while at other times only one of them is used. 'Maxwell's equations' is not uncommon but less frequent than 'Maxwell equations'. A microbiologist will speak of the 'Petri dish' and not or only very rarely of 'Petri's dish' (Section 5.3). The gas law named after Robert Boyle is almost always called 'Boyle's law' whereas references to the 'Boyle law' are uncommon. The non-possessive form is more convenient or even necessary for eponyms with more than one name. Chemical engineers speak of the 'Haber–Bosch process' for the production of ammonia from air ($3H_2 + N_2 \rightarrow 2NH_3$) named after Fritz Haber and Carl Bosch. They never speak of 'Haber's and Bosch's process' and only occasionally of 'the process of Haber and Bosch'.

A small group of scientists have been eponymously honoured not only with particular theories or discoveries but also with broader world views or concepts by adding the suffix -ism to their names. As we speak of Stalinism and McCarthyism in twentieth-century political history, so historians of science and ideas speak of Aristotelianism, Copernicanism, Newtonianism, Darwinism, Mendelism, and Freudianism. These are primarily labels indicating that the ideas of the persons in question were influential beyond science in the more concrete or narrow sense. The labels are not necessarily honorific, such as illustrated by the term *Lysenkoism* referring to the largely non-scientific ideas of the notorious Trofim Lysenko, an agricultural scientist in Stalin's Soviet Union.[37] Lysenko was anything but a great scientist, but for a period of two decades he was an influential as well as controversial figure in both biology and politics. Nor was the eighteenth-century Austrian physician Franz Anton Mesmer a great scientist, and yet he gave name to the term *Mesmerism* (Section 2.1). Not all eponyms ending on -ism belong to the mentioned category. The term *Daltonism* does not denote a particular conception of the physical world inspired by John Dalton's atomic theory, but is a medical term for a defectiveness of the eye (Section 1.3).

---

[35] Collins English Dictionary, https://www.collinsdictionary.com/dictionary/english/eponym.
[36] Anderson (1996). The issue of possessive versus non-possessive eponyms has attracted much attention in the medical world. See Abel (2014).
[37] Joravsky (1986).

A follower or adherent of the influential Swiss physician called Paracelsus and his philosophy of nature (*Paracelsianism*) was known as a 'Paracelsian' in the sixteenth and seventeenth centuries.[38] In the same period, a natural philosopher or other person who accepted Copernicus' cosmology in the seventeenth century was a 'Copernican' and similarly we have in later eras names such as 'Newtonian', 'Freudian', and 'Mendelian'. The same names are used in adjectival form for the theories and discoveries associated with the scientists: Paracelsian medicine, Newtonian mechanics, Freudian psychology, Mendelian laws, etc. Several widely known medical eponyms belong to this class, such as the 'fallopian tubes' connecting the uterus to the ovaries and so named after the sixteenth-century Italian anatomist Gabriel Fallopius.

While in these cases the name needs a referent (either a person or a subject) there are also a few eponymous words with the *-an* suffix that stand alone. When physicists and mathematicians speak about a *Lagrangian* or a *Hamiltonian* they are not referring to followers of the Italian-French mathematician Joseph-Louis Lagrange or the Irish mathematical physicist William Hamilton. No, they speak of a well-defined quantity associated with particular physical problems and theories. For example, 'the Hamiltonian of this system is so-and-so' and 'the Lagrangian is so-and-so for this interaction'. There are no similar phrases for Newtonian, Darwinian, Euclidean, etc.

The total number of eponyms in science is unknown but dauntingly large. In medicine alone the number is estimated to be on the order of 20,000 which suggests that at least 50,000 eponyms have been in circulation through history.[39] Many of those used in the past have disappeared only to be replaced by new eponyms. In this respect this group of words shares the destiny of words in general whether associated with science or not. As the Roman poet Horace (Horatius) put it in his *Ars Poetica* (The Poetic Art) from around AD 20: 'Many a word long disused will revive and many now high in esteem will fade if custom wills it, in those powers lie the arbitrament, the rule and the standard of language.'[40]

The number and distribution of eponyms have been studied within the framework of so-called scientometrics, a sociological research area dealing with quantitative analyses of science and scientific literature.[41] The term means 'measurement of science' and dates in its Russian version from 1969, the same year that the roughly equivalent 'bibliometrics' was coined. However, the scientometric approach seems to be of limited relevance to how eponyms have evolved through the history of science and especially so in the period before 1900 where reliable bibliometric data are largely absent.

As argued by Marco Beretta, an Italian historian of science, eponyms can be found even in the ancient world where they were used to pay tribute to worthy philosophers, doctors, and patrons of science. The Hippocratic oath of medical ethics is named after the great physician Hippocrates living around 430 BC. Eponymic references to the oath can be found as early as in the first century AD. In his *Naturalis Historia* Pliny

---

[38] The name Paracelsus was a pseudonym. His real name was Theophrastus von Hohenheim or in full Phillipus Aureolus Theophrastus Bombastus von Hohenheim.

[39] Beretta (2019).

[40] Cited in Igea (2013), p. 972.

[41] Scientometric studies of eponymy include Cabanac (2014) and Schubert, Glänzel, and Schubert (2022).

the Elder wrote of 'a sort of ambition, as it were, of adopting plants by bestowing upon them one's name ... so desirable a thing did it appear to have made the discovery of some new plant, and thus to have contributed greatly to the benefit of mankind.'[42]

During the Renaissance, the coining of eponyms gained momentum in astronomy, medicine, pharmacy, and other areas of science or proto-science. Galileo's naming of the four Jupiter moons (Section 6.2) is one example and another is the prominent Jesuit astronomer Giovanni Riccioli's names for the features of the Moon. In his *Almagestum Novum* (The New Almagest) published in 1651, Riccioli advocated the geocentric cosmology of Tycho Brahe according to which the other planets revolved around the Sun. Introducing the eponymous lunar nomenclature that is still in use today, Riccioli coined a large number of eponyms in honour of past astronomers. Although himself an anti-Copernican, he named a large lunar crater after Copernicus (another and equally prominent crater he called 'Tycho'). Most of Riccioli's names refer to ancient astronomers, but several of them also honour astronomers who had recently passed away such as Galileo and Kepler. The French priest and natural philosopher Pierre Gassendi appears on his list, and he was still alive. Riccioli even includes himself as a name for a lunar crater, a rare example of a self-suggested eponym.[43]

Over the next two centuries eponymies proliferated in all sciences and fields of technology with the zenith appearing in the nineteenth century. While there are only very few self-suggested eponyms in the history of science, they appear more frequently in the history of technology where inventors have sometimes named marketable products after themselves. For example, when the French artist Louis Daguerre in 1839 invented one of the earliest forms of photography, he called the process *daguerreotype* and the pictures were sold under the same name. In 1907 the Belgian-American chemist Leo Baekeland invented the world's first commercial plastic, which he called *Bakelite*.

The frequent use of eponyms in scientific terminology may seem to collide with the generally accepted aim of describing nature in neutral and objective terms. It may even seem an aberration from the norms of science by its emphasis on the individual and neglect of social and other impersonal factors. In an influential paper of 1957, the sociologist of science Robert Merton highlighted eponymy as an important element in the reward system of science. 'Heading the list of the immensely varied forms of recognition long in use is eponomy', he wrote. Merton described the phenomenon as 'the most enduring and perhaps most prestigious kind of recognition institutionalized in science' and said that in this way 'scientists leave their signatures indelibly in history; their names enter into all the scientific languages of the world.'[44] Given the large number of eponyms and the fact that many of them are long forgotten, this is surely an exaggeration.

Merton further suggested that eponyms may be ordered in a hierarchy, some of them being more prestigious than others. To have a historical epoch named after

---

[42] Beretta (2019), p. 225.

[43] For lunar nomenclature through the ages, see Whitaker (1999) and Altschuler and Ballasteros (2019), pp. 35–44. Of the 1,586 named craters on the Moon, only 28 are in honour of a woman. Among those so honoured are Lise Meitner, Emmy Noether, and Marie Curie, none of whom were astronomers.

[44] Merton (1973), pp. 298–300.

a person (the Newtonian or Darwinian era) or to be credited as the 'father' of a particular science (Georges Cuvier as the founding father of palaeontology, etc.) is a much greater recognition than to have a name associated with one of the numerous objects, units, or theories in science. While there are at least a hundred 'fathers' of scientific and medical fields, there are but a handful of 'mothers' and these are rarely called so.

The second half of the twentieth century witnessed a drastic change from the traditional mode of science conducted by a single scientist to large and often very large teams of scientists. Since eponyms are by definition personal labels they do not easily go together with the increasing collectivization of science. The scientists making up a large research team or collective including several hundred or even thousand members tend to become anonymous. As pointed out by Donald Beaver in a paper of 1976: 'Multiple authorship of research contributions either makes it difficult to distribute eponymic credit fairly, or leads to eponymous stature of none, when faced with unwieldy combinations of names.'[45]

Eponymies are associated not only with experimental discoveries but also with important scientific results variously labelled laws, principles, theories, and rules. Of these categories, 'law' is traditionally considered the most elevated and prestigious. Bibliometric data indicate that whereas the cumulative number of named laws grew steadily during the nineteenth century, from about 1915 this component decreased drastically (Figure 1.2). On the other hand, the number of eponymous theories continued to grow, possibly because scientists in the later period preferred to call what formerly would have been 'laws' as 'theories'. The theories of relativity and quantum physics are no less fundamental than the laws of thermodynamics, but they are rarely referred to as laws. The theory of global plate tectonics marked a revolution in the earth sciences (Section 1.6), but there is no law of plate tectonics. To quote Beaver, alluding to Shakespeare's *Romeo and Juliet*: 'To a modern scientist, theories, principles and rules smell just as sweet as laws; anonymous ones smell sweeter.'

Although the number of new eponyms have not followed the growth of science in the post-World War II era, there is no indication that the old tradition of celebrating scientific progress with eponyms is fading away. It is only in relative terms that there are fewer of them in modern science. A recent study based on 8,636 biomedical eponyms concludes that 'eponyms are now used more than at any point in the past', but also that 'they have been losing market share to more scientific terms for the last two decades.'[46]

Can we be sure that if a scientist is eponymously credited with a discovery or something else, then the discovery was actually made by this scientist? Although the answer is definitely no, it does not justify what the statistician Stephen Stigler in 1980 elevated to a law named after himself. What is known as 'Stigler's law of eponymy' states in its simplest form that 'no scientific discovery is named after its original discoverer.'[47] As Stigler was aware, this is a gross exaggeration—Kepler did, in fact, discover the

---

[45] Beaver (1976), p. 91.

[46] Thomas (2016).

[47] Stigler (1980). Scientific eponyms do not always relate to discoveries or inventions. For example, there are many eponymous units of science, but units are conventions agreed upon and not things that can be discovered.

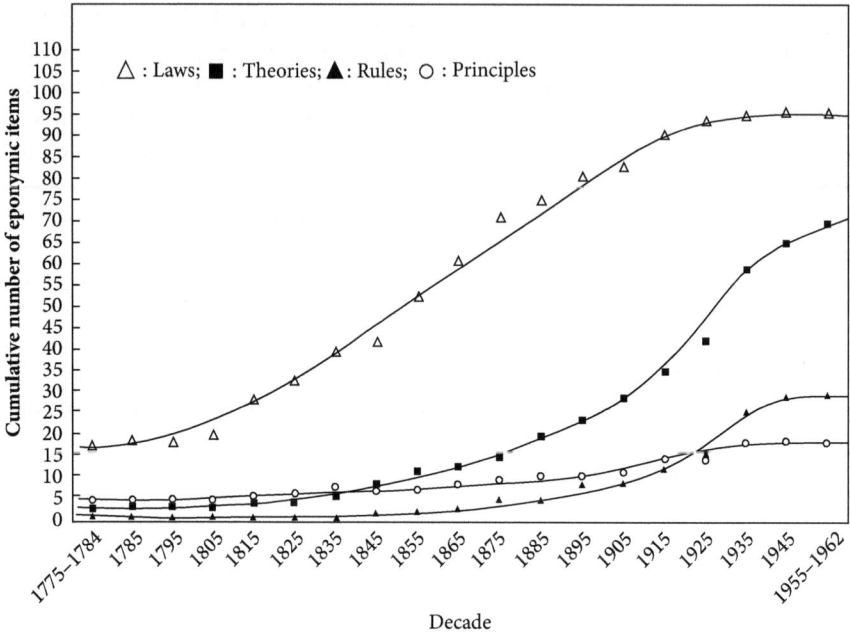

Figure 1.2 The historical cumulation of particular eponymic categories from about 1775 to 1960.

Reproduced from Beaver (1976), *Social Studies of Science* 6: 89–98.

planetary laws named after him—but he used it to emphasize that in a great many cases eponyms are misleading or at least inaccurate as far as priority is concerned. Like Merton, Stigler considered eponyms as rewards of extraordinary prestige that secure the scientists in question 'a kind of intellectual immortality'. As mentioned, this is in my view to place the notion of eponymy on a pedestal where it does not belong. In many cases eponyms are not taken very seriously by either the scientific community or the public at large. They are sometimes proposed and adopted for no good reason and with no intention of corresponding to the authentic discovery history.

Eponyms have often been controversial for priority reasons, for practical reasons, and for political and ideological reasons. The name of planet Neptune is an eponym referring to a Roman god and not to its discoverer. All the same, the choice of name caused considerable debate in the mid-nineteenth century (Section 6.2), and later the eponym Hertz for the unit of frequency was considered problematic in Nazi Germany because Heinrich Hertz was Jewish (Section 2.3). In a letter of 1849, Charles Darwin commented critically on the custom of denoting new biological species by a name including that of the discoverer or first describer:

> I have been led of late to reflect much on the subject of naming & I have come to a fixed opinion that the plan of the first describer's name being appended for perpetuity to species has been the greatest curse to natural History. ... I feel sure as long as

species-mongers have their vanity tickled by seeing their own names appended to a species, because they first miserably described it, in two or three lines, we shall have the same *vast* amount of bad work as at present.[48]

Citing Darwin's critique, the American entomologist James Needham complained in a paper of 1930 that names of species, whether eponymous or not, were often monstruously long, non-pronounceable, and anything but simple.[49] As an extreme example he referred to the proposed name for a little amphipod crustacean found in Lake Baikal,

Brachyuropushkydermatogammarus grewlingii mnemonotus Dybowski,

so named by and after the Polish naturalist Benedykt Dybowski. As Needham saw it, and he was probably not alone, the name was simply ridiculous. He proposed to replace Dybowski's name with the simpler non-eponymous term *Gammarus mnemonotus*.

On a more general level, the American physicist Duane Roller pointed out in 1947 that eponymies give a wrong impression of how science develops: 'In placing emphasis on the work of a very few individuals, important factors are left out of account, such as the socio-economic and, today especially, the fact that a great many individuals may have contributed to a particular discovery.' Moreover, Roller anticipated Stigler's law: 'When proper names are retained, one certainly should make sure that the recognition is given to the real discoverer, as given by priority of publication. . . . National pride seems frequently to have an effect on the name associated in different countries with the same discovery.'[50]

More recently objections to the eponym naming tradition have resurfaced and intensified, either with respect to particular eponyms or, more broadly, to the tradition itself. Objecting to the eponymous recognition of former Nazi doctors, in 2007 two British medical scientists concluded that the entire system of eponymy ought to be shelved:

Eponyms lack accuracy, lead to confusion, and hamper scientific discussion in a globalised world. Instead of using eponyms, we should use our interest in medical history to provide fair and truthful accounts of scientific discoveries and to dissect individual contributions. We call on the editors of medical journals and textbooks to abandon the use of eponyms.[51]

Recently a group of biologists have proposed a similar taxonomic reform concerning the vast amount of eponyms in the zoological and botanical sciences. From a contemporary perspective, they argue, the traditional eponymic system

---

[48] Letter to H. E. Strickland, 29 January 1849. Darwin Correspondence Project, https://www.darwinproject.ac.uk/letter/DCP-LETT-1215.xml
[49] Needham (1930).
[50] Roller (1947).
[51] Woywodt and Matteson (2007). For arguments for and against medical eponyms, see Abel (2014).

is potentially problematic, as many of those honoured are strongly associated with the social ills and negative legacy of imperialism, racism and slavery. Moreover, 19th-century and early 20th-century taxonomy was largely dominated by white men who, by and large, honoured other men (funders, colleagues, collectors and so on) of their own nationality, ethnicity, race and social status.[52]

The authors consequently conclude that naming species in honour of real people is unnecessary and should be avoided.

Political and ideological motives also played an important role in the recent attempt to change the name of NASA's celebrated James Webb Space Telescope to a more politically correct name. Webb, who ran the space agency from 1961 to 1968, was accused of discriminating against gay and lesbian employees and therefore to sin against the moral norms of a later period. Could a multibillion-dollar scientific instrument be eponymously named after such a dubious person? In October 2022, the venerable Royal Astronomical Society founded in 1820 declared that because Webb had engaged in 'unacceptable behaviour', 'The RAS now expects authors submitting scientific papers to its journals to use the JWST acronym rather than the full name of the observatory.'[53] That is, the hidden eponym was OK, but not the full one. After having investigated Webb's moral habitus, NASA decided to keep the name and in December the Royal Astronomical Society retracted its ban.[54]

The German physicist Johannes Stark, a recipient of the 1919 Nobel Prize, turned into an ardent Nazi and advocate of non-Jewish science. In 1970 the International Astronomical Union, apparently unaware of his commitment to the Nazi cause, decided to name a lunar crater after him. Then, fifty years later and now referring to his unacceptable political views, the name was deleted.[55] On the other hand, physicists happily continue to speak of the spectroscopic Stark effect for which he was awarded the Nobel Prize. They also continue to refer to the Schrödinger equation despite Schrödinger's reputation as a sexual predator and possibly a paedophile. Entomologists are familiar with a species of a blind cave beetle which since 1937 has been called *Anophthalmus hitleri* after Adolf Hitler. The name was proposed by Oskar Scheibel, an Austrian collector. Despite suggestions of renaming the small insect, this is still its official name. 'A species of any other name would sound as sweet', a biologist recently paraphrased Shakespeare.[56]

Metaphors are almost as abundant in science as they are in art and fiction literature, but they function differently. As regards the latter case, it has been said: 'The poetic metaphor sports a penumbra of further metaphors and implications, which may themselves clash with the conventional usage and the tacit knowledge of the reader, be flagrantly contradictory with one another, and fly in the face of previous

---

[52] Guedes et al. (2023). On the Western dominance of eponymies in the naming of biological species, see also DuBay, Palmer, and Piland (2020).

[53] Witze (2022). https://ras.ac.uk/news-and-press/news/ras-and-jwst

[54] Politically and ethically motivated renaming of at least some eponyms in science is advocated in Herbert et al. (2022).

[55] https://www.prospectmagazine.co.uk/ideas/technology/40359/astronomers-unknowingly-dedicated-moon-craters-to-nazis.-will-the-next-historical-reckoning-be-at-cosmic-level

[56] Smith (2022). 'What's in a name? That which we call a rose by any other name would smell as sweet.' *Romeo and Juliet*, Act II, Scene II.

comparisons in the same text. Far from being considered an error, this is part of the calculated impact of poetic language.'[57] Scientific metaphors are clearly very different as their primary function is to open up for new avenues of research that establish connections between different domains of nature. They have to be intelligible, disciplined, and consistent both internally and with existing background knowledge. For this reason, when incorporated into scientific thinking the initially vague and polysemic metaphor fades and loses much of its metaphoric nature.

Thomas Blount, a seventeenth-century English lexicographer, defined in his *Glossographia Anglicana Nova* a metaphor as follows: 'A Trope in Rhetorick, by which we put a strange and remote Word, for a proper Word by reason of its resemblance with the thing of which we speak; as smiling Meadows, a youthful Summer, &c.'[58] In general, metaphors can be described as a semantic and rhetorical connection between an object (or something else) which is principal and literal and an object which is subsidiary and figurative. The latter is a metaphor for the former. People in the seventeenth century were impressed by the mechanical clock and in order to emphasize the regularity and divine order of the heavenly world they described it metaphorically as a 'clockwork universe' (Section 6.1). Of course, they did not mean that the universe actually was a clock, but that it works *as if* it were.

Linguists and other writers disagree of how to define the concept of metaphor, more precisely. While some definitions are broad, including analogies, similies, and metonymies, others are more restrictive. The version given in *Collins English Dictionary* belongs to the first class: 'A *metaphor* is an imaginative way of describing something by referring to something else which is the same in a particular way. For example, if you want to say that someone is very shy and frightened of things, you might say that they are a mouse.' Incidentally, another metaphoric reference to the small rodent is the ubiquitous computer mouse (Section 5.2).

In the eighteenth century, light was often conceived as tiny particles expelled from the luminous body. Within the framework of the so-called projectile theory the hypothetical light particles were routinely likened to particles of matter such as cannon balls moving at the extreme speed of light. However, Benjamin Franklin did not find the widely used military metaphor to be convincing. As he wrote in a letter of 1752: 'I am not satisfy'd with the Doctrine that supposes Particles of Matter, call'd Light, continually driven off from the Sun's Surface, with a Swiftness so prodigious! Must not the smallest Particles conceivable, have, with such a Motion, a Force exceeding that of a 24 pounder discharg'd from a Cannon?'[59] Notice the term 'discharge' which here appears in a ballistic context and which Franklin and others also used metaphorically in their studies of electricity (Sections 2.1 and 2.2).

In his early thinking about the electromagnetic ether Maxwell invented a metaphor in the form of a mechanical analogy sometimes called the 'idle wheel' or 'molecular vortex' model. Although this pictorial model was literally wrong it played a crucial

---

[57] Mirowski (1989), p. 278. Pulaczewa (1999).

[58] Blount (1707), available as Google Book. The dictionary first published in 1656 appeared in several editions. It defined about 11,000 hard and unusual words. The slightly later dictionary *The New World of Words* compiled by the author Edward Phillips copied many of its entries from Blount's *Glossographia*. Phillips (1720) can also be found as a Google Book.

[59] Quoted in Cantor (1987).

heuristic role in the considerations that led him to the discovery of the electromagnetic field equations. In his paper 'On Physical Lines of Force' from 1861, Maxwell noted about the mechanical model:

> I do not bring it forward as a mode of connexion existing in nature, or even as that which I would willingly assent to as an electrical hypothesis. It is, however, a mode of connexion which is mechanically conceivable, and easily investigated, and it serves to bring out the actual mechanical connexions between the known electro-magnetic phenomena; so that I venture to say that any one who understands the provisional and temporary character of this hypothesis, will find himself rather helped than hindered by it in his search after the true interpretation of the phenomena.[60]

In a later address to the British Association, Maxwell discussed the relationship between physical analogies and what he called, for the first time, *scientific metaphors*. Although the latter were primarily illustrations that carried no explanatory force, he suggested that they played an important part in science and were more than just a play with words: 'The figure of speech or of thought by which we transfer the language and ideas of a familiar science to one with which we are less acquainted may be called Scientific Metaphor.'[61] As an example he mentioned that everyday words such as velocity, momentum, and force have acquired precise meanings in the theory of dynamics. Maxwell elaborated:

> These generalized forms of elementary ideas may be called metaphorical terms in the sense in which every abstract term is metaphorical. The characteristic of a truly scientific system of metaphors is that each term in its metaphorical use retains all the formal relations to the other terms of the system which it had in its original use. The method is then truly scientific—that is, not only a legitimate product of science, but capable of generating science in its turn.

Thus, Maxwell suggested what is perhaps the major function of metaphors in science, namely that they deepen the insight of the researcher and help him or her to discover things in nature. Analogies and metaphors sometimes facilitate the transfer of a well-understood scientific model to new domains. 'We describe by metaphor because metaphor enables us to advance our thinking from the known to the unknown', says Frieda Stahl, a physicist and educator.[62] In general, a good metaphor is an eminently useful mental picture that serves as more than just a pedagogical device. According to physicist Alan Lightman, 'metaphors in modern science carry a greater burden than metaphors in literature or history or art'. Philosopher Susan Haack agrees: 'The role of metaphors in scientific inquiry is much larger than their role in literary inquiry.'[63]

However, there is no agreement as to how metaphors work in science, more precisely, or if they are of value in scientific communications at all. While few scholars

---

[60] Maxwell (1965), part I, p. 486. See the detailed analysis in Cat (2001).
[61] Address given to the 1870 British Association meeting in Liverpool. Maxwell (1965), part II, p. 227.
[62] Stahl (1987).
[63] Lightman (1989), p. 97. Haack (2019), p. 2064. See also Crease (2000).

deny that metaphors are valuable as a heuristic tool, it is an open and much-discussed question whether they are also of constitutive value for scientific reasoning. The versatile French physicist and chemist Pierre Duhem is today mostly remembered for his contributions to the history and philosophy of science. Although he did not use the term 'metaphor' in his influential monograph *The Aim and Structure of Physical Theory*, he wrote extensively of 'analogy'. The terms metaphor and analogy are related but different. A metaphor is a figure of speech that directly compares two things for rhetorical effect. An analogy too shows that two things are alike, but here the goal is explanatory, namely to make a point about the comparison.

'The history of physics shows us', Duhem stated, 'that the search for analogies between two distinct categories of phenomena has perhaps been the surest and most fruitful method of all the procedures put in play in the construction of physical theories.'[64] Duhem was particularly impressed by the mathematically identical equations that sometimes turn up in different abstract theories of, for example, heat and electricity. 'Analogies consist in bringing together two abstract systems; either one of them already known serves to help us guess the form of the other not yet known, or both being formulated, they clarify each other.' While Duhem thus praised analogies (or metaphors) to be a source of inspiration as well as an 'infinitely valuable' method of theory construction, he warned that analogies should not be confounded with models.

Since the seventeenth century there has been a tradition arguing that metaphorical language is merely decorative and aimed at popularization, whereas it and similar rhetorical devices should be shunned in serious science and other rational discourses. Francis Bacon was fully aware that words are more than just words and that many words and phrases are counterproductive by guiding the scientific mind in a wrong direction. In his *Novum Organum* from 1620, he said:

> Men believe that their reason is lord over the words, but it happens, too, that words exercise a reciprocal and reactionary power on our intellect. Words, as a Tartar's bow, shoot back upon the understanding of the wisest, and mightily entangle and pervert the judgment.[65]

Several of the key figures in the scientific revolution, including Bacon and Robert Boyle, argued for a revision of the English language to make it more fit for the new science. Some of them even suggested to pattern the desired purified language on the algebraic symbols used by mathematicians throughout Europe. When Boyle reported some of his experiments, he apologized that he had not followed the 'succinct way of writing' but instead 'delivered things, to make them more clear, in such a multitude of words, that I now seem even to myself to have in divers places been guilty of verbosity'.[66]

In his influential *Essay Concerning Human Understanding* from 1690, the philosopher John Locke argued that words are nothing but signs of the ideas of the speaker.

---

[64] Duhem (1974), pp. 95–7, French original of 1906.
[65] Quoted from Müller (1862), p. 11.
[66] Quoted in Shapin and Schaffer (1985), p. 63.

He offered his low opinion of metaphors or what he called 'figurative language and literary allusion'. According to Locke: 'If we want to speak of things as they are, we must allow that all the art of rhetoric (except for order and clearness)—all the artificial and figurative application of words that eloquence has invented—serve only to insinuate wrong ideas, move the passions, and thereby mislead the judgement; and so indeed are perfect cheats.'[67] Many of the members of the early Royal Society shared Locke's view that scientific communication must avoid ambiguities of language and represent meanings without distortion. In his *History of the Royal Society* from 1667, Thomas Sprat formulated the planned reform of academic language as the intention

> to reject all the amplifications, digressions, and swellings of style: to return back to the primitive purity, and shortness, when men deliver'd so many *things*, almost in an equal number of *words*. They have exacted from all their members, a close, naked, natural way of speaking; positive expressions clear senses; a native easiness: bringing all things as near the Mathematical plainness, as they can: and preferring the language of Artisans, Countrymen, and Merchants, before that, of Wits or Scholars.[68]

However, the suggestions of the linguistic purists were largely rhetorical and not followed up to any extent in their actual writings in the *Philosophical Transactions* and elsewhere. In these writings there was no shortage of verbosity, metaphors, and other 'wrong ideas'.

As argued by Miles MacLeod, British natural philosophy during the scientific revolution relied upon a reconceptualization of language, a 'linguistic revolution' which rejected the traditional belief that words and objects of nature were essentially equivalent and of equal status. Contrary to this view, the empirical philosophers of the new age considered the meaning of words to be fixed by convention only and with no intrinsic correspondence to the actual fabric of nature. Language was arbitrary and conventional, they claimed, nothing but a means for rational communication. It followed that there was no natural connection between words and ideas and that the chosen vocabulary was not directly involved in the construction of knowledge. As Locke said:

> *Words*, by long and familiar use, as has been said, come to excite in men certain *Ideas* so constantly and readily, that they are apt to suppose a natural connexion between them. But ... every man has so inviolable a liberty, to make Words stand for what *Ideas* he pleases, that no one hath the Power to make others have the same *Ideas* in their minds that he has, when they use the same Words that he does.[69]

At least in some respects, Isaac Newton seems to have shared the ideas about a pure scientific language as expounded by Locke, Sprat, and other members of the Royal Society. He was keenly aware that his scholarly works in mechanics and optics could be misunderstood if presented in a popular non-mathematical manner relying

---

[67] Locke (1836), p. 372 available as Google Book. Book III, par. 34. See also Jones (1932).
[68] Sprat (2003), p. 112.
[69] Locke (1836), p. 294. Book III, par. 8. See MacLeod (2016) for the British linguistic revolution.

on analogies and metaphoric phrases. Language of this kind was acceptable and unavoidable in interpreting Scripture but not in interpreting nature. In a letter to Thomas Burnet, a theologian and author of cosmogonical tracts, Newton explained that Moses' account of creation was not intended to be technically correct and yet it was, in a sense, true:

> As Moses I do not think his description of the creation either Philosophical or feigned, but that he described realities in a language artificially adapted to the sense of the vulgar [common people]. Thus ... when he tells us God placed those lights in the firmament, he speaks I suppose of their apparent not of their real place, his business being not to correct the vulgar notions in matters philosophical [but] to adapt a description of the creation as handsomly as he could to the sense & capacity of the vulgar.[70]

In the same letter he defended Moses' use of 'figurative expressions' although admitting that they were 'more poetical than natural'. Newton's concern with finding unambiguous non-metaphoric words for his theory of universal gravitation has been analysed in a paper by John M. Coetzee, the South African author and recipient of the 2003 Nobel Prize in Literature. As Coetzee points out, since the days of Newton the general trend in science has been towards a language purged of metaphoric content. And yet he doubts 'whether a metaphor-free language in which anything significant or new can be said is attainable'.[71]

Newton repeatedly described his theory of gravitation by means of terms such as 'attraction' and 'attract', a usage which his great contemporaries Leibniz and Huygens found highly objectionable. To them, they were nonscientific and occult words that belonged to a previous dark age. Huygens had his own theory of gravity which rested on Cartesian principles and did not assume any mutual influence between two bodies. 'Attraction' was not yet a technical term used to describe, for example, the behaviour of magnets and electrical bodies. In an account in *Philosophical Transactions* of 1671 of Erasmus Bartholin's discovery of double refraction it was said about a crystal that 'this substance is Electrical, attracting (to speak with the Vulgar,) when heated, straw, Feathers, &c.'.[72]

A century later, neither physicists nor philosophers had any problem with the words which no longer carried with them the original vulgar and animistic connotation. To quote Coetzee, it is a general feature that 'words that are originally metaphorical soon "become frozen" or "die" in their new senses and are thereafter no longer metaphorical because they are no longer felt to be metaphorical'.

Far from eschewing the 'amplifications, digressions, and swelling of style' so vehemently criticized by Sprat, scientific authors in the Enlightenment era indulged in them. Examples from two of the era's most prominent scientists illustrate how far they had moved away from the former ideal of a pure and naked ideal of scientific

---

[70] Letter of 13 January 1681. See https://www.newtonproject.ox.ac.uk/view/texts/normalized/THEM00253.
[71] Coetzee (1982), p. 4.
[72] *Philosophical Transactions* no. 67 (1671), p. 2041.

discourse. In his influential work *Les Époques de la Nature* published in 1778, the French naturalist Georges-Louis Leclerc (Comte de Buffon) argued eloquently that the geological past of the Earth could be revealed in much the same way as the history of human societies:

> In civil history, one consults documents, studies old medals, deciphers antique inscriptions, to determine the epochs of human revolution and to establish the dates of moral events. Likewise, in natural history, one must rummage through the Earth's archives, pull ancient monuments from the entrails of the Earth, reassemble their remains, and put together in a body of evidence all of the indications of physical change that can allow us to reach back into the different ages of Nature. ... Nature, being contemporary with matter, space, and time, its history is that of all substances, of all places, of all ages.[73]

Buffon's metaphorical description of the hidden layers of the Earth as 'archives' became popular in later evolution geology and palaeontology. In the twentieth century the archive metaphor turned up in sciences as diverse as cosmology and climatology. By drilling up ice cores from Greenland and the Antarctic and analysing their composition, scientists have established 'icy archives' from which a 'climate archive' of past temperatures can be derived.[74]

Eleven years after Buffon published his great work the famous German-born British astronomer William Herschel contemplated how the slow evolution of the heavenly nebulae could be recognized by means of observation. He suggested that an evolutionary picture of the universe could be obtained by collecting a large amount of data from different parts of it, some far away and others closer to the Earth. Herschel expressed the method in a language no less eloquent, no less rhetorical, and no less metaphorical than Buffon's:

> The heavens ... are now seen to resemble a luxuriant garden, which contains the greatest variety of productions, in different flourishing beds; and one advantage we may at least reap from it is, that we can, as it were, extend the range of our experience to an immense duration. For, to continue the simile I have borrowed from the vegetable kingdom, is it not almost the same thing, whether we live successively to witness the germination, blooming, foliage, fecundity, fading, withering, and corruption of a plant, or whether a vast number of specimens, selected from every stage through which the plant passes in the course of its existence, be brought at once to our view?[75]

In Darwin's *Origin of Species* from 1859 metaphors appeared abundantly, not only as convenient manners of speaking but also as necessary parts of his entire argument for evolution by natural selection. As Darwin was fully aware, the basic claims of his theory were hidden from immediate perception and would be visible only over

---

[73] Buffon (2018), p. 3.
[74] Lorius et al. (1992).
[75] Herschel (1789), p. 226.

an immense span of time—not unlike what was the case with Herschel's nebulae. For this reason, metaphorical and analogical language was indispensable. When he introduced the famous phrase 'struggle for existence', he explicitly presented it as a metaphor:

> I use the term Struggle for Existence in a large and metaphorical sense, including dependence of one being on another, and including (which is more important) not only the life of the individual, but success in leaving progeny. . . . As the missletoe is disseminated by birds, its existence depends on birds; and it may metaphorically be said to struggle with other fruit-bearing plants, in order to tempt birds to devour and thus disseminate its seeds rather than those of other plants. In these several senses, which pass into each other, I use for convenience sake the general term of struggle for existence.[76]

Such explicit references to metaphors are unusual in scientific texts. Elsewhere in *Origin* Darwin referred to how naturalists spoke of 'the skull as formed by metamorphosed vertebrae [and] the jaws of crabs as metamorphosed legs'. Although this was admittedly a manner of speaking, a metaphor, in this case Darwin thought that it was more than just that. It was a metaphor that came close to reality:

> Naturalists, however, use such language only in a metaphorical sense: they are far from meaning that during a long course of descent, primordial organs of any kind—vertebrae in the one case and legs in the other—have actually been modified into skulls or jaws. Yet so strong is the appearance of a modification of this nature having occurred, that naturalists can hardly avoid employing language having this plain signification. On my view these terms may be used literally.

Metaphors occur no less frequently in chemistry than in the other sciences. Mary Jo Nye gives several examples of metaphoric language in the history of chemistry, as when the French chemist Auguste Laurent in the mid-nineteenth century described the action of atoms and molecules in terms such as 'copulation', 'birth', 'attack', and 'marriages of convenience'. As she points out, this kind of language should not be taken too seriously: 'The metaphors are intended to be illuminating but not rigorous. The language is playful and whimsical, or simply picturesque and striking to the memory.'[77] Laurent was involved in an extended dispute concerning the structure of organic compounds in which one of the other participants was the leading German chemist Hermann Kolbe. At one stage during the dispute, Kolbe made use of a metaphor with sociopolitical roots:

> In my conception, the constitution of a chemical compound is like that of a well-organised constitutional state, with one sovereign and a number of subordinate members standing nearer or farther from him, which is organised such that in place

---

[76] Darwin (1964), pp. 61–3. Pramling (2008). Another key phrase of Darwinism, 'survival of the fittest', was coined by the philosopher Herbert Spencer in 1864 and later adopted by Darwin.

[77] Nye (1993), pp. 78–9. For a more philosophical perspective, see Mahootian (2015).

of a given individual, a group consisting of various individuals of equal rank can function.[78]

Metaphors are often used to argue against or sometimes to ridicule a theory by presenting (or misrepresenting) it by means of as-if similes. Fred Hoyle's labelling of the explosive origin of the universe as a 'big bang' is an example of this kind of rhetoric (Section 6.5). When the American chemist Gilbert N. Lewis in 1916 suggested a new model of the 'cubic atom' with stationary electrons, it was received unfavourably by many scientists. According to Lewis, a pair of electrons could belong to the outer shells of two different atoms and thereby function as a chemical bond between them (e.g. H:Cl and H:O:H). The Polish-American chemist Kasimir Fajans later described the theory in terms of metaphorical rhetoric:

> Saying that each of two atoms can attain closed electron shells by sharing a pair of electrons is equivalent to saying that husband and wife, by having a total of two dollars in a joint account and each having six dollars in individual bank accounts, have eight dollars apiece![79]

Today it is generally agreed that metaphors are useful if far from indispensable in creative scientific work. They inevitably carry with them potential misunderstandings and, moreover, it cannot be known in advance whether a metaphor is helpful or if it is a blind alley that turns out to be counterproductive. In later chapters we shall meet a large number of science metaphors of various kinds.

Given that a scientific report, apart from its factual content, is also aimed at convincing its readers of its soundness and importance it is hardly surprising that rhetorical elements almost invariably are parts of the science literature. Since the 1970s the 'rhetorical approach' to texts and oral addresses has been cultivated by many historians, sociologists, and rhetoricians.[80] According to some of these scholars, not only are scientific texts permeated by rhetoric, even the so-called scientific facts about nature are rhetorically constructed. Most philosophers of science, not to speak of the scientists themselves, believe that this most radical claim is unconvincing. What is more, the amount and significance of rhetorical devices depends on the period and scientific discipline in question. Whereas rhetoric is abundant in Darwin's Origins from 1859, it is harder to find in Einstein's 1915 article 'Zur allgemeinen Relativitätstheorie' in which he introduced the general theory of relativity.

The claim that the experienced world is a rhetorical construction has a counterpart in some linguistic theories suggesting that our perceptions and thought processes are not only influenced by the spoken language but are determined by it. In a weak version, 'As language changes so too do the structure and content of knowledge.'[81] The stronger version is that language shapes our reality or what we perceive as such. This is often called the 'Whorfian hypothesis' after the American linguist and chemical

---

[78] Quoted in Rocke (2010), p. 178.
[79] Quoted in Brock (1993), p. 477.
[80] E.g. Gross (1990) and Bazerman (1988).
[81] Roberts (1991), p. 200.

engineer Benjamin Lee Whorf, who stated it around 1940. An alternative name is the 'linguistic relativity hypothesis' which alludes to the consequence that all phenomena are relative to equally valid linguistic frames of reference (see also Section 5.3).[82] It goes without saying that Whorf's hypothesis is controversial. In his paper on Newton's language, Coetzee used the opportunity to repudiate the narrow version of the Whorfian hypothesis: 'The case of gravitational attraction does not at all demonstrate what Whorf asserts about Newtonian cosmology as a system, namely that the key concepts of the cosmology emerge smoothly from or fit smoothly into, the structures of Newton's own language(s).'[83]

At the end of this section I want to call attention to an example from early nuclear physics which illustrates the close but sometimes problematic connection between metaphor and model.

When physicists in the 1930s struggled to understand the structure of atomic nuclei they were forced to rely on analogies and using a metaphorical language. The most successful of the nuclear models was the 'liquid drop' model originally conceived by George Gamow but usually associated with the improved model introduced by Niels Bohr in 1936. As early as 1929, at a time when physicists conceived the atomic nucleus as a compact collection of alpha particles, protons, and electrons, Gamow suggested that it might be understood in analogy with a drop of water. It was the idea of surface tension that led him to the liquid drop model and its associated metaphoric name. As the name suggests, the nucleus—which by 1936 was known to consist of protons and neutrons—was likened to a drop of water consisting of a large number of molecules. The physicists simply transferred the terms and concepts from the well-known water drops to the poorly understood atomic nuclei. Gamow thought of the nucleus to be 'in a way similar to a water-drop held together by surface tension' and Bohr spoke of the emittance of particles from the nucleus as 'evaporation'.[84] He and other physicists suggested that the compact atomic nuclei could be ascribed a 'temperature'.

The language of the liquid drop model whether in Gamow's or Bohr's version was throughout metaphoric as it referred to phenomena known from classical physics and everyday experience. In the event the comparison between a water drop and an atomic nucleus proved fruitful. It was on the basis of the liquid drop model that Bohr in early 1939 provided the first qualitative explanation of uranium fission.

1.3 Atoms and molecules

Importantly, many (but far from all) names used in science carry epistemic and social connotations with them, and as such they are more than just neutral linguistic reference terms. This is one reason why the choice of names is often controversial and may arise passions not only among the scientists but also in the public arena. The

---

[82] https://plato.stanford.edu/entries/linguistics/whorfianism.html. Whorf's hypothesis is also known as the Sapir–Whorf hypothesis, where the first name refers to the American linguist and anthropologist Edward Sapir. The double eponym is widely considered a misnomer as the statements of the two linguists differed in important respects.

[83] Coetzee (1982), p. 11.

[84] Quoted in Stuewer (1994). For Gamow's analogy, see Little (2008).

importance of names or words in the historiography of science was pointed out in a pioneering study of the term *scientist* made by Sydney Ross, a Scottish-American colloid chemist, bibliophile, and historian of science (Section 1.4). According to Ross:

> The history of a word is never solely a matter of etymology: the need for a new word is socially determined, right at the start, and any subsequent changes of denotation, as well as the cluster of connotations surrounding it, are also in response to demands from society. The word cannot be isolated from its historical background; indeed, some key words offer a concise and suggestive clue to the historian or sociologist.[85]

More than half a century earlier, Clifford Allbutt, a professor of medicine (or 'physic') at Cambridge University, made a similar point in a manual for the writing of scientific papers, namely that semantics is no less important than etymology:

> It is a common and tiresome error to suppose that the meaning of a word is to be governed by its etymology. Even on its invention a word may be derived awkwardly or ineptly; but, however apt in origin, words must grow or drift with things they signify, and thus become endowed with ever new and cumulative content. ... Indeed, when words do not develop thus we may suspect that thought is stationary or declining.[86]

This observation is as true today as it was in 1905. As expressed metaphorically by a biologist, 'Words propagate in the linguistic world like organisms in the biosphere and, like organisms, they are subject to the forces of natural selection.'[87] According to Peter Bowler, a historian of biology, the complex history of the term 'evolution' provides 'a good illustration of the care which must be taken in studying the origin of even the most commonly used words.'[88] The Latin *evolutio* refers to an act of unrolling or unfolding and was originally applied by embryologists to denote how the embryo developed from pre-existing parts. In the eighteenth century 'evolution' was used in astronomy and geology as indicating a kind of progressive natural process. The meaning and popularity of the term changed significantly with Darwin's theory of natural selection, but even after the Darwinian theory had been broadly accepted, the term 'evolution' was used by different writers in a variety of different senses.

The erroneous view referred to by Allbutt—that the meaning of a word is located in its etymology—was popular in ancient Greece and the early Middle Ages when it was generally believed that a word could be ascribed a true meaning independent of its common use. But Plato knew better. In his dialogue *Cratylus* from about 360 BC, Socrates discusses the origin of names and the connection between language and thought. While one of Socrates' two interlocutors (Hermogenes) argues that names are arbitrary and conventional, the other (Cratylus) believes they are natural in origin. Plato lets Socrates say: 'How to learn and make discoveries about the things that are is probably too large a topic for you or me. But we should be content to have agreed

---

[85] Ross (1962), p. 65. See also Miller (2017).
[86] Allbutt (1905), pp. 90–3.
[87] Zacharias (2020), p. 345.
[88] Bowler (1975). *OED*, entries 'evolution' and 'evolutionary'.

that it is far better to investigate them and learn about them through themselves than to do so through their names.'[89]

Since many things are named arbitrarily and not in accordance with their 'nature', not all things can be etymologized. Moreover, the belief that the proper meaning of a term can be uncovered by its etymology obviously disregards that the meaning of words constantly changes over time. It is known as the 'etymological fallacy'.[90] If a word has long since developed new connotations, it makes no sense to argue from its etymon that its original meaning is still valid. An extreme version of the fallacy was expounded in a large encyclopaedia called *Etymologiae* written about AD 630 by the Spanish Archbishop Isidore of Seville. According to Isidore, the study of the origins of words would lead to truths about nature, history, and much more. He thought that the royal road to knowledge was by way of words and their origin:

The knowledge of a word's etymology often has an indispensable usefulness for interpreting the word, for when you have seen whence a word has originated, you understand its force more quickly. Indeed, one's insight into anything is clearer when its etymology is known.

Nonetheless, Isidore was well aware that the etymological approach has its limits:

Not all words were established by the ancients from nature; some were established by whim, just as we sometimes give names to our slaves and possessions according to what tickles our fancy. Hence it is the case that etymologies are not to be found for all words, because some things received names not according to their innate qualities, but by the caprice of human will.[91]

To illustrate the semantic change of a word more concretely, consider first the *atom* of ancient Greek philosophy, a term which can be translated as an indivisible or uncut particle. The atom is an *a-tomos*, with *a* for 'absence of' or 'without' and the Greek *tomos* (τόμος) meaning 'to cut' or 'to slice'. An atom is something which cannot be cut into smaller slices or pieces. We have the second part of the word in, for example, *tomography*, the art of producing images of sections of a body. A *microtome* is a specialized instrument used for cutting thin slices of a block of tissue or some other material. The word *diatom(s)* for an important group of microscopic algae is divided as *dia-tom* and not *di-atom*. The name is Greek with *dia* meaning 'through'; that is, a diatom looks like an organism 'cut through'. Confusingly, in chemistry the adjective 'diatomic' has been used since the mid-nineteenth century for molecules consisting of two atoms (such as oxygen O_2 and carbon monoxide CO). While the noun 'diatom' belongs to biology, 'diatomic' refers to chemistry.

The basic postulate of Leucippus, Democritus, Lucretius, and other adherents of ancient atomistic philosophy was that all matter consists of qualitatively similar

[89] Cooper (1997), p. 154.
[90] Kolb (2019).
[91] Barney et al. (2006), p. 55. For an online edition of *Etymologiae* in English translation, see https://sfponline.org/Uploads/2002/st%20isidore%20in%20english.pdf

minimal parts or *atoms* moving incessantly and randomly in a void. According to Lucretius' *De Rerum Natura* from about 50 BC, atoms could only combine into a finite variety of ways. He used the alphabet as a metaphor, stating that atoms are like letters: just as the letters of the alphabet can create a large number of different words, so the atoms can combine to form a large number of different objects. The name 'atom' and the concept associated with it was in a modified version revived by Daniel Sennert, Pierre Gassendi, and others in the seventeenth century, but the idea was turned into a scientifically fruitful atomic theory only by John Dalton in the early years of the nineteenth century.

Dalton's atoms—he also called them 'ultimate particles', 'simple elementary particles', or just 'particles'—were indivisible, but in contrast to earlier thinkers he suggested that they differed qualitatively and quantitatively from one chemical element to another. Moreover, he used the term atom not only for the particles of elements such as oxygen and iron but also, to the confusion of modern readers, for those of chemical compounds. Thus, in his *New System of Chemical Philosophy* published in 1808 he wrote of 'an atom of water', 'an atom of alcohol', and 'an atom of sugar'. Translated into present nomenclature, he stated the compositions of the three atoms—what we would call molecules—to be HO, C_3H, and C_4O_2H.

Apart from his atomic theory, Dalton is eponymously associated with a couple of other terms. He arrived to his idea of atoms through studies of gases, which in 1802 resulted in what is known as Dalton's gas law, namely that the total pressure of a mixed gas equals the sum of the partial pressures of the individual gases. In the case of atmospheric air, its pressure is the sum of that of oxygen and nitrogen (argon had not yet been discovered). The 'dalton' with lower case d is an atomic mass unit defined as 1/12 of a neutral carbon-12 atom. The dalton unit was originally proposed in 1924 by an American chemist, but at the time based on oxygen and without taking its isotopes into consideration.[92] That is, an 'atom of oxygen' weighed 16 daltons. The name but not the definition was eventually adopted by the International Union of Pure and Applied Chemistry (IUPAC) in 1993 and endorsed by its sister organization IUPAP (International Union of Pure and Applied Physics) in 2005.

In a different context Dalton's name turns up also in the history of the medical sciences. 'Had Dalton not conceived an atomic theory his name might have been remembered even more widely than it is', writes Frank Greenaway, a historian of science. He explains:

> The word 'Daltonism' is still used by Continental scientists for 'Colour-blindness' and had British scientists not preferred to identify Dalton rather with atomic theory the word might have come into general use in this country as well. It is certainly less committing that the unhappy word 'colour-blindness' which was shown, early on, to be misleading.[93]

Experimenting on himself Dalton realized that he suffered from what he called 'anomalous vision', a particular kind of colour blindness or rather defectiveness where

---

[92] Tanner (1924).
[93] Greenaway (1966), p. 98.

blue colours are perceived as red and vice versa. In 1798 he published a paper on the subject, the first of its kind. The term *Daltonism* appeared in print 1827 and continued to be widely used in many European countries. However, in England the eponym largely disappeared after Dalton's death in 1844.

The British surgeon and ophthalmologist James Dixon disliked the term, but not because he did not want to honour Dalton with a name. On the contrary, he found 'Daltonism' to be an unworthy eponym for the admired natural philosopher. Besides, he preferred a name derived from Greek in accordance with the long-established tradition of nomenclature. His own suggestion for this kind of colour blindness was *acritochromacy*, meaning something like inability to distinguish between colours. According to Dixon, this name 'supplies the meaning we are in want of, and is tolerably pronounceable'. Perhaps it is, but his neologism, although used from time to time, was unsuccessful (it is included in *OED*, but as an obsolete word). In a book of 1859, Dixon argued why 'Daltonism' should not be part of the English language:

> Of all the unfortunate invention of pathological nomenclature, the word *Daltonism* ... seems to me one of the worst. It seems an indignity to the memory of such a man to connect his name in this way with a physical defect. We would surely wish to remember our great men for their mental excellences and not for their bodily imperfections. ... Dalton should be immortalized as the propounder of the 'atomic theory', not as the man who mistook red for blue.[94]

The word *molecule* is a contraction of the Latin *moles* for mass and the diminutive suffix -*culus*, which is also Latin. It thus means 'very small mass'. The term had on occasions been used since about 1650 but only in the general sense of extremely minute parts of matter more or less synonymous with atoms. In his treatise *Elements of Natural Philosophy* first published posthumously in 1720, John Locke wrote about sensible matter that it consisted of 'unconceivable small bodies, or atoms, out of whose various combinations bigger *molleculæ* are made'.[95]

Dalton did not use the term which was only introduced in chemical theory by Amedeo Avogadro in an important but initially overlooked paper of 1811 in *Journal de Physique*. His use of 'molecule' was novel and eventually successful, but since he did not coin the word it was not a proper neologism. Moreover, and again to the confusion of modern as well as contemporary readers, he wrote exclusively of molecules whereas he altogether avoided the term atom. This he replaced by either 'elementary molecule' or 'half-molecule'. For example: 'The integral molecule of water will be composed of a half-molecule of oxygen with one molecule, or, what is the same thing, two half-molecules of hydrogen.'[96] The molecule of an element he called a 'constituent molecule'. Throughout much of the nineteenth century, the words atom and molecule were given different meanings and sometimes mixed up.

Of course, these words were also used outside chemical and physical theory, either figuratively or in the context of other sciences. It may be of interest to note that

---

[94] Dixon (1859), p. 278.
[95] Locke (1758), p. 110. For a complete history of the concept of molecule, see Kubbinga (2001).
[96] Leicester and Klickstein (1952), p. 234.

although Shakespeare did not use the term 'atom', he did use a close derivative on a few occasions. 'It is as easy to count atomies as to resolve the propositions of a lover', we read in Shakespeare's play *As You Like It* from 1599. And in the slightly earlier *Romeo and Juliet* there is a passage on an imaginary coach of miniscule size 'drawn with a team of little atomi'. About half a century after Shakespeare, the verb *atomize* came into use in the meaning of destroying something or dividing it into smaller fragments. If a perfume or some other liquid is transformed into a fine spray, it is atomized by means of an *atomizer*, a noun known in English only from about 1865 (*OED*). The tiny vapour droplets produced by an atomizer are small indeed, but they are far from being of an atomic or molecular scale.

At the time of Dalton there already was a tradition in natural history to refer to tiny elementary organic bodies as molecules. The important concept of *Brownian motion* eponymously refers to the Scottish botanist Robert Brown who in 1828 published his observations of the motion of microscopic bodies immersed in fluids. Strictly speaking, Brown did not discover the phenomenon named after him and the eponym may thus be an example of Stigler's law of eponymy. Observations of a similar kind but restricted to particles of organic substances had been reported by John Needham and several other naturalists in the eighteenth century. On the other hand, Brown was the first to demonstrate that the irregular motion was exhibited also by inorganic particles and thus was not vital in origin. His article in *Philosophical Magazine* carried the title: 'A Brief Account of Microscopical Observations Made in the Months of June, July and August, 1827, on the Particles Contained in the Pollen of Plants; and on the General Existence of Active Molecules in Organic and Inorganic Bodies'. Clearly, what Brown referred to as molecules were something very different from the molecules of Avogadro and later chemists.

If students of chemistry and physics are unaware of Avogadro's introduction of molecule in the scientific vocabulary, as they most likely are, they do know about the law and constant to which the Italian chemist has laid name. 'Avogadro's law' states in the elementary textbook version that equal volumes of different gases contain the same number of molecules, while 'Avogadro's number' N denotes the number of molecules in 24 litres of a gas at normal temperature and pressure. In his paper of 1811, Avogadro did indeed present his law, but there was no mention of the number, which was only discussed and eventually determined to be approximately $N = 6 \times 10^{23}$ much later. In a paper of 1865 the Austrian physicist and chemist Josef Loschmidt showed how the quantity, or rather the equivalent number for one cubic centimetre, could be estimated but without actually doing so. The Avogadro-Loschmidt number first appeared numerically in a condensed version of Loschmidt's paper from the same year but written by an unknown author.[97] The given value was 8.7×10^{17} molecules/cm^3 or $N = 2.1 \times 10^{22}$.

There is little doubt that the eponym Avogadro's number lacks historical justification and that 'Loschmidt's number', as sometimes used in German-language publications, is more justified in this respect. The attribution of the number or constant to Avogadro seems to have originated with the French physicist and Nobel laureate Jean Perrin, who in 1909 advocated 'Avogadro's constant'. According to

---

[97] Hawthorne (1970). Jensen (2007a).

Perrin, 'This invariable number N is a universal constant, which may appropriately be designated *Avogadro's Constant*'.[98] Neither this term nor Avogadro's number appears in the earlier literature. The name is an example of Stigler's law discussed in Section 1.2.[99]

As far as Avogadro's law or hypothesis is concerned, in 1814 it was independently restated by André Marie Ampère better known for his fundamental contributions to electrical theory. For this reason, the law was sometimes known as the Avogadro–Ampère gas law, a term which can still be found in French literature in particular. The law had wide-ranging implications since it implied that elemental gases were diatomic, contrary to what Dalton claimed. Thus, in modern notation, Dalton's H, O, and N were replaced with H_2, O_2, and N_2. Likewise, the synthesis of water was no longer $H + O \rightarrow HO$ but $2H_2 + O_2 \rightarrow 2H_2O$. Despite Avogadro's reasoning, his hypothesis was generally rejected or ignored, and the same was the case with Ampère's version of it. When Jean Baptiste André Dumas in 1826 revived the Avogadro–Ampère law, he was no more successful.

One of the reasons for the greatly delayed acceptance was undoubtedly the ambiguous and confusing terminology.[100] In addition to the terms atom and molecule, chemical authors in the period 1810–60 used a variety of other loosely defined terms including 'ultimate particles', 'composed atoms', 'corpuscles', and 'simplest parts'. As mentioned, Avogadro wrote of half-molecules in the meaning of atoms, and Dumas even used the contradictory word 'half-atom'. He later suggested 'chemical molecule' and 'physical molecule' to distinguish between what we call atoms and molecules. The terminological mess only disappeared in 1860 when a large number of European chemists (plus one from Mexico) convened at an important congress in Karlsruhe, Germany, and there reconsidered the question of atomic weights in relation to Avogadro's law.

About forty years later it was realized that the atom, despite its name, is divisible as it consists of even smaller particles. Joseph John (J. J.) Thomson's atomic model from the early years of the twentieth century pictured the atom as consisting of a large number of electrons (which he called 'corpuscles', see Section 2.2) moving inside a frictionless and positively charged sphere. Thomson's contemporary Oliver Lodge described the model in metaphoric terms:

The bulk of the atom may be composed of an indivisible unit of positive electricity, constituting a presumably spherical mass or 'jelly', in the midst of which an electrically equivalent number of point electrons are as it were 'sown'; these electrons probably distributing themselves in rings ... revolving in regular orbits about the centre of the jelly.[101]

Later elementary presentations adopted another culinary metaphor, namely the popular 'plum-pudding', which still appears frequently in introductory physics textbooks

[98] Perrin (1910), p. 10, a translation of a French memoir from 1909.
[99] Crepeau (2009).
[100] Ihde (1964), p. 121. Weiner (1943).
[101] Lodge (1906), p. 148.

and on numerous websites. However, Thomson never used the metaphor, which first turned up in an anonymous review article from 1906.[102] Given that Thomson's electrons moved at high speeds around the centre of the positive charge it is a misleading image. After all, plums or raisins in a pudding do not move significantly. And yet, despite being misleading, the image has stuck. As noted in Section 1.2, metaphors are useful but always inaccurate, and in some cases they are grossly distorting representations of reality.

Whereas scientists and the general public associate the term 'atomism' with the view that matter consists of smallest material particles called atoms, to philosophers it may have a different connotation. In 1918 the British philosopher Bertrand Russell published a work with the title 'Philosophy of Logical Atomism' in which he argued that all philosophical discourse about the world can be broken down to certain 'atomic facts'. He also spoke of 'logical atoms' and 'molecular propositions'. Russell coined the term 'logical atomism' in 1911 and it was later adopted and discussed by other philosophers in the analytical tradition. 'The logic which I shall advocate is atomistic', he wrote, 'as opposed to the monistic logic of the people who more or less follow Hegel.' As Russell pointed out, his atoms were completely different from those considered by the physicists and chemists: 'The atom I wish to arrive at is the atom of logical analysis, not the atom of physical analysis.'[103]

At the time when Russell wrote about his logical atoms, Ernest Rutherford introduced his nuclear model of the physical atom although originally without using the word 'nucleus' for the massive central part of the atom. Curiously, one can find '(atomic) nucleus' three years before Rutherford's atom. In an elementary textbook from 1908, the Danish chemist Sophus M. Jørgensen wrote about the atom that it is 'now considered to be a nucleus of positive electricity, around which negative electrons rotate . . . like the planets in the solar system'.[104] However, it was only with Niels Bohr's extension of Rutherford's model into an atomic theory that physicists began speaking of the nucleus of the atom. *Nucleus*, which means 'little nut' in Latin, had for long been used in other areas of science, for instance to denote the central part of a living cell or the bright core of a comet. Newton wrote in *Principia* about the tails of comets that they were fine vapours emitted by 'the head or nucleus of the comet'. Much later, biochemists used *nucle-* in words such as nucleoprotein and nucleic acid, terms that date from about 1900.

Within a few years the Rutherford–Bohr nuclear model became the accepted view and at around 1930 'nuclear physics' emerged as a new and exciting branch of physics. The term only appeared in the scientific literature in the late 1920s. To most people the distinction between atom and nucleus was and still is of no significance, something which is reflected in the unfortunate current usage of terms such as 'atomic energy' and 'atomic bombs'. These are clearly misnomers since conventional energy sources based on coal, oil, or gas, and also conventional chemical explosives such as TNT, are 'atomic' as well. After all, the energy is due to chemical processes involving

---

[102] Hon and Goldstein (2013).
[103] Russell (2010), pp. 2–3.
[104] Jørgensen (1908), p. 26. Planetary atomic models were proposed prior to Rutherford's nuclear atom. Jørgensen most likely referred to one of those models.

electrons, which are no less part of the atom than are the nuclei. The unfortunate usage was sanctioned at an early stage; witness that the United States created its Atomic Energy Commission (AEC) in 1946 and the United Nations' International Atomic Energy Agency (IAEA) dates from 1957. Whereas atomic bomb is still more frequently used than 'nuclear bomb', today 'nuclear energy' is more frequent than atomic energy (however, the more generally used term is 'nuclear power'). 'Atomic reactor' instead of 'nuclear reactor' was sometimes used in the past, but it was never popular and is now obsolete.[105]

1.4 The origin of 'scientist'

Although the term *science* entered the English language in the Middle Ages, it was originally with a meaning quite different from the present one. Derived from the Latin word *scientia* meaning knowledge, the word denoted the art of deriving certain knowledge demonstratively in the sense of Aristotle. Science was associated with necessity and deductive logic, not with experiment or observation. It was only with the scientific revolution in the early seventeenth century that science became experimental or natural philosophy, terms which for a long period of time were used to designate what we would today call science or natural science. Still in 1704 John Harris defined in his *Lexicon Technicum* science as 'Knowledge founded upon, or acquir'd by clear, certain, and self-evident Principles.' As regards the subject of 'Physicks, or Natural Philosophy', he wrote that it is 'the Speculative Knowledge of all Natural Bodies, ... and of their proper Natures, Constitutions, Powers, and Operations'.[106]

When the Royal Society was founded in 1660, its motto was chosen to be *Nullius in verba*, which is often misread as 'nothing in words', allegedly an indication that the founding fathers of the society wanted to emphasize how insignificant words were in the new empirical study of nature. However, a much better translation of the Latin phrase is 'on the word of no one', meaning that rhetorical knowledge claims about nature as based solely on the reading of Aristotle and other ancient authorities should be disregarded. According to Clive Sutton, a British science educator and historian of science:

> I suspect that all readers are potentially affected by the negative ring in the N of 'nullius' and are likely to feel it in association with Verba which in other contexts can just mean 'words'. (How N-ish our negative words are: not, never, nothing, no, null, nil.) One picks up, therefore, an anti-word feeling. To readers unlearned in Latin the motto suggests strongly a distrust of words; it looks and sounds like 'Nothing in words', but if someone had chosen the anti-word idea for an heraldic device they would probably have written *Nihil in verbis*.[107]

---

[105] For the frequency of these and other terms as given by their occurrence per million words in modern written English, see the *OED*.

[106] Harris (1704). Available online as Google Book. The definitions of 'science', 'physicks', and 'natural philosophy' were almost the same in Blount (1707), who defined a 'naturalist' as 'one skill'd in Natural Philosophy'.

[107] Sutton (1992), p. 39.

Following Bacon, the new generation of natural philosophers agreed to pay more attention to their own observations than to accepted wisdom, but of course their observations must be reported in words. Nevertheless, as Sutton points out, 'the anti-word tradition comes down to us very strongly, and no scientist or science teacher wants to be caught dabbling with "mere words"'.[108]

One of the oldest scientific journals, and the longest-running in the world, was founded by the German-born Henry Oldenburg, secretary of the Royal Society, in the spring of 1665. Characteristically, it was called (and is still called) *Philosophical Transactions*. The lengthy subtitle indicated what 'philosophical' referred to: *Giving Some Accompt* [Account] *of the Present Undertakings, Studies, and Labours of the Ingenious in Many Considerable Parts of the World*. The 'ingenious' were the scientists in the new empirical tradition except that the word scientist did not yet exist. They also described themselves as 'virtuosi' or sometimes 'Christian virtuosi'. The latter term entered the title of a book published by Boyle, in which he argued that the study of nature was a religious duty (*The Christian Virtuoso*, 1690). The Italian word 'virtuoso' used for one who excelled in music or some other art was based on the Latin *virtus* meaning 'goodness' or 'moral strength'.

Twenty-two years after the foundation of *Philosophical Transactions*, Isaac Newton published his famous *Principia* the full title of which was *Philosophiæ Naturalis Principia Mathematica*, that is, Mathematical Principles of Natural Philosophy. The word science did not appear in the English translation of 1729 based on the third edition of 1726 and the word physics only once. As a third and later example of (natural) philosophy in the sense of science, consider John Dalton's important textbook in which he based chemistry on a new theory of atoms. The book published in two parts in 1808 and 1810 carried the title *A New System of Chemical Philosophy*. Two years later, Humphry Davy published his *Elements of Chemical Philosophy*.

Clearly, the term 'philosophy' was still in the early nineteenth century used in a meaning unrecognizable to most professional philosophers of a later era. The designation of science as philosophy was common at the time when two of the premier British journals of physics and chemistry, apart from *Philosophical Transactions*, were *Philosophical Magazine* and *Annals of Philosophy* founded in 1798 and 1813, respectively. The subtitle of the latter journal was *Magazine of Chemistry, Mineralogy, Mechanics, Natural History, Agriculture and the Arts,* topics which from our present point of view are decidedly un- or non-philosophical.

By the mid-nineteenth century, *science* as a common name for the empirical and theoretical studies of nature had won wide but not full acceptance. The semantic evolution of science in English can be followed through the editions of the *Encyclopaedia Britannica*, which in its first edition of 1771 defined the term as a philosophical doctrine 'deduced from self-evident and certain principles, by a regular demonstration'.[109] In the sixth edition of 1823, science was treated in more detail but still as part of a more general philosophical concept. Science, the reader was informed, could mean 'The knowledge of things, their constitutions, properties, and operations: this, in a little more enlarged sense of the word, may be called Φυσικη [physics] or natural

---

[108] Sutton (1994). See also Dear (1985).
[109] Quotations from Roelli (2021), pp. 24–5.

science'. It could also mean 'mechanics, or the application of the powers of natural agents to the use of life'. In the 1911 edition the word had become narrowed down to essentially its modern meaning:

> In general usage a more restricted meaning has been adopted, which differentiated 'science' from other branches of accurate knowledge. For our purpose, science may be defined as ordered knowledge of natural phenomena and of the relation between them; thus it is a short term for 'natural science', and as such is used here technically in conformity with a general modern convention.

Still at the time of the founding of the British Association for the Advancement of Science in 1831 the term was used almost synonymously with philosophy. The definition of science as a single term for the occupation of men of science, and as distinct from particular disciplines such as anatomy and geology, was much discussed in British cultural circles. The *Monthly Repository* brought in 1834 an article with the title 'On the Application of the Terms Poetry, Science and Philosophy' in which the anonymous author discussed the similarities and differences between the three fields. 'There is no reason', he said,

> why the term Science should be confined to mathematics or physics. Every collection of general propositions, on any subject, comprehending all that is known concerning it, arranged with a view to communicate information in a synthetic Form, and ... to state principles and exhibit results, may surely be called a Science. Why, for example, should there not be the Science of metaphysics, of morals, of jurisprudence, or of political economy, as well as of astronomy, mechanics, and chemistry.[110]

During the first half of the nineteenth century the terms philosophy and science were often used interchangeably with, for example, 'experimental philosophy' being largely a synonym of 'experimental science'.

Just at the time when the anonymous author wondered about the meaning of science in *Monthly Repository* a new name was coined for the cultivator or practitioner of natural science. At its first three meetings the British Association addressed its activities to 'friends of science', 'cultivators of science', and 'interpreters of science', but not yet to 'scientists'. The neologism *scientist* was proposed by the English polymath, philosopher, and theologian William Whewell (Figure 1.3), first in an anonymous review of Mary Somerville's *On the Connexion of the Physical Sciences* that appeared in the March 1834 issue of *Quarterly Review*. Referring to the meetings of the British Association, Whewell noted the lack of a common word for those occupied with the natural sciences. '*Philosophers* was felt to be too wide and too lofty a term ... *savans* was rather assuming, besides being French instead of English; some ingenious gentleman [Whewell] proposed that, by analogy with *artist*, they might form *scientist*.'[111]

---

[110] *Monthly Repository of Theology and General Literature* **6** (1834): 323–31. On the status and definition of science, see Yeo (1993).

[111] Quoted in Ross (1962), p. 72, reprinted in Ross (1991), pp. 1–39. Danielson (2001).

Figure 1.3 Portrait of William Whewell
(1794–1866).
Courtesy of Wikipedia Commons, Wellcome Collection
Gallery.

As a possible alternative Whewell mused that perhaps the Britons could translate the name used by the British Association's sister organization in Germany, the Gesellschaft deutscher Naturforscher und Ärtzte founded in 1822. 'But it was not found easy to discover an English equivalent for *natur-forscher*. The process of examination which it implies might suggest such undignified compounds as *nature-poker*, or *nature-peeper*, for these *naturae curiosi*; but these were indignantly rejected.' More seriously, and with greater impact, Whewell repeated his suggestion in his major work *The Philosophy of the Inductive Sciences* published six years later:

The terminations *ize* (rather than *ise*), *ism*, and *ist*, are applied to words of all origins: thus we have to *pulverize*, to *colonize*, *Witticism*, *Heathenism*, *Journalist*, *Tobacconist*. Hence we may make such words when they are wanted. As we cannot use *physician* for a cultivator of physics, I have called him a *Physicist*. We need very much a name to describe a cultivator of science in general. I should incline to call him a *Scientist*. Thus we might say, that as an Artist is a Musician, Painter, or Poet, a Scientist is a Mathematician, Physicist, or Naturalist.[112]

---

[112] Whewell (1840), p. cxiii.

In contrast to the author of the article in *Monthly Repository*, Whewell's concept of science was narrower as it excluded theology, economy, and the humanities. Whewell was not the only one who at the time looked for English words for cultivators of the various sciences. The French-American eccentric polymath Constantine Rafinesque, for a period professor of botany in Kentucky, likewise contemplated of how to relate the names of the sciences to those cultivating them. In a book published shortly after his death in 1840, he wrote:

> Those who apply themselves to any of these sciences are called Chemists, Botanists, Zoologists, and so on, the termination being generally in IST; but there are some Anomalies, we say Astronomer, Geographer, &c, and thus when the sciences end in *nomy* or *graphy* the term of the men must be in ER—the term of Physists and Cosmist for natural philosopher and natural historian, ought to be introduced; the French say *Physician*, now applied like *Physics* to the medical profession and science by us, erroneously but of too long standing to be altered.[113]

Rafinesque did not suggest a word for the cultivator of science in general, as Whewell did. His term 'physist' was close to 'physicist', but it was a protologism not used by other authors and not included in *OED*.

One might expect that Whewell's name scientist immediately caught on, but this was far from the case. For quite a long time the term was either ignored or resisted, and for considerably longer in Britain than in America. Ngram data show that by 1880 the term philosopher appeared in English-language books about eight times more frequently than scientist and 1.4 times more frequently than 'chemist'. Fifty years later philosopher was still more common than scientist, but now 'chemist' was twice as common as philosopher.

In France, where *savant* referred to scholars in general, there was no obvious equivalent to scientist and not even to natural philosopher. The situation in Germany was different again. The term *Naturwissenschaft*, which is today the equivalent of 'science', existed in the eighteenth century but with a broader meaning and without being much used until the following century. *Wissenschaft* included (and still includes) all academic knowledge and *Wissenschaftler* corresponded to both the English scientist and scholar. For the investigator of nature, the German word was not 'Naturwissenschaftler' but *Naturforscher*, a term with approximately the same meaning as Whewell's scientist. While it took several decades before science in its present meaning became a standard term in English, Naturwissenschaft was widely used in Germany by the 1830s.[114]

Many British natural philosophers found the new term scientist to be ignoble, even offensive, as it reduced the status of science from a vocation to a kind of profession or business. Instead of being called scientists, they preferred 'men of science' in analogy to the respected 'men of letters'. The man of science could also be called

---

[113] Rafinesque (1840), p. 12, accessible as Google Book. In the third edition of *On the Origin of Species*, Darwin referred to Rafinesque's ideas of evolution. He is known not only as a naturalist but also as a pioneer of American comparative linguistics. See Belyi (1997) for his early contributions to this area.

[114] See details in Phillips (2012).

a 'scientific worker'. In 1894, the prominent evolutionist and anthropologist Thomas Huxley criticized an American science magazine for using the term scientist, which he considered a vulgar Americanism no better than 'electrocution'. Much in favour of 'man of science' instead of 'natural philosopher', Huxley associated the undignified scientist with a culturally impoverished language.[115] About ten years later, in a manual for the writing of scientific papers, the term only occurred once, indicating that it was not generally accepted. '"Scientist" seems to me as proper as "artist" or "naturalist", and better than "orientalist"', the physician Clifford Allbutt opined.[116]

As scientist was a controversial term, so the older term men of science or its equivalent scientific men raised objections in some circles. Because, could these phrases be rightfully reserved for the new class of natural scientists? The English artist, writer, and cultural critic John Ruskin thought it was a 'ridiculous notion'. In a work of 1878, he protested:

> It has become the permitted fashion among modern mathematicians, chemists, and apothecaries, to call themselves 'scientific men', as opposed to theologians, poets, and artists. ... [But] there is a science of Morals, a science of History, a science of Grammar, a science of Music, and a science of Painting; and all these are quite beyond comparison higher fields for human intellect, and require accuracies of intenser observation, than either chemistry, electricity, or geology.[117]

In America, the term scientist had to a large extent replaced man of science by the 1880s if not without some reservations. John Billings, a librarian and surgeon, gave in 1886 an address to the Philosophical Society of Washington, DC, in which he reflected on the duties of scientific men. Well aware of the popular word scientist—'a coinage of the newspaper recorder'—he found it to be too comprehensive as a designator for the true cultivator of science which he insisted on calling scientific man or man of science. 'The suggestion which the word [scientist] conveys to my mind is rather that of one whom the public suppose to be a wise man, whether he is so or not, of one who claims to be scientific.'[118]

The name continued to be controversial in Britain for a surprisingly long time and much longer than in the United States. Noting that William Thomson preferred 'naturalist' over scientist and physicist, in the introduction to the first volume of *Electromagnetic Theory* published in 1894 Oliver Heaviside pointed out that there were different kinds of naturalists. He mused that those dealing with animate nature might be called 'organists' or 'vitalists', whereas 'materialists' was a suitable name for the physicists and others investigating inanimate nature. 'Buffon, Cuvier, Darwin, were typical vitalists. Newton, Faraday, Maxwell, were typical materialists. All were naturalists. For my part I always admired the old-fashioned term "natural philosopher" [but] ... There are no natural philosophers now-a-days.'[119]

---

[115] White (2002), p. 1. The term 'electrocution', a condensation of 'electrical execution', was an American neologism first appearing in 1889 just before the first use of the electric chair.

[116] Allbutt (1905), p. 39.

[117] Quoted in Ross (1962), p. 70.

[118] Billings (1886), p. 541.

[119] Heaviside (1951), pp. 1–2.

As late as 1924, there appeared in the journal *Nature* an extensive debate concerning the journal's policy of avoiding scientist.[120] The debate was spurred by a letter from the Cambridge physicist Norman Campbell, who argued in favour of scientist as a label for 'the discovery that there is something common in the intellectual attitude of all the sciences and other branches of learning'. As to the commonly used man of science, he considered it a poor substitute which was not only artificial but also 'offensive to feminists'.

However, Sir Richard Gregory, a British astronomer who served as editor of *Nature*, maintained that scientist was an unfortunate and unnecessary word.[121] While some of the correspondents, such as the astrophysicist Herbert Dingle, agreed with Campbell that there was no justification for rejecting scientist in *Nature* or elsewhere, other prominent British men of science expressed their dislike of the term. Henry Armstrong, a respected chemistry professor of the conservative school, was among the antagonists. 'If I had ever favoured the term—I hate it—I should cease from using it', he wrote, adding that the term had become 'a word of evil import in the public ear'.[122] What he had in mind was more vehemently expressed by the eminent zoologist Edwin Ray Lankester, according to whom scientist was a disgracing term which often referred to charlatans:

I hope *Nature* will continue to refuse the word 'scientist' ... The eminent scientist Barney Bunkum is already flourishing in the United States and in English newspapers. I think *we* must be content to be anatomists, zoologists, geologists, electricians, engineers, mathematicians, naturalists (the last is a term I like and wish to use). We cannot find a name to join all the followers of different branches of science, but I think that 'rationalist' rightly designates those who, in their inquiries into the unknown in whatever field, strictly follow the scientific method.[123]

The Scottish biologist D'Arcy W. Thompson, a pioneer of mathematical biology and the author of the famous work *On Growth and Form*, largely agreed with Lankester. 'Most men of science would surely rather be called so than be dubbed scientist', he wrote, commenting that 'The widely used term "Christian Scientist" has helped to make matters worse.'[124] Armstrong likened scientist to words such as theosophist, telepathist, and thaumaturgist (a performer of miracles). He suggested *sciencer* as a euphonious alternative, but the word never caught on. Yet another of the critics found scientist objectionable on formal and euphonious grounds: 'It has a destructive effect in a sentence, and when spoken the last syllables must be gobbled. "Naturalist" may be gobbled fairly easily; few people notice it; but "scientist" is difficult.'[125]

---

[120] Baldwin (2015), pp. 6–9. Ross (1991), pp. 33–5.

[121] Campbell (1924). Contributors to the discussion in *Nature* 1924–5 included H. Wildon Carr, H. W. Fowler, J. H. Fowler, D'Arcy W. Thompson, W. J. Sedgefield, O. Lodge, R. W. Chambers, I. Gollancz, E. Ray Lankester, C. Allbutt, H. Dingle, R. Paget, R. Fessenden, and J. W. Williamson.

[122] *Nature* **115** (1925): 85.

[123] *Nature* **114** (1924): 823.

[124] *Nature* **114** (1924): 824. So-called Christian Science denoted the religious belief system of the Church of Christ founded in the United States in the late nineteenth century.

[125] *Nature* **115** (1925): 50.

There were several reasons for the discomfort with the term scientist. Some felt that it was an Americanism and others that it was a clumsy hybrid of Latin and Greek. However, most of the discomfort related to the social and cultural implications of the word and not to its form. It took more than a decade before scientist became the accepted British term for a person doing scientific research. Nonetheless, the word did catch on. The frequency in books in ppm (parts per million words) was only 0.13 in 1860, but it increased drastically to 9.6 in 1910, to 28 in 1940, and to 43 in 1970 (*OED*). Thus, it belongs to the 2,000 most common words in modern written English.

Whewell not only introduced scientist but also the word *physicist* for 'a cultivator of physics'. As he noted, physician was already occupied as a name for a medical doctor. Whereas Faraday approved of scientist, for linguistic reasons he objected to physicist. In a letter to Whewell of 20 May 1840, he expressed his dislike: '*Physicist* is both to my mouth and ears so awkward that I think I shall never be able to use it. The equivalent of three separate sounds of *i* in one word is too much.'[126] William Thomson alias Lord Kelvin sided with Faraday. Referring to the definition in Johnson's *Dictionary* of naturalist as 'a person well versed in Natural Philosophy', he asserted:

> Armed with his authority, chemists, electricians, astronomers and mathematicians may surely claim to be admitted along with merely descriptive investigators of nature to the honourable and convenient title of Naturalists, and refuse to accept so un-English, unpleasing, and meaningless a variation from old usage as *physicist*.[127]

Although the word 'technologist' for a person specializing in technology existed at Whewell's time, it was rarely used and seems to have been uncontroversial. *Technology* is rooted in the Greek word *techne* (τέχνη) commonly transliterated as 'art' or 'craft'. According to the ancients it and the word *technologia* (which can be found in Aristotle) denoted not only the activity of producing something non-natural but also a particular kind of knowledge. In something like its modern meaning technology was first defined in late-eighteenth-century Germany, where it principally referred to the academic or scientific study of arts and crafts. For Johann Beckmann, a German economist and chemist who is credited for coining 'technology' in 1780, it was a branch of knowledge dealing with various kinds of arts and applied sciences. Although the term was well known in the nineteenth century it only became common in the post-World War II period.

The related word 'technician' for a person doing practical work in the laboratory, for example, came into common use only around the mid-twentieth century. It was earlier used in a broader sense and sometimes with the implication that the technician lacked creativity ('just a technician'). Whether called technicians or not, assistants and operators performing manual tasks in the laboratory have always been an integral if often overlooked part of experimental science.[128]

---

[126] Ross (1961), p. 216.
[127] Ross (1962), p. 73.
[128] Shapin (1989).

1.5 Lost sciences

Both physicist and physician are derived from the Greek word *physis* (φύσις) commonly translated as 'nature'. Originally *physis* referred to both the practice of medicine and the study of nature: the practitioners of the first art were the physicians, those of the latter the physicists. According to Walter Flood's glossary of scientific words from 1960:

> The Greek word *physikos* ... means 'natural, pertaining to nature, produced by nature'. It has given rise to a range of English words with rather divergent meanings. The adjective *physical* means natural (as in physical geography), material (as in physical properties) and bodily (as in physical exercise). A *physician* (who is skilled in the medical arts) is really one who has a knowledge of nature. *Physics* is that branch of science which deals with the general properties of matter and with the various forms of energy.[129]

In a manuscript on the classification of the sciences from 1872, Maxwell pointed out that although the Greek 'physical' was an exact equivalent of the Latin 'natural', in the course of time the meaning of the two words had separated:

> Natural science is now understood to refer to the study of organized bodies and their development while Physical Science investigates those phenomena primarily which are observed in things without life. ... What is commonly called Physical Science occupies a position between the abstract sciences of arithmetic algebra and geometry and the morphological and biological sciences.[130]

The word *physis* is also the root of *physiology*, a term which originally referred to natural philosophy in general and carried many of the same connotations as physics.[131] John Harris' *Lexicon Technicum* from 1704 explained: 'PHYSIOLOGY, *Physicks*, or Natural Philosophy, is the Science of Natural Bodies, and their various *Affections*, *Motions*, and *Operations*.'

Some thirty years later the lexicographer Benjamin Martin published his *Philosophical Grammar* subtitled *Experimented Physiology or Natural Philosophy*. Martin equated 'physicks' with 'physiology'. In the seventeenth century a physiologist did not necessarily study the organs of living bodies. The related word *physic* (or more often *physick*) was widely used from the Middle Ages well into the Enlightenment to denote the art of healing also known as medicine or the practitioner of this art, that is, a medical doctor. According to Blount's *Glossographia Anglicana*, 'physick' meant 'the Art of curing Diseases, or Medicines prepared for that purpose'.[132] In 1540 Henry VIII founded the Regius Professorship of Physic at Cambridge University, a chair which still exists under this archaic name. Its occupants have always been physicians, never physicists.

---

[129] Flood (1960), p. 145.
[130] Maxwell (1995), p. 776.
[131] Welch (2009).
[132] Harris (1704). Martin (1735). Blount (1707).

The contemporaneous tradition known as *natural magic* was another branch of renaissance natural philosophy which does not exist any longer. To a modern reader the term may appear to be an oxymoron, for is not magic most unnatural and unscientific? It is, but only because we associate the word with a meaning different from that of the past. While magic as ordinarily understood is indeed foreign to science, in the late Renaissance and early modern period it might signify an alternative way to obtain knowledge of the secrets of nature and to do so, at least in part, by means of observations and experiments.[133] Many historians of science argue that this occult tradition played a major role in the development of the experimental method and that leaders of the scientific revolution such as Kepler, Gassendi, Boyle, and even the great Newton, were influenced by it. They were not only scientists but also *magi* (not magicians!) who to some degree shared the aims and methods of natural magic.

The Italian natural philosopher Giambattista della Porta wrote in 1558 a book in Latin called *Magia Naturalis*, which over the next century came in several extended editions and in 1658 was translated into English as *Natural Magick*. As indicated by its subtitle, it was more about physics and other empirical sciences than about magic as currently understood: 'Wherein are Set Forth all the Riches and Delights of the Natural Sciences'. The translation of *Magia* as 'magic' (or *magick* in old English spelling) can give the false impression that it is about magic as we understand the word today. The term is better translated as 'mastery', as the goal of Porta and his fellow *magi* was to comprehend and control the occult qualities of nature in order to make use of them. Among the topics in Porta's work were magnetism, distillation, transmutation of metals, pneumatic experiments, and—not unimportant—how to beautify women. His chapters on physical experiments dealt mostly with topics that according to our standards belong to medicine and pharmacy.

Much later the term 'natural magic' was revived in Victorian Britain, but then in the meaning of entertaining and surprising experiments for the public. A little earlier, in 1832, the eminent physicist David Brewster published the first edition of a popular book with the title *Letters on Natural Magic*.[134] It principally dealt with optical illusions and instruments, acoustics, machines to produce sounds and speech, mechanical automata, and wonders of the chemical laboratory. Seven years later, Henry Talbot, a pioneer in photography, described his new art—what he called 'photogenic drawing'—as a marvellous example of natural magic. 'It is a little bit of magic realized:—of natural magic', he wrote in the *Literary Gazette*.[135]

Several of the names of the chief scientific disciplines that we recognize today can be found centuries ago, such as is the case with chemistry, astronomy, mathematics, anatomy, and botany. Examples of sciences named more recently, for about two hundred years ago, include biology and geology (Section 1.6). Genetics as a name dates from 1907 and the latecomer molecular biology is from 1938.[136] As far as it concerns the group of older names for scientific disciplines, it would be a grave mistake to believe that because a name has survived for long, the meaning of it has remained

---

[133] Henry (2000).

[134] Available online as https://www.gutenberg.org/ebooks/51645.

[135] Nickel (2002).

[136] On the baptism of these and other disciplinary terms in twentieth-century biology, see Powell et al. (2007).

essentially unchanged. Although Aristotle wrote treatises with the titles physics and meteorology, in Latin called *Physica* and *Meteorologica*, they had little in common with the sciences as known later on. The term *meteorology* directly derived from Aristotle's classic is worth a digression.

From a modern point of view 'meteorology' may seem to be an embarrassing misnomer insofar that it apparently means the study of meteors (meteor-ology), something that meteorologists do not occupy themselves with but are happy to leave to the astronomers. The word 'meteor' derives from the Greek *meteoros* (μετέωρος) with the vague meaning of anything high in the air or something elevated. According to Aristotle, meteorology was a discourse about the elevated natural phenomena—the 'meteors'—in the sublunary world. They included all kinds of transient atmospheric phenomena, not only what we call meteors but also comets, the Milky Way, lightning, the aurora borealis, and more. Importantly, meteorology in the Aristotelian sense was *not* concerned with the weather.[137] The close connection between meteorology and weather, so obvious today, was a result of a reconceptualization of meteorology dating from about 1800.

The meteors of the past—up to the late eighteenth century—were typically considered products of terrestrial vapours interacting with the higher strata of air. There were different kinds of meteors and only one of them corresponded to the later concept of a meteor or shooting star. According to a dictionary of 1704:

METEORS ... appear high in the Air, and they are either Fiery, Airy, or Watery. *Fiery Meteors*, are such as consist of a Fat, Sulphuerous kindled Smoak, whereof there are several kinds such as *Ignis Fatuus, Trabs, Ignis Pyramidalis, Draco Volans, Capra Saltans, Thunder* and *Lightning*, etc. *Airy Meteors*, are such as consists of Flatuous and Spirituous Exhalations, as *Winds* etc. *Watery Meteors*, consist of Vapours, or Watry Particles.[138]

At about the same time Thomas Blount used the term *meteoroscopy* to denote 'the part of Astronomy that treats of sublime heavenly Bodies, distance of Stars, &c'. He summarily defined meteorology as 'a Discourse of Meteors' such as did Samuel Johnson in his later dictionary. Only with the later limitation of 'meteors' to denote shooting-star objects, and the recognition that these come from outside the atmosphere, did the term meteorology become a misnomer. *Aerology* might seem to be a better term, but it never replaced meteorology and to the limited extent it was used it was for the study of the atmosphere as a whole. In about this sense it appeared in Martin's *Philosophical Grammar*, a scientific dictionary from 1735 written in the form of a dialogue.

Martin defined aerology as the study of 'the Nature of the *Atmosphere*, or Region of *Air*, and all the *Phænomena* thereto belonging'.[139] 'Meteorography' was one branch of aerology and so was 'anemography', which 'signifies a philosophical *Description* of *Winds* in general'. The Greek word *anemos* (άνεμος) for the winds appears in the still

---

[137] Janković (2000), pp. 16–22, a history of meteorology, airs, and weather in British history until the early nineteenth century.

[138] Harris (1704). See also Blount (1707) and Johnson (1755). For the history of the study of meteors and meteorites, what is called meteoritics, see Burke (1986).

[139] Martin (1735), available as a Google Book.

commonly used 'anemometer', an instrument that measures the speed and direction of wind and goes back to about 1600 (it also appears in the flowering plant anemone or 'windflower'). With a word construction of his own, a protologism, Martin further included among the atmospheric studies what he called *phantasmatography*, a field which 'treats of the *celestial Appearances*, or such *Phænomena*, as exist only in Vision, and not Corporally'. What he had in mind were atmospheric apparitions such as rainbows, halos, and parhelia (also known as mock suns). These and other halo-like phenomena are caused by diffraction by ice crystals in the upper atmosphere, an explanation which only dates from the mid-nineteenth century.

Today, meteorology is often defined simply as the study of the *atmosphere*, a term that is compounded by two Greek words, atmos (ἀτμός, vapour) and sphere (σφαίρα). It may be thought to be classical but is not, as the origin of the term is to be found in The Netherlands in the late Renaissance and not in ancient Greece. *Atmosphaera* was coined as a neologism in 1608 by Willebrord Snellius (better known for Snell's law of refraction), who needed a new name for what his compatriot Simon Stevin had called *dampcloot* in Dutch.[140] Snellius' name soon became popular, appearing first in English (as 'Atmo-sphæra') in John Wilkins' Galileo-inspired *The Discovery of a World in the Moone* from 1638. Many of the first natural philosophers who adopted the term 'atmosphere' used it in the context of the new anti-Aristotelian experimental science that rejected earlier scholastic traditions. Gassendi and Boyle were among those who promoted the word and suggested how the concept of air related to that of the atmosphere. Harris' *Lexicon Technicum* from 1704 included a long entry on Boyle's experiments on airs, defining the atmosphere as 'the lower Part of the Region of the Air or Ether, with which our Earth is encompassed all round; and up into which the Vapours are carried, either by Reflection from the Sun's heat, or by being forced up by the Subterranean Fire.'

As there are old names for branches of science that still exist, so there are names and corresponding concepts that became obsolete and have disappeared from the scientific vocabulary (such as did Martin's phantasmatography). Modern physics consists of a confusingly large number of subdisciplines, but among them are no longer 'mosaic physics' and 'cosmical physics', to mention but two examples of what may be called lost sciences.

Natural philosophers in the first half of the seventeenth century were eager to develop a new and pious physics as an alternative to the established one based upon the ideas of the heathen Aristotle. The result was what some of them called *mosaic physics* or philosophy, a physical and cosmological world view based on a literal reading of the Old Testament and Genesis in particular.[141] Among the mosaic physicists was the Norwegian-Danish astronomer and theologian Cort Aslaksen, a pupil of Tycho Brahe and the author of *Physica et Ethica Mosaica* from 1613. Another was the Czech philosopher John Amos Comenius better known as a pioneer pedagogue. Mosaic physics had almost nothing in common with what came to count as physics and yet, for half a century or so it was highly regarded and thought to be

---

[140] See Martin (2015) for the origin of the name atmosphere and its role in early modern natural philosophy.
[141] Blair (2000).

consonant with the essence of scientific physics alias Christian philosophy. By the end of the seventeenth century, the project of mosaic physics had been largely abandoned except that parts of it lived on in the form of so-called *natural theology*—the influential belief that the study of nature can discern some of God's attributes.

The once so popular discipline or research area called *cosmical physics* is of a much later date and has in common with mosaic physics only that it no longer exists. Or rather, although the individual sciences of which it was a conglomerate still exist, cosmical physics as an umbrella discipline does not. On 27 December 1904, *The Times* described this new branch of science as promising a bright future:

'Cosmical Physics' is a department of science which is rapidly growing in favour. As a subsection of the British Association, for instance, it is scarcely three years old. ... Once born it has justified its existence by growing rapidly and healthily, and has apparently fascinated the readers of papers and addresses in the main section.

What for a few decades was known as cosmical physics (in German *kosmische Physik*) was an ambitious interdisciplinary science that flourished in the fin-de-siècle period from about 1890 to 1915. In brief, it was an attempt at synthesizing and coordinating those parts of astronomy and the earth sciences that could be understood from the perspective of laboratory physics and the laws of nature. The central research problems of cosmical physics were typically taken to be terrestrial phenomena in relation to the solar system, including geomagnetic storms, the aurora borealis, meteorites, oceanography, and atmospheric electricity. Cosmology and aspects of astrophysics might also be included, such as suggested by a few cosmical physicists.[142]

During the first decade of the twentieth century, cosmical physics obtained semi-official recognition as a scientific category, if not as a proper discipline.[143] For example, it entered the classification systems of leading German abstract journals as a subdiscipline of physics. In 1902 the British Association for the Advancement of Science decided to establish a new subsection on 'Astronomy and Cosmical Physics' under its Section A dealing with the mathematical and physical sciences. As chairman of the subsection served the distinguished physicist Arthur Schuster, who at the time specialized in physical meteorology. Other leading cosmical physicists were the Austrian meteorologist Wilhelm Trabert, the Norwegian physicist Kristian Birkeland, and the Swedish physical chemist and Nobel laureate Svante Arrhenius.

Arrhenius and Trabert both wrote massive textbooks with the same title, *Lehrbuch der kosmischen Physik*, the first in 1903 and the latter in 1911. Yet another textbook with the same title was first published by Johann Müller in 1856 with several new editions, the latest in 1894. The famous innovator of the earth sciences Alfred Wegener may also be counted among the loose group of cosmical physicists, at least for a while. In an undated letter to his father-in-law, the meteorologist Wladimir Köppen, he wrote:

---

[142] Details in Kragh (2013).
[143] On the criteria of discipline formation, see Good (2000).

Recently I have often wondered whether 'cosmical physics' might have more justification as a university subject than such a specialized subject as meteorology.... It would be necessary that cosmical physics is represented at each university where physics is taught, and that cosmical physics is obligatory at the physics exams for high school teachers. Should thoughts like these get across, then the huge public of physics students would be won for the cause.[144]

Despite its promising start, cosmical physics remained a fragile construction with almost no institutional setting. It barely survived World War I and then only in a weak state that included only a single university chair, namely one occupied by Felix Exner at the University of Innsbruck. When the International Union for Geodesy and Geophysics was founded in 1919 it comprised six earth sciences including meteorology, seismology, and terrestrial magnetism and electricity, but not cosmical physics which was also not part of other of the unions of the newly established International Research Council.

Although cosmical physics disappeared from the landscape of science at about 1920, the term lived on for a while, but then associated with meanings different from those of the original project. For example, in 1926 James Jeans wrote on 'Recent Developments of Cosmical Physics' in *Nature*, and in 1929 Lord Rayleigh (Robert John Strutt) discussed 'Problems of Cosmical Physics' in an article in *Science*. In both cases they conceived the term as approximately equivalent to astrophysics. They did not refer to the older tradition in cosmical physics with its aim of presenting a synthetic perspective on the earth sciences and their relations to solar and planetary physics.

Arrhenius' dedicated occupation with cosmical physics was in part inspired by a slightly earlier work on the effect of changes in the amount of atmospheric carbon dioxide on the Earth's climate. In a now-classic paper in *Philosophical Magazine* of 1896 he developed a model that causally connected variations of carbon dioxide to climate changes and to the Ice Ages in particular.[145] He spoke of the mechanism by using 'hothouse' as a metaphor but only used the term once and then with a reference to an earlier work by the French physicist Joseph Fourier. Arrhenius did not employ the terms 'greenhouse effect' and 'greenhouse gas(es)' with which his theory came to be identified. It has been known for a long time that 'greenhouse effect' is in fact a misnomer, but it has not prevented the popularity of the term. A greenhouse gets warm because the warm air cannot escape it, but this is not the way that carbon dioxide and other greenhouse gases cause the temperature of the Earth to rise.

If Arrhenius did not coin the term 'greenhouse effect', who did? The best offer seems to be the British theoretical physicist John Henry Poynting, who in a paper of 1907 investigated the temperatures of the planets: 'The "greenhouse effect" may perhaps be understood more easily if we first consider the case of a greenhouse with horizontal roof of extent so large compared to its height above the ground that the effect of the edges may be neglected.'[146] Poynting's work was general and theoretical

---

[144] Wegener (1960), p. 72. Greene (2015), pp. 54–9.
[145] Crawford (1996), pp. 145–56.
[146] Poynting (1907), p. 750.

in nature and not related to the Earth in particular. He did not cite Arrhenius or other scientists working in the tradition of cosmical physics. Nor did Arrhenius cite Poynting's paper. When Poynting is known today it is mainly because of the eponymous 'Poynting vector', referring to the direction and magnitude of electromagnetic energy flux. The terms greenhouse effect and greenhouse gas only became widely used in the 1960s and at first in astronomical contexts such as in discussions of the Venus atmosphere.

While chemists and other scientists often refer to simple chemical substances by their formulae—e.g. O_3 instead of ozone and HCl instead of hydrogen chloride—such usage is practically unknown in everyday language. There is one notable exception, though, namely the climate sinner number one, carbon dioxide. In today's conversation and mass media the compound often appears as just CO_2 or more commonly CO2 (pronounced see-oh-too) rather than carbon dioxide. Although CO_2 is a symbol and not a word, it is used as a word, a noun with no plural form. *OED* does not, as a rule, include chemical formulae, but in 2021 it made an exception for CO_2 because of its prominent role in discussions about climate change. In modern English the word CO_2 appears as often as carbon dioxide, namely with a frequency of about seven words per million words. No other chemical compound has changed from symbol to word in a similar manner.[147] No one, not even chemists, would say about a liqueur that it contains 43% of C_2H_5OH (see-too-aytch-five-oh-aytch). So much for carbon dioxide.

Phrenology is another and very different example of a lost science or rather one that for a time aspired to be a science, and for a period was widely if far from universally accepted as one, but then lost its scientific status. Like alchemy and astrology, it was degraded from a science to a pseudoscience or perhaps, in this case, a proto-neuroscience. The root of the term phrenology is the Greek word *phren* (φρήν) meaning 'mind' or 'soul' and it can thus be translated as the science of the mind. The root word can also be found in the adjective *schizophrenic* (a split mind) and is related to 'frenzy' and 'frenetic' in the sense of 'manic', say in the expression 'he was frenetically active'.

Pseudo (ψεύδω) means 'false' in Greek. With regard to *pseudoscience* (or the hyphenated pseudo-science), the word only entered the English language in the late eighteenth century, when we meet authors characterizing alchemy, for example, with this derogatory term. *OED* refers to an example from 1796. About half a century later, *St. James's Magazine* included an article on 'The Pseudo-Sciences', indicating that the plural version of the term was well known at the time. The author commented on the name:

> The term pseudo-science is hybrid, and therefore objectionable. *Pseudognosy* would be better etymology, but the unlearned might be apt to association with it the idea of a *dog's nose*, and thus, instead of taking 'the eel of science by the tail', take the cur of science by the snout; so that all things considered we had better adopt the current term *pseudo-sciences*. A *pseudo*, false, or bastard science, may be defined a science the object of which is chimerical, and therefore unattainable.[148]

---

[147] *OED* also includes H_2O (aytch-too-oh, water) as a word, but mentions that it is rarely used and then typically in humorous contexts.

[148] 'The pseudo-sciences', *St. James's Magazine*, February 1842, pp. 162–71.

As examples of pseudosciences, the article mentioned the usual suspects alchemy, astrology, and homoeopathy, but not phrenology. Nothing more was heard of the term 'pseudognosy', which is derived from Greek with the last part *gnosy* meaning knowledge or investigation. Among the words with the same termination are prognosis, diagnosis, and geognosy, the latter an obsolete name for geology well known in the mid nineteenth century (see Section 1.6). In late 1953, the famous chemist Irving Langmuir, a Nobel laureate of 1932, coined a new and largely synonymous word for pseudoscience. Without referring to this name, he spoke of *pathological science*.[149] There is no doubt that according to Langmuir's criteria, phrenology belonged to the class of pathological sciences.

In a nutshell, phrenologists in the early nineteenth century studied the contours on the skull based on the hypothesis that they indicate a person's traits of characters and mental faculties. By carefully observing and mapping an individual's skull they obtained data which they ascribed to a number of 'organs' in the brain. Each of the different organs was ascribed a special mental function. The characteristics of the organs, and their size in particular, supposedly determined the psychological constitution of the individual. Thus, in contrast to the traditional introspective approach phrenology was an empirically based method which aimed at explaining psychology in terms of physiognomic studies.

It should be added that at the time phrenology was introduced, the term *psychology* was still relatively new and more popular on the Continent than in Great Britain. Although derived from the Greek word *psyche* (ψυχή) for soul or spirit, it is a late Renaissance invention not known in ancient Greece. The Latin *psychologia* first appeared in print in the sixteenth century as the philosophy or science of the human soul and during the Enlightenment it gained wide currency in Germany and France. In Britain, on the other hand, the word and its content remained controversial well into the nineteenth century. The influential philosopher Herbert Spencer's textbook *Principles of Psychology* from 1855 was instrumental in making the term acceptable. As a later psychologist put it: 'The battle for the term "psychology" was won, even though to the average Englishman it for long retained an "exotic" flavour, and to the psychologists themselves its definition was destined to be the cause of many controversies in the future.'[150]

Although phrenology goes back to the Austrian physician Franz Joseph Gall in about 1800, he did not use the term but instead spoke of his new science as *organology* or *craniology*, sometimes *cranioscopy*. It was his student and assistant Johann Spurzheim who gave the name 'phrenology' to the new science and successfully promoted and popularized it. Scientists and philosophers of a materialist inclination were fascinated by the possibility that mental processes might one day be explained on a purely biological basis. Although phrenology apparently constituted a more scientific and experimental approach to the study of the mind, it remained controversial in the medical profession. Nonetheless, the study of the skull became very popular and was in the early Victorian period widely considered a respectable science institutionalized in, for example, the Phrenological Society of Edinburgh whose members

---

[149] Langmuir (1989).
[150] Quoted in Lapointe (1970).

included many prominent intellectuals and medical doctors. Like other fields of science, phrenology had its own journals, such as *Phrenological Journal and Miscellany* founded in 1823 and *American Phrenological Journal* from 1838. However, by the 1840s the Gall–Spurzheim theory was increasingly discredited and within a decade or so it came to be regarded as nothing but a pseudoscience.

1.6 From the history of the earth sciences

Nicolaus Steno in Denmark and Robert Hooke in England are often credited as the founding fathers of scientific *geology*, and yet none of them ever used the term. Despite being Greek (discourse about the Earth) it was unknown in the Greek–Roman period and only coined in the mid-seventeenth century.[151] Early modern natural philosophers delighted in inventing new terms based on composite Greek words and generally produced a large number of scientifically relevant neologisms. One of them was *geologia*.

The first occurrence in print of geology in something like its modern sense, that is, the scientific study of the Earth, appeared in a book from 1657 written by the Norwegian priest Mikkel Pedersen Escholt with the title *Geologia Norvegica* (Figure 1.4). The book is remarkable not only by including the term geology but also by being the first scientific work ever published in Norway, a country which at the time was part of Denmark. Despite its Latin title, it was written in Danish and only became known outside Scandinavia when it appeared with the same title in an English translation of 1663. Although it introduced geology as a term in natural philosophy, the term only appeared once, namely in the title, and not in the text. Besides, although the author described the mountainous Norwegian nature, the book was essentially in the popular physico-theological tradition and not scientific in a modern sense. Its lengthy subtitle referred to the 'Physical, Historical, and Theological Grounds and Reasons Concerning the Causes and Significations of Earthquakes'.

At the end of the seventeenth century there appeared several other books with *Geologia* as the title. One of them, written by the English clergyman Erasmus Warren in 1690, shared with Escholt's book that the term was found only on the title page. Strictly speaking, the word geology appeared in print some two centuries earlier, but then in a sense entirely different from how it was later used, namely for the non-theological study of law. In a book called *Philobiblon* completed about 1345 and published as early as 1473, Richard de Bury, Bishop of Durham, wrote that this study 'we may call by a special term *Geologia* or the earthly science [which is not] to be properly numbered among the sciences'.[152] The word only appeared once.

It may be relevant at this stage to briefly compare the appearance of the word geology with that of 'biology', which—perhaps surprisingly—arrived at the scene

---

[151] Dean (1979). See also Howarth (2020).
[152] *Philobiblon* is available online as https://www.philobiblon.com/philobiblon.shtml. Adams (1954), p. 166.

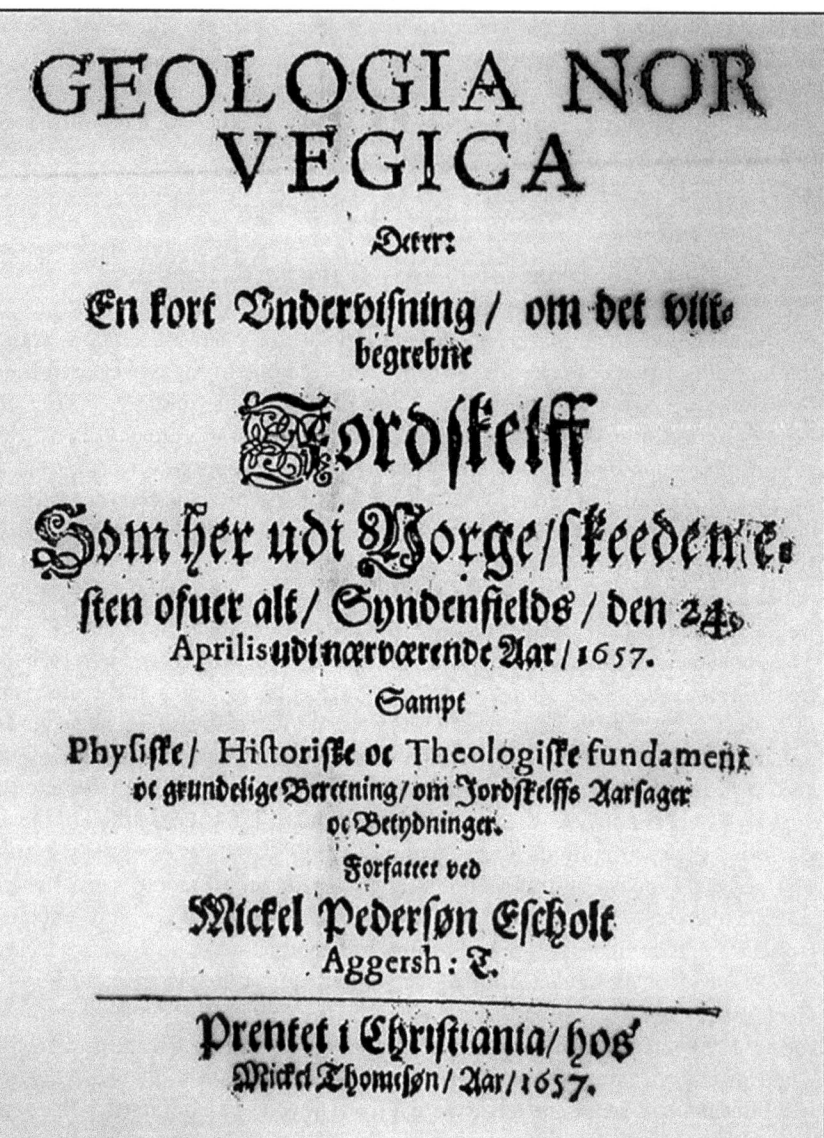

Figure 1.4 Escholt's *Geologia Norvegica* printed in Christiania (Oslo) 1657, the first publication with the term 'geology' as the science of the Earth.

more than a century later. It is generally agreed that *biology* as the study of living organisms and life in general dates from about 1800 when it was included in the German physiologist Gottfried Treviranus' *Biologie oder Philosophie der lebenden Natur* (Biology, or the Philosophy of the Living Nature) and also in a book by Jean-Baptiste Lamarck, the leading French naturalist. Interestingly, Lamarck used the name in his one and only contribution to geology, a work from 1802 with the title *Hydrogéologie*. It is less well known that biology entered in the subtitle of a book from 1766 by the German author Michael Hanov.[153] This book is one more example of a new scientific term occurring only in a title, and in this case also in the running head of a chapter, and not in the text. Hanov also introduced the word 'bionomy', which did not catch on any better than the later proposal 'geonomy' did. His book seems to have been uninfluential as far as the term biology is concerned.

In his *History of the Inductive Sciences* from 1837, Whewell discussed botany, zoology, physiology, and comparative anatomy without referring to biology. Three years later, in his *Philosophy of the Inductive Sciences*, he made up for the omission, now introducing biology as a generic term for all the life sciences and also referring to 'philosophy of biology'. Charles Darwin was a 'naturalist' and not a 'biologist'. The latter word for a practitioner of biology or natural history only appeared at around 1830, in the same period when physicist was coined (Section 1.3), but for a long time it was little used. Tellingly, the statutes of the Nobel Foundation from 1900 referred to a prize in 'the domain of physiology or medicine' and not to biology. Although many biologists have been awarded the coveted prize, to this date none of the prizes have formally been in biology. It should be mentioned, if only as a curiosity, that 'biologist' can be found earlier with a different connotation. According to the *OED*, the author of a biography, what we call a biographer, was sometimes called a biologist.

In 1877, the French anthropologist Charles Letourneau published a book which in its English translation carried the title *Biology*. After noting that the word biology 'is far from bearing in the scientific vocabulary a completely settled import', he took it to mean 'all the great facts and great laws of life, or nearly what is usually understood by "General Physiology", when this denomination is applied to the two organic kingdoms'.[154] In the nearly 500 pages of Letourneau's book 'biological' appears much more frequently than biology, and biologist does not appear at all. Thomas Huxley gave the same year a lecture in which he too found it necessary to justify the name of the new science and distinguish it from the older 'natural history'. 'All clear thinkers and lovers of consistent nomenclature', he said, 'have [now] substituted for the old confusing name of natural history, which has conveyed so many meanings, the term biology, to denote the whole of the sciences which deal with living things, whether they be animals or whether they be plants'. Huxley's definition of biology was essentially the same as Letourneau's.[155] Still in the early years of the twentieth century, natural history appeared more frequently in books than biology. It took until about 1970 before biologist became more popular than naturalist (Google Ngram).

The following table gives an indication of the popularity or visibility of words describing experts or researchers in various fields of science in the years 1800, 1880,

---

[153] McLaughlin (2002).
[154] Letourneau (1878), p. v. The two organic kingdoms were botany and zoology.
[155] Huxley (1877).

1960, and 2010. The numbers adopted from Ngram (*OED*) are the frequencies per million words in the Google Book corpus.

	1800	1880	1960	2010
scientist	—	3.7	42	36
researcher	—	—	8.4	45
engineer	6.1	20	35	23
astronomer	5.1	5.6	4.1	3.6
chemist	6.1	6.9	6.7	3.7
physicist	—	1.8	6.2	4.8
geologist	1.4	5.3	3.6	1.9
naturalist	7.3	7.5	3.5	2.9
biologist	—	0.8	3.3	3.7
physiologist	1.4	2.7	1.7	0.8
sociologist	—	0.4	7.0	5.5
meteorologist	—	0.3	0.6	0.4
psychologist	—	1.1	15	13

Mineralogy, the science of minerals, is in its hybrid Latin–Greek form 'mineralogia' a little older than geology (geologica). The term 'mineral' is from the medieval Latin *minerale* and refers directly to what is taken out from a mine. It was in use well before mineralogia first appeared in print in 1636 in a voluminous compilation with this title. The author was Bernardo Cesi, an Italian Jesuit scholar who at the time had passed away.[156] Cesi stated that he coined the word *mineralogia* because he found it to be pleasant and useful. However, the 626-page-long book was in the older Aristotelian tradition and made no impact on the new generation of natural scientists. The British physician John Webster published in 1671 a book titled *Metallographia* in which he briefly and critically referred to Cesi's tome on 'minerologie' (as he called it) and thereby introduced the word in English.

Curiously, the term 'mineralogist' turned up in English at an earlier date, namely in the polymath and writer Thomas Browne's *Pseudodoxia Epidemica* from 1646, where it appears twice together with a number of other neologisms including 'electricity', 'ambidextrous', 'hallucination', 'locomotion', and 'computer'.[157] Like so many other works in the period, the short title of Browne's book was in Latin but the longer title and the entire text was in the vernacular.

It took longer for mineralogy to take on than what was the case with geology. This only happened when the Swedish chemist Johan Gottschalk Wallerius in 1747 published his *Mineralogia eller Mineralriket* (Mineralogy or the Mineral Kingdom) which three years later was translated into German and from this edition into French in 1753. In all likelihood, Wallerius was unacquainted with Cesi's earlier work and thus reinvented the term. While Ceci's mineralogia was a protologism, Wallerius' was a neologism which was quickly adopted by the Enlightenment scientific community.

---

[156] Mottana (2017).
[157] Online edition: http://penelope.uchicago.edu/pseudodoxia/pseudodoxia.shtml.

The word appeared in Samuel Johnson's *Dictionary* from 1755 as 'The doctrine of minerals', and 'mineralogist' (with a reference to Browne) as 'One who discourses on minerals'.

Before proceeding with the history of the word geology there is another word that merits attention, namely *fossil*. The German naturalist Georg Bauer better known as Agricola published in 1546 an important work called *On the Nature of Fossils* and about two decades later the Swiss physician, philologist, and polymath Conrad Gessner (or Gesner) completed his book *On Fossil Objects* (both were in Latin). The subtitle of Gessner's work was *A Book on Fossil Objects, Chiefly Stones and Gems, Their Shapes and Appearances*. The point is that during the Renaissance and a long time thereafter the term fossil had a broader and substantially different meaning than it has today. In his *Glossographia Anglicana*, Blount said that 'all Bodies whatever that are dug out of the Earth are by Naturalists commonly called by the general Name of Fossils'.

The word is based on the Latin *fossilis*, which means anything dug up from the Earth whether metals, minerals, or stones of organic origin. Thus, in the early classification fossils belonged to metallurgy, mineralogy, petrology, and palaeontology, and not exclusively to the latter science. As the historian of science Martin Rudwick points out in a pioneering study of the history of palaeontology: 'The word "fossil" has changed its meaning radically since Gesner's day. ... This change in the meaning of the word "fossil" is far more than a trivial point of etymology: it is a clue to the first major problem in the history of palaeontology.'[158]

The Renaissance meaning of fossil was common still in the late eighteenth century when the term did not only refer to petrified organic remains. There are reminiscences of the old tradition even in today's language. When we speak of 'fossil fuels' we refer to coal, oil, and natural gas, substances which have nothing in common with the petrified plants and animals found in stones. We may also speak more figuratively and often in a derogatory sense of an old person as a fossil, or the term may be used for anyone who fails to absorb new ideas. Yet another connotation is to characterize any physical body or signal from the far past that has survived to the present as a fossil. For example, cosmologists often refer to the background radiation formed billions of years ago as a fossil of the Big Bang (Section 6.5).

The word geology only spread slowly in Great Britain and even more slowly in Germany and most other parts of continental Europe. The slow acceptance may be illustrated by the Swiss naturalist Jean-André de Luc, who in a work of 1778 mentioned the word only to reject it because it was so little known: 'By *Cosmologie* I mean here the knowledge of the earth only, and not that of the universe. *Geologie* would, in this sense, have been the proper word, but I am afraid to employ it, because it is not in use.'[159] As indicated by de Luc, 'cosmology' was at the time known as a term for the study of the universe or world, but 'cosmos' could also refer to the planetary system or even to one of its members, the Earth (see also Section 6.4). When Ptolemy's famous *Geographia* was first translated into Latin in 1406, it carried the title *Cosmographia*. In modern parlance 'world' sometimes denote the universe at large but more often

---

[158] Rudwick (1976), p. 1.
[159] Dean (1979), p. 39.

the referent is the Earth, as in 'the world championship in football' or 'the world's richest man'.

In England, geologia became geology in the 1730s. A quarter of a century later, the term appeared in Samuel Johnson's *Dictionary of the English Language* as 'The doctrine of the earth; the knowledge and nature of the earth'.[160] A fossil was simply defined as 'that which is dug out from the earth' including 'fossil or rock salt'. Johnson's dictionary did not include an entry on either biology or physics, and chemistry only appeared under 'chymistry'. The term geology became moderately common some forty years later and it is also in this later period that derivatives such as 'geological' and 'geologist' made their entrance in the English language. James Hutton made frequent use of the two terms in his *Theory of the Earth* from 1795, where geology also appears. With the establishment in 1807 of the Geological Society of London the words received official recognition and soon became popular.

In the German and Scandinavian countries geology met competition from the word 'geognosy' promoted by the influential German mineralogist Abraham Werner since about 1790. Werner's *geognosy* was not identical to geology though, for it differed from the latter science by being restricted to factual knowledge of the Earth's crust such as revealed by analysis of the minerals in rocks. It was not about *geogony*, a term used for speculations concerning the origin of the Earth and which for a period was used intermittently with *cosmogony*. Whereas in English- and French-speaking countries geology came to be used almost exclusively, in German and related languages geognosy was the preferred term, although its exact meaning was a matter of dispute. These national differences reflected different cultural and national styles of science.[161] Werner and the scientists sharing his ideas—the 'geognosts'—did not avoid the term geology, which they typically conceived as an umbrella term for several subdisciplines of which geognosy was one and 'geogony' another.

Yet another word that Werner promoted in his system of the Earth and possibly coined was the not easily pronounceable *oryctognosy* (German: Oryktognosie) with which he referred to the identification and classification of minerals or fossils. For the art of measuring minerals Werner and his followers spoke of *oryctometry*. The first of these terms occurred fairly frequently until the mid-nineteenth century, after which it declined and practically vanished at about 1900. As an American geologist later commented, 'The word is a specimen of polysyllabic cacophony which fortunately gradually disappeared'.[162] That the term was unwelcome also in England in the early nineteenth century is indicated by an anonymous book review in the *Quarterly Journal of Science, Literature, and the Arts* of a popular book on mineralogy:

> Mineralogy and geology are quite sufficient without oryctognosy—a lately introduced and bad term, derived from ορυσσω, *fodio*, and γινωσκω, *nosco*, meaning, if it mean [*sic*] any thing, the science of digging, and consequently as applicable 'to potatoes and carrots', as to metals and stones. We notice this, because it may mislead the young reader, in writing for whom more than common care should be taken not to

---

[160] Johnson (1755). Online as https://johnsonsdictionaryonline.com/
[161] Klemun (2015).
[162] Adams (1954), p. 216.

introduce terms calculated to give false ideas ... which, if etymology has any thing to do with the *application* of a term, is eminently the case here, for by no possible construction can oryctognosy (etymologically) be made to signify 'a knowledge of minerals by their external character'.[163]

Werner and his pupils were much concerned with the terminology and taxonomy of minerals. Incidentally, the term taxonomy was only coined in 1813 when the French botanist Augustin de Candolle based it on Greek words for 'order' (*taxi*) and 'law' (*nomos*). According to Ludwig August Emmerlin, one of Werner's assistants,

> It is necessary in every art or science to have words or names to designate the things of which we treat. They serve particularly to fix the different concrete ideas, to express ourselves intelligibly, and to communicate our thoughts.[164]

Among the rules of nomenclature recommended by Werner's school of mineralogy was that 'The names ought to be taken from one language only', and another was that 'We should write and pronounce the words according to their true signification and etymology ... long names are difficult to pronounce and inconvenient to write and remember'. As far as eponyms were concerned, they should be used 'only when a philosopher has been the first to make known a mineral'. The latter recommendation caused the Irish chemist and mineralogist Richard Chenevix to object: 'With this proviso it seems difficult to determine who may claim the right to the honour of transferring his name to a mineral ... Should the discoverer not happen to be an author or professor, it appears that he has no right to the honour ... The folly and vulgarity of such a system should be evident.'

By the 1820s geology had become a very popular science, with the result that the term was used also as a verb. As gentlemen (and ladies too) could *botanize*, so they could *geologize*. In Darwin's *Voyage of the Beagle* published in 1839, there is an entry from 1 June 1832 in which he writes, 'I took a long ride, in order to geologize some of the surrounding hills'.[165] Some sciences, but far from all of them, can and have been used in this way if with rather different meanings. The oldest may be 'anatomize', a verb appearing in about 1500 in the sense of conducting a dissection or anatomy as it was often called. The 'anatomical theatre' of the late Renaissance was a lecture hall where the professor dissected or anatomized corpses in front of students and visitors. *Anatomy* (ana-tomy) is linguistically related to 'atom' insofar that the first term means 'cutting up' and the second 'cannot be cut' (Section 1.3).

The verb 'astronomize' has been used, but only rarely. To 'mathematize' a theory or a set of empirical data usually means that it is turned into a formal mathematical system. One may say, for example, that Maxwell mathematized Faraday's ideas about electromagnetism or that Ronald Fisher mathematized genetics. Sciences such

---

[163] *Quarterly Journal of Science, Literature, and the Arts*, no. 27 (1823). The reviewed book was *Conversations on Mineralogy* written by Delvalle Lowry and published in 1822.

[164] Quoted from an English translation in Chenevix (1811), which is also the source of the other quotations.

[165] http://darwin-online.org.uk/content/frameset? itemID=F1925&viewtype=text&pageseq=434

as physics, chemistry, and biology do not invite verbs of this kind, although 'biologize' has been used from time to time. In 1906 an American psychologist wrote about the 'biologizing psychologist' and said about a trend in modern psychology 'that—if I may use the expression—it "biologizes" consciousness'.[166] I have never come across 'physicize' and definitely not 'chemistrize'.

With regard to the term anatomy it should be pointed out that it, like so many other words, was used and still is used in a variety of nonscientific meanings. Thus, the term may refer to a piece-by-piece examination of something in general. The English author Robert Burton published in 1621 a book with the title *The Anatomy of Melancholy*. In 1959 the American movie called *Anatomy of a Murder* appeared with James Stewart in the leading role. I once wrote a paper titled 'Anatomy of a Priority Conflict' in which I examined in detail the priority conflicts surrounding the discovery of element 72 in the periodic table (Section 5.4).

According to a broadly accepted view in the eighteenth century, the crust of the Earth was formed by a series of catastrophes in the distant past with the biblical deluge assumed to be the last and most catastrophic. Terms such as 'diluvial' and 'diluvium' widely used in the nineteenth century were derived from the Latin word for deluge or flood. As an alternative some natural philosophers hypothesized an Earth of immense age where the transformations had occurred continuously with a strength corresponding to those presently observed. The controversy between the two rival views of the formation of the Earth's surface—the first called *catastrophism* and the second *uniformitarianism*—constitutes an important and well-researched chapter in the history of geology.[167]

Charles Lyell's famous *Principles of Geology* published in three volumes 1830–3 argued forcefully against catastrophism and for the uniformitarian model of the Earth. The two key terms, catastrophism and uniformitarianism, first appeared in the *Quarterly Review* of 1832, in a critical yet laudatory review of the second volume of Lyell's book written anonymously by William Whewell. It was he and not Lyell who coined the words or, to be precise, Whewell wrote about 'the Uniformitarians and the Catastrophists' who had different answers to the question: 'Have the changes which lead us from one geological state to another been, on a long average, uniform in their intensity, or have they consisted of epochs of paroxysmal and catastrophic action, interspersed between periods of comparative tranquillity?'[168]

As mentioned, Whewell was an erudite wordsmith who suggested several of the technical terms still used in mineralogy, geology, and electrochemistry (for his contributions to the vocabulary of the latter field, see Section 2.2). The *Oxford English Dictionary* has no less than 628 quotations to his work. In his *History of the Inductive Sciences* from 1837, he noted 'how powerful technical phraseology is, for the perpetuation either of truth or error'.[169] Like most of his contemporaries, Whewell believed that in general scientific terms should be in Greek or Latin, for other reasons because such terms 'are readily understood over the whole lettered world'. On the other hand,

---

[166] Bentley (1906). *OED* includes 'biologize' but not 'physicize'.
[167] E.g. Hallam (1989), pp. 30–64.
[168] *Quarterly Review* **47** (1832): 103–32, p. 126.
[169] Whewell (1837), vol. 1, p. 60.

he also admitted the advantage of indigenous terms, which were 'intelligible much more clearly and visibly than those borrowed from any other source'. Likewise, he expressed a relaxed attitude to etymology and philological correctness. 'We are not to be too scrupulous about the etymology of scientific terms', he said.[170]

Although not an original scientist himself, Whewell contributed several scientific papers on a variety of topics including mathematics, geology, mineralogy, and crystallography. He was interested in the study of tides in particular, a subject for which he proposed the term *tidology*.[171] This was not a success, but others of his neologisms were. In his correspondence with Lyell, Whewell expressed much interest in the nomenclature related to the new uniformitarian geology such as the names for the geological epochs. Among the neologisms included in *Principles of Geology* are the still-used *Pleistocene*, *Pliocene*, and *Miocene*, but originally Lyell intended to designate the epochs with the longer and not very elegant termination -*synchronous*, e.g. *Pleiosynchronous*.

The currently accepted -*cene* termination was suggested by Whewell, who in a letter to Lyell of 31 January 1831 wrote: 'The termination *synchronous* seems to me to be long, harsh, and inappropriate.'[172] Whewell first suggested -*neous* as a better ending: 'Do you like this? They are shorter than yours. ... The words meioneous and pleioneous (or whatever termination you take), might be spelled mioneous, plioneous, &c., and would be so according to the Latin and old English spelling, in which the Greek εɩ become *i*.' In a postscript to the letter, Whewell came up with yet another idea:

It has occurred to me that καινός is a better word than νέος, and I propose for your four terms, 1 acene, 2 eocene, 3 miocene, 4 pliocene. The termination *cene* is right, as in epicene; αɩ becoming *e* as well οɩ. For Eocene you might say spaniocene, but I like your *eo* better. Is not this shortest and best?

Whewell continued to think of Lyell's geological names. In a letter of 19 February 1831, he wrote: 'I ... will try if I can hit on anything else to make your nomenclature more tidy. ... You have been very meritorious in making geological words English in their form; I delight in your plesio*saurs*, and pterodac*tyls*.' For the record, the mentioned geological epochs are today defined in terms of their ages from the present: Pleistocene (11,700 years to 2.6 million years), Pliocene (2.6 to 5.3 million years), Miocene (5.3 to 23 million years), and Eocene (34 to 56 million years).

Acknowledging the assistance of Whewell, in the third volume of *Principles of Geology* Lyell added a glossary of geological terms. He had wished, he said, to avoid 'the numerous foreign diphthongs, barbarous terminations, and Latin plurals, which have been so plentifully introduced in late years into our scientific language'. Moreover, he dissociated himself from the unscholarly mode of 'fabricating Greek derivatives and compounds, many of the latter being a bastard offspring of Greek and Latin.'[173]

---

[170] Whewell (1847), vol. 2, pp. 539, 540, and 530. A revised edition of Whewell (1840).
[171] Reidy (1996).
[172] Todhunter, vol. 2 (1876), pp. 110–12.
[173] Lyell (1830–1833), vol. 3, p. 53. Online as http://www.esp.org/books/lyell/principles/facsimile/

At about the same time that Whewell and Lyell discussed geological names, another prominent British natural philosopher and polymath, John Herschel, reflected on the general issue of nomenclature in science:

> The imposition of a name on any subject of contemplation, be it a material object, a phenomenon of nature, or a group of facts and relations, looked upon in a peculiar point of view, is an epoch in its history of great importance. It not only enables us readily to refer to it in conversation or writing, without circumlocution, but, what is of more consequence, it gives it a recognized existence in our own minds, as a matter for separate and peculiar consideration.... Nomenclature, then, is, in itself, undoubtedly an important part of science, as it prevents our being lost in a wilderness of particulars, and involved in inextricable confusion.[174]

Herschel was convinced of the necessity for a rational nomenclature, but he also recognized that any system of nomenclature must be flexible as it depends on the limited state of knowledge at any given time:

> Any one may give an arbitrary name to a thing, merely to be able to talk of it; but, to give a name which shall at once refer it to a place in a system, we must know its properties; and we must *have* a system, large enough, and regular enough, to receive it in a place which belongs to it, and to no other. It appears, therefore, doubtful whether it is desirable, for the essential purposes of science, that extreme refinement in systematic nomenclature should be insisted on. ... All nomenclature must be a balance of difficulties, and a good, short, *unmeaning* name, which has once obtained a footing in usage, is preferable to almost any other.

In Herschel's *Preliminary Discourse on the Study of Natural Philosophy*, from where the quotations are taken, there were numerous references to 'geologist(s)' and 'chemist(s)' but none to 'scientist(s)', 'physicist(s)', or 'biologist(s)'. He readily admitted geology as a genuine science, saying that, 'geology, in the magnitude and sublimity of the objects of which it treats, undoubtedly ranks, in the scale of the sciences, next to astronomy'.

With the rapid growth of geology in the nineteenth century followed the formation of several interdisciplinary research areas combining geology with some other science. Although chemistry had always been important to mineralogy and vice versa, it took until 1838 before the Swiss-German chemist Christian Schönbein coined the term *geochemistry* (Geochemie). But it takes more than a name to establish a new research field and in the case of geochemistry the route was long and difficult.[175] Tellingly, when modern geochemists speak of the 'fathers of geochemistry' they refer to scientists not yet born when Schönbein introduced the term.[176] It was only when geology began to move away from pure natural history that geologists recognized

---

[174] Herschel (1851), pp. 139–40. First edition 1831. Available online: https://www.gutenberg.org/ebooks/54897
[175] Kragh (2000) and references therein.
[176] Fairbridge (1999).

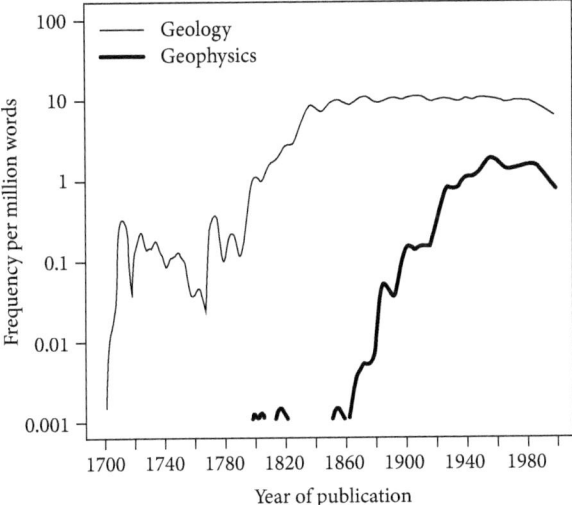

Figure 1.5 Semilogarithmic plot of the frequency of the terms 'geology' and 'geophysics' in English, French, German, Italian, and Spanish ca. 1700–2000. Data from Google Ngram database.

Reproduced from Howarth (2020), *Earth Sciences History* 39: 1–27.

chemists as potential partners. Yet it was a matter of debate which of the two sciences, geology or chemistry, should be the senior partner in the new interdisciplinary branch of science. Should it be 'geological chemistry' or 'chemical geology'? Well into the twentieth century these two terms were more common than 'geochemistry'.

As pointed out by Gregory Good, an American historian of science, the appearance of a name can be a critical indicator of the emergence of a new research field. And yet, the appearance and use of a name 'indicates only a desire to designate a new field, not success in doing so'.[177] This is illustrated by many cases, including those of geochemistry and geophysics. The word *geophysics* (Geophysik in German) was first used in 1834 but only appeared sporadically in print during the next forty years, when more or less equivalent names such as 'terrestrial physics', 'physics of the Earth', and 'physical geography' were more used. Geophysics became better known and soon widely accepted after the German periodical *Beiträge zur Geophysik* (Contributions to Geophysics) began publication in 1887 (Figure 1.5). Like geochemistry, geophysics had to justify its existence in relation to two larger and substantially different scientific disciplines. With the creation of the American Geophysical Union in 1919 geophysics obtained its first institutional basis.

The popularity of the terms can be illustrated by the number of times they appeared in the journal *Nature* between 1869 and 1900:

---

[177] Good (2000).

Geology: 7,291; Physics: 6,415; Chemistry: 7,984; Geophysics: 17; Geochemistry: 8; Chemical Geology: 25; Geological Chemistry: 6; Physical Geology: 220; Geological Physics: 8.

Before proceeding with the geological sciences, I shall briefly refer to the biological counterpart of geophysics, that is, *biophysics*, a name that may have been coined by the versatile English mathematician Karl Pearson. In *The Grammar of Science* published in 1892, Pearson discussed how the life sciences related to the inorganic sciences, and in this context he suggested that a new interdisciplinary science was needed, namely one that brought biology under the wings of physics:

> This branch of science which endeavours to show that the facts of *Biology—of Morphology, Embryology*, and *Physiology*—constitute particular cases of general physical laws has been termed *Ætiology*. It would perhaps be better to call it *Bio-physics*. This science does not appear to have advanced very far at present, but it not improbably has an important future.[178]

In a footnote he added about aetiology: 'From the Greek αἴτιον, a cause. The name does not seem very aptly chosen, especially as it has a very definite meaning of older origin in medical practice.' Indeed, aetiology (or etiology) had for long been used in medicine to denote the study of the cause or causes of diseases. It took several decades before Pearson's neologism became generally accepted as the name for a new biological subdiscipline. In a paper of 1956, the Nobel laureate in physiology Archibald V. Hill compared the scientific status of biophysics with that of biochemistry, noting that the first and younger branch of science had 'only recently acquired a name and personality.'[179] Through most of the twentieth century, the word biochemistry has been about ten times as common in English writing as biophysics (Google Ngram). Also in the research literature biochemistry appears much more frequently than biophysics.

After the acceptance in the 1970s of global plate tectonics as a unified theory of the Earth, the term geology has, to some extent, been replaced by the broader concepts of 'earth science(s)' or 'geoscience(s)'. These terms have now gained wide currency, but none of them are new. For example, in the preface to the 1929 edition of his seminal book on continental drift Alfred Wegener stated: 'Scientists still do not appear to understand sufficiently that all earth sciences must contribute evidence towards unveiling the state of our planet in earlier times.'[180] Wegener thought of his theory as an earth science and not as a new geological theory in the traditional sense. Since the 1970s many institutions and university departments formerly labelled with the terms geology and geological have changed to earth science or geoscience. All modern geologists are earth scientists, but not all earth scientists are geologists.

Wegener is famous for his daring theory of continental drift which he first proposed in 1912 and three years later expounded in greater detail in a monograph with the title *Die Entstehung der Kontinente und Ozeane*. He did not use the term

---

[178] Pearson (1900), p. 528.
[179] Hill (1956).
[180] Wegener (1966), p. vii.

(continental) 'drift' but the German *Verschiebung* meaning displacement. Although Wegener never referred to either 'drift theory' or 'continental drift', in a paper of 1919 he used the German term *Trift* instead of the more cumbersome *Verschiebung*. The *Trift* or drift metaphor alluded to driving or floating of timber and had previously been used for ocean currents. As noted by a later seismologist and science writer: 'It implied a gentleness of movement, an effortless subtlety as possessed by the wisps of smoke or clouds; nothing that was connected with the bulky, clumsy or resistant.'[181]

The second German edition was critically and anonymously reviewed in *Nature* in February 1922 under the heading 'Wegener's Displacement Theory'[182] and when the third edition of the book was translated into English two years later, 'Verschiebung' was correctly translated as 'displacement'. Nonetheless, it was the misleading label continental drift which stuck and still sticks. This non-Wegenarian term may have been coined by the British geologist Philip Lake in a review essay of early 1923, where he used it only in the headline of the essay.[183] The first *OED* quotation to Wegener's theory is from the 1926 edition, where the geology professor John Gregory authorized the expression continental drift. Several other British and American scientists adopted the term in the early 1920s, among them the mathematician and geophysicist Harold Jeffreys who strongly opposed Wegener's hypothesis. In November 1926 an American symposium on 'The Problem of Continental Drift' took place. Wegener did not attend but contributed to the proceedings volume with a paper on his 'displacement theory'. When the paper appeared in print two years later, its title referred to continental drift, the phrase then universally adopted for Wegener's theory.[184]

Although practically dead for more than two decades, Wegener's theory of continental drift was revived in the 1960s when it was incorporated in the more comprehensive theory known as global plate tectonics. By the mid-1960s the crucial idea of 'sea floor spreading', a name due to the American marine geologist Robert Dietz, was generally accepted but the new theory of the Earth still had no name.[185] In a much-cited paper of 1967 Dan McKenzie and Robert Parker proposed the name 'paving stone theory' which did not catch on and was soon forgotten.[186] The label 'global tectonics' dates from 1968 and 'plate tectonics' first appeared in 1969. 'Tectonics' was a term previously used by geologists but the combination with 'global' or 'plate' was new. The term is derived from the Greek *tektōn* (τέκτων) meaning builder or constructor. *Architektōn* is a chief or arch-builder, an architect.

With the rise of plate tectonics new words were coined or reintroduced such as 'transform fault', 'tectonophysics', and 'tectonostratigraphy', the latter a somewhat monstrous word not easier to pronounce than Werner's oryctognosy. In 1964 a new journal called *Tectonophysics* was founded. The rapid growth of plate tectonics required agreement on new terms and concepts, which in 1972 led to the formation

---

[181] Wood (1985), p. 68.
[182] *Nature* **111** (1922): 202–3.
[183] Lake (1923). An abridged version with the same title appeared in the more widely read *Nature* **111** (1923): 226–8. See also Chateau-Smith (2022).
[184] Greene (2015), p. 521. Newman (1995).
[185] Dietz (1961).
[186] McKenzie and Parker (1967).

of a working group under the Commission on Geodynamics on international ter-minology.[187] The term 'tectonophysics' had been used earlier. In 1940 the American Geophysical Union established a Section of Tectonophysics for the purpose of promoting research in the physical structure of the Earth.

In a paper to the American Philosophical Society of 1968, J. Tuzo Wilson, one of the chief architects of plate tectonics, reflected on the recent development of the earth sciences and how they related to Thomas Kuhn's new ideas of paradigms, crises, and revolutions in science. Arguing that continental drift and plate tectonics were indeed revolutions in Kuhn's sense, he ended his paper with an appeal: 'We should start afresh and combine all investigations into the study of a mobile earth-structure under the name of one new science—geonomy.'[188] The term *geonomy* as a terrestrial coun-terpart to astronomy is rarely used, but it goes back long before Wilson. In his 1840 classification of all sciences (Section 1.4), Rafinesque defined geonomy as the 'science of terrestrial bodies, with 2 great branches, *Geognosy*, the Earth itself, *Somognosy*, the bodies it contains.'[189] The word also appeared in the title of a book published in 1858 by the American geologist James Grimes, according to whom geonomy was 'a science which relates to the physical laws of the earth, and includes all the essential facts of geology and physical geography.'[190]

Presumably unaware of Grimes' book, Wilson reinvented the word. A few years later it appeared in a monograph of 1975 titled *A Textbook on Geonomy*, but that was about all.[191] It seems that the word is no longer used. The term *geodynamics* fared better. Although it had popped up on a few occasions in the scientific literature before World War II, it became widely used only in the late 1960s. The British math-ematician Augustus Love published in 1911 an important book with the title *Some Problems in Geodynamics*, but—following the old tradition of Escholt and Warren in the seventeenth century—the term did not appear in the body of the book. 'Geo-dynamics' received official recognition in 1968 and in 1984 the first issue of *Journal of Geodynamics* appeared. Web of Science lists about 11,700 papers referring to the term.

If geophysics and geochemistry, why not geoastronomy? A few scientists in the nineteenth century dreamt of integrating geology and astronomy into a wider mul-tidisciplinary science. One of them was the American geologist Alexander Winchell who in 1883 coined a word for the dream. 'I have attempted', he wrote,

> to take the reader over the system of evidences from which he may thus reason in laying the foundation of a science which, from one point of view, may be styled the geology of the stars; and, from another, the astronomy of the earth. It is the science of Comparative Geology. It is Astrogeology.[192]

---

[187] Dennis and Atwater (1974).

[188] Wilson (1968). Chateau-Smith (2022) argues that an earlier revolution in the earth sciences, Louis Agassiz's theory of glaciation, 'exemplifies a Kuhnian paradigm shift, revealed by a change in vocabulary, with a noticeable increase in new words'.

[189] Rafinesque (1840), p. 6.

[190] Grimes (1858), p. 2.

[191] Jacobs (1975).

[192] Winchell (1883), p. vii.

Winchell's 'astrogeology' was no more successful than Grimes' geonomy, but in contrast to the latter word it reappeared and even became popular a century later.

Modern earth sciences are not only closely related to physics and chemistry but also to biology and astronomy. What is today called 'planetary geology' or sometimes astrogeology are thriving fields of research originating in the 1980s. Planetary geology is a discipline belonging to the broader category called 'planetary science' and is mainly concerned with the geology of the planets and their moons, asteroids, comets, and meteorites. Although the *geo-* prefix typically and etymologically suggests topics relating to the Earth, scientists speak of planetary geology for historical and convenience reasons. Indeed, from a linguistic point of view, this word and also astrogeology are curious and strictly speaking oxymoronic words given that *ge-* in geology refers to the Earth. 'Astrobiology', another popular research area, is in this respect a consistent combination of syllables as the word *astron* (star) does not exclude 'life' (Section 6.3).

Finally, the modern astrophysical research area known as *asteroseismology* is one more example of the connections between the earth sciences, in this case seismology, and the astronomical sciences. For astero- and helioseismology and the names associated with this branch of science, see Section 6.2. At about the same time that asteroseismology developed into a new science, the elusive neutrinos entered geophysics (for the neutrino, see Section 3.4). The nascent field of *neutrino geology* took its beginning in the mid-1980s when physicists realized that information about the inner Earth might be obtained by measuring the neutrinos (actually antineutrinos) emitted by naturally occurring beta-radioactive sources such as K-40, U-238, and Th-232. Neutrino geology first appeared in the scientific literature in a Russian paper of 1984.[193] The first detection of *geoneutrinos*, as they are called, was reported in 2005 and today neutrino geology is a burgeoning subfield of science integrating geophysics, astrophysics, and elementary particle physics. Another and even more recent branch of this subfield is *muography*, a technique using muons from the cosmic rays to generate images of, for example, volcanoes and glaciers. The neologism 'muography', an abbreviation of 'muon radiography', was coined by the Japanese geophysicist Hiroyuki Tanaka in a paper of 2009 analysing the eruption of volcanoes.

Like in other branches of science, in the earth sciences there is and have always been grand but heterodox, and often eccentric, theories that run counter to what is considered mainstream science. The old view of 'mother Earth' as a purposefully designed self-regulating organism has long been discarded and yet it was reintroduced at about the same time that global plate tectonics became widely accepted. According to the English independent scientist James Lovelock, life did not simply adapt to the physical and chemical conditions of the Earth, but these conditions are themselves parts of a larger system where life is no less important for the surface of the Earth than the surface is for life. What is relevant in the present context is Lovelock's name for his controversial hypothesis, which he called *Gaia* in a letter to *Nature* of 1972. He wrote of the Earth as were it itself a living entity, a creature:

---

[193] Askar'yan (1984).

Such a large creature, even if only hypothetical, with the powerful capacity to homeo-
stat the planetary environment needs a name; I am indebted to Mr. William Golding
for suggesting the use of the Greek personification of mother Earth, 'Gaia'.[194]

Lovelock found fashionable metaphors such as 'Spaceship Earth' to be 'misleading
and unnecessary as a replacement for the older concept of the Earth as a very large
living creature, Gaia, several giga-years old who has moulded the surface, the oceans,
and the air to suit her'.[195]

In Greek mythology, the goddess Gaia was the personification of the Earth and
the ancestral mother of all life. The root of the same word appears as *geo* in words
such as geology and geography. Lovelock's coining soon became a household word,
popular as well as controversial. As Lovelock and his adherents used the term, Gaia
denoted an object, the living Earth. The eponymous name was not itself new as it
had been used since antiquity, but its use in geoscience was. As an alternative and
more scientific name for the Gaia hypothesis, Lovelock suggested in 1984 the term
geophysiology, which was a neologism and not merely a protologism as in a proposal
made almost a century earlier by the English geographer Halford Mackinder. Lovelock called attention to Hutton's old theory of the Earth from 1788 which 'likened the
Earth to a superorganism and recommended physiology as the science for its investigation'.[196] He apparently considered the geophysiological Gaia theory to be a modern
revival of Hutton's long forgotten ideas. As indicated by the only thirty references in
Web of Science, the term geophysiology failed to catch on.

[194] Lovelock (1972). The British novelist William Golding, best known for his debut novel *Lord of the Flies* from 1954, was Lovelock's neighbour in Bowerchalke, a village in southern England.

[195] Giga-years = billion (10^9) years. The Spaceship Earth metaphor dates from the mid-1960s and was popularized by the American architect Richard Buckminster Fuller in a book of 1969 with the title *Operating Manual for Spaceship Earth*. For Buckminster Fuller and chemical nomenclature, see Section 5.2.

[196] Lovelock (1989). For Mackinder's use of 'geophysiology' in 1895, see *OED*.

2

Electricity and electromagnetism

The electrical sciences and their many-sided applications are well suited to illustrate how concepts and names, and the relations between them, have changed over time. As new electrical phenomena were discovered and new instruments designed, the scientists involved recognized that a new vocabulary was required. In the early period it resulted in words such as 'Leyden jar', 'galvanic currents', and Volta's 'electric organ' soon to become a 'battery'. Naturally, the terminology reflected what the scientists knew at the time and not our present understanding of the phenomena. From our perspective the chosen words may appear to be confusing and inconsistent, but they made sense at the time. If we read the texts in which Faraday introduced the word 'ion' we will scarcely recognize that it has any connection to our present usage of the term.

With the discovery of electromagnetism, the electrical sciences expanded to many other branches of science, among them chemistry, astronomy, medicine, and the earth sciences. Electromagnetic theory formed the backbone of the all-important electrical technology which thoroughly changed society. In the cases of the terms 'transistor' and 'ion' we are able to follow in detail from primary sources how they were coined and which alternatives were suggested. Eponymies turn up repeatedly in the history of electricity, many of them relating to the units for electrical and magnetic quantities. Units such as ohm, volt, and ampere are well known, but it is not generally known that the decibel unit, often associated with acoustics, is also of electric origin. Nor is it obvious that this unit is a hidden eponym honouring the Scottish-American inventor of the telephone.[1]

The discovery of superconductivity in 1911 was not only of the utmost scientific importance, it also marked the introduction of the *super-* prefix in names of modern physics. Section 2.5 elaborates on this class of names and related science prefixes such as *ultra-*, *hyper-*, and *meta-*. In Section 2.1 two well-known words with background in the electrical sciences are singled out for closer inspection. One of the words is 'plasma' and the other is 'transistor'. While the first word had been used earlier, but in a medical and not a physical context, the second was a genuine neologism.

2.1 Electricity before electromagnetism

Despite some notable work in the seventeenth century by William Gilbert, Otto von Guericke, and others, electricity as a science only took off in the following century.

---

[1] On scientific units and their names, see Jerrard and McNeill (1992).

It has been called the greatest scientific invention of the Enlightenment.[2] The name electricity derives from the Greek word for amber (elektron, ἤλεκτρον), and something like it, namely the adjective *electric*, was first used in English by Francis Bacon around 1620. In the early modern period it was often spelled 'electrick' and used as a noun. We have a reminiscence of this meaning in the term *dielectric* coined by Faraday on the suggestion of Whewell for an insulating transparent material such as glass. The term electricity was originally the property of behaving like an electric(k), that is, possessing the power of attracting light bodies. It was in this sense that Thomas Browne introduced the English word in his *Pseudodoxia Epidemica* from 1646.[3]

During much of the eighteenth century natural philosophers discussed the nature of electricity in terms of hypothetical *imponderabilia* designated 'fluids' or sometimes 'effluvia'. 'Electrical virtue' was another word in common use for what would later be called charge. As the name 'imponderable' suggests, these fluids, effluvia, or virtues were substances with no weight (*pondus* = weight in Latin).

According to the dualistic two-fluid theory there were two varieties of electricity, one associated with wax or amber and the other with glass, and the two would cancel each other in neutral bodies. This hypothesis was favoured by the French chemist Charles François Dufay (or du Fay), who in 1734 introduced the terms *resinous* and *vitreous* for the two electric fluids, the first relating to wax and the other to glass: 'There are two distinct electricities, very different from one another; one of which I call *vitreous electricity*, and the other *resinous electricity*. ... The characteristick of these two electricities is, that the body of the *vitreous electricity*, for example, repels all such as are of the same electricity; and on the contrary, attracts all those of *resinous electricity*.'[4] The two terms that Dufay derived from Latin (*vitrum* = glass and *resinosus* = resin) continued to be used for more than a century, but they are now obsolete and have long been replaced by the positive–negative (+/−) terminology. By 1734 none of the adjectives were new as they had earlier been applied in contexts other than electricity. For example, they both appeared in Browne's *Pseudodoxia Epidemica*.

In contrast to Dufay's view, Benjamin Franklin proposed around 1749 a single-fluid theory in which the two states of electrification were due to the addition or subtraction of one fluid. The first state he called *positive* and identified it with vitreous electricity; the other *negative* state he identified with resinous electricity. Since the two set of terms built on different conceptions of the nature of electricity, it was more than just a replacement of one nomenclature with another. With Franklin's terminology, but not with Dufay's, one could charge a body either positively or negatively. In 1858, an American writer turned the dispute between the two theories of electricity into poetry:

> He [Franklin] taught the doctrine of Du Faye was wrong,
> And made of twain Electrons, one again.
> Showed how two natures to the sprite belong;

---

[2] Fara (2002), p. 2. For the history of electricity in the seventeenth and eighteenth centuries, see also Assis (2010).
[3] Heathcote (1967).
[4] Dufay (1734).

He taught the doctrine of Du Faye was wrong,
Which all his mad caprices may explain,
As in excess, or need, each in its turn may reign.[5]

In his massive *History and Present state of Electricity* from 1767 the English polymath Joseph Priestly suggested that all electrical phenomena could be explained by the one-fluid theory as well as by the more popular two-fluid theory. He summarized the first theory as follows: 'According to this theory, all the operations of electricity depend upon one fluid *sui generis*, extremely subtile and elastic, dispersed through the pores of all bodies; by which the particles of it are as strongly attracted, as they are repelled by one another.' Priestley introduced a whole vocabulary of terms related to the study of electricity. Many of them were derivatives of *electric* including electrical, electrify, electrification, and electrician. Others were combined terms such as electric fluid, electric circuit, electric light, electric fire, electrical battery, communicative electricity, medical electricity, and conductor of electricity. With regard to the terms used in electrical research, Priestley ascribed many of them to the British natural philosopher and clergyman John Theophilus Desaguliers:

> To Dr. Desaguliers we are indebted for some *technical terms* which have been extremely useful to all electricians … He first applied the term *conductor* to that body to which the excited tube conveys its electricity; which term has since been extended to all bodies that are capable of receiving that virtue.[6]

The discussions concerning the two theories of electricity were closely connected with two of the period's most important electrical instruments, the *Leyden jar* or bottle and the *electrophorus*. As to the first instrument it is usually if not quite correctly credited Pieter van Musschenbroek, a professor at the University of Leyden (Leiden), who made the invention in 1745. The name jar or bottle, as well as the design of the instrument, directly reflected the current conception of electricity, namely that it was a fluid which could be stored in a bottle.

The word for this original form of a condenser or capacitor was adopted early on and is one of the few cases in which a scientific instrument or apparatus is named after a location and not eponymously after a person. When it comes to chemical elements, for example, such naming is common (copper, ytterbium, francium, berkelium, etc., Section 5.4), but not so in other branches of science. The Leyden jar is what linguists call a *toponym*, a word named after a city, country, or some other geographical location. The electrophorus, a simple instrument that separated positive and negative charges, was very important to Alessandro Volta, who improved it and coined its name from Greek. It can be translated as 'electricity bearer' just as the earlier substance phosphorus can be translated as 'light bearer' (Section 5.4).

Franklin and other electricians in the period experimented with charging or discharging Leyden jars connected in series. For such a device Franklin used as early as 1749, in a letter to a friend, the term 'electrical battery'. He used the term twice,

---

[5] Richards (1858), p. 39. Note the author's surprising use of 'Electrons'. See also Section 2.2.
[6] Priestley (1794), pp. 438 and 60.

first when reporting on some new experiments: 'Upon this we made what we called an *electrical-battery*, consisting of eleven panes of large sash-glass, arm'd with thin leaden plates, pasted on each side, placed vertically, and supported at two inches distance on silk cords'.[7] How could he had known that more than half a century later this term would be almost monopolized by Volta's invention of a battery consisting of electrochemical cells?

Franklin's term, which later appeared in print, was the first use of battery in an electrical context. As indicated by his 1749 letter, it is likely that he was inspired by the commonly known battery of guns. Batteries of Leyden jars were well known to electricians in the second half of the century and they appeared under this name in many manuals and treatises on electricity. In experiments from the 1780s the Dutch natural philosopher Martinus van Marum connected a large electrostatic generator to a battery of no less than 135 Leyden jars.

It may seem strange to call Franklin an electrician, a word which today refers to people installing, maintaining, and repairing electrical equipment, but in the past the term was also used for scientists doing research in electricity and related fields. As mentioned, this is how Priestley employed the term. Still in the nineteenth century electrician was not only a term used for telegraph engineers and other practical men. For example, the trade journal *The Electrician* established in 1861 included long articles of a theoretical nature that principally appealed to physicists rather than engineers. It was in this journal that Oliver Heaviside published many of his important and mathematically abstruse papers later collected in his book *Electromagnetic Theory*.

As static electricity was applied for medical purposes in the form of 'electrotherapy', so magnetism found its way into the practice of medicine. The Austrian physician Franz Anton Mesmer wrote his 1766 doctoral thesis on the influence of the planets on the human body but without subscribing to traditional medical astrology. About a decade later he claimed to have discovered a subtle magnetic fluid that was everywhere and penetrated the human body. Sometimes, he said, the passage of the fluid was blocked with the result that people got sick, but fortunately Mesmer's magnetic cures were able to reinstate the natural state of the magnetic fluid. After having allegedly cured patients with nervous diseases by means of strokes of iron magnets, Mesmer concluded that there existed an *animal magnetism* different from the one studied by the physicists.

The medical establishment considered Mesmer to be nothing but a charlatan, a view confirmed by a scientific commission which examined the magnetic cures. Although Mesmer's alternative system was controversial, it was also popular and survived well into the nineteenth century. A patient treated with animal magnetism was 'mesmerized' and the whole system known as 'Mesmerism'. Those performing the art were called 'magnetizers' or sometimes 'mesmerists'. It took until the mid-nineteenth century before mesmerism disappeared, then no longer as a medical science but as a pseudoscience.

Luigi Galvani's 'animal electricity' fared better than Mesmer's animal magnetism, in particular because the former concept proved to be very important to Alessandro

---

[7] http://www.benjamin-franklin-history.org/franklin-collison-april-29-1749/

Volta's famous construction of the battery or pile in 1800. In presenting his apparatus to the Royal Society, Volta referred to the importance of terminology: 'New instruments should be given names, depending not only on their form, but also on the effects, on the principle on which they are based.'[8] Without suggesting his own name to be associated with the invention, he preferred *artificial electric organ* but also used a couple of other names such as *column apparatus* and *electro-motive apparatus*. The word battery did appear in his article in *Philosophical Transactions*, but Volta used it only in connection with a series of Leyden jars. In contrast to these batteries, he emphasized, the new device had the advantage that it did not need to be charged and recharged by an external machine.

Volta's seemingly strange term artificial electric organ makes sense only in the light of how he arrived at his device. In the process that brought him from animal to metallic electricity the study of electric fishes, such as the torpedo also known as the electric ray, played a crucial role. He saw a clear analogy between the electric organ of the torpedo and his column of metal discs separated by a humid material. In his address to the Royal Society he wrote of the 'natural electric organ' belonging to the fish. 'To what electricity then, or to what instrument ought the organ of the torpedo or electric eel, &c. to be compared?' he asked. The answer was to be found in his demonstration

> that conductors are also, in certain cases, exciters of electricity in the case of the mutual contact of those of different kinds, &c. in that apparatus which I have named the *artificial electric organ*, and which being at bottom the same as the natural organ of the torpedo, resembles it also in its form, as I have clearly advanced.[9]

During the early years of the new century Volta's admired instrument was referred to by different names. One of those widely used was *pile* or 'electric pile', terms which were introduced in 1800 by the British chemist and inventor William Nicholson. One hundred and forty-two years later, another 'pile' was constructed in Chicago, much bigger and more powerful than Volta's. Enrico Fermi and his collaborators stacked layers of graphite and uranium on one another with the aim of creating a self-sustaining chain reaction in the world's first nuclear reactor. Without thinking of Volta's old electrical pile they called the arrangement a pile, a name which 'was very convenient and useful at the time, because it did not reveal the purpose of the structure.'[10]

Apart from pile, others of the early terms were 'voltaic battery' and the much-used 'galvanic battery' which were due to Humphry Davy in late 1800. Given that Galvani did not construct a battery and that his ideas of electricity differed substantially from Volta's, the latter term is peculiar and almost oxymoronic. The same is the case with the commonly used term 'galvanic current'. But Galvani's name was popular and widely associated with the new kind of electricity called 'galvanism'. Scientists specializing in research with Volta's pile, such as the German chemist Johann Wilhelm Ritter, were sometimes called 'galvanists' (*OED*) but never 'voltaists'.

---

[8] Pancaldi (2003), pp. 246–8.
[9] Volta (1800), p. 311.
[10] Glasstone (1958), p. 427. Lockard (1950).

Whewell later objected, not unreasonably, to the widespread association of electricity with Galvani's name. As he wrote, 'In the discovery of what is termed galvanism, Volta's office was of a higher and more philosophical kind than that of Galvani; and I have, on this account, urged the propriety of employing the term *voltaic*, rather than *galvanic* electricity.'[11] For the frictional electricity produced by electrostatic machines he proposed the eponymous 'franklinic electricity' (after Benjamin Franklin) as a term corresponding to voltaic electricity. Whewell's protologism *franklinic* disappeared shortly after it was introduced.

The name of Galvani also appeared in the verb 'galvanize' and that in two different meanings. First, to galvanize something referred to a technology in which a metal (such as zinc) was coated by means of an electrochemical process to another metal (such as iron). Second, the term was and still is used metaphorically, namely in the sense of causing someone into action, to greatly stimulate someone, or to make someone very excited. In this metaphorical sense it is equivalent to 'electrify', a term alluding to an electric shock and used synonymously with galvanize since the late eighteenth century. A later and today much-used electrical metaphor is an expression of the kind 'I'm exhausted and need to recharge my batteries', which first appeared in the 1890s.

In her novel *Villette* from 1853, the British author Charlotte Brontë wrote about a person that 'her approach always galvanized him to a new and spasmodic life', and also, 'I conceived an electric chord of sympathy between them, a fine chain of mutual understanding.'[12] At about the same time, the American poet Walt Whitman wrote a long poem which in 1867 appeared with the title 'I Sing the Body Electric'. The same kind of dual meaning is found in the verb mesmerize relating to Mesmer's system of animal magnetism. To mesmerize a person may refer to the person being fascinated, spellbound, captivated, or 'hypnotized'. The Greek *hypnos* (Ὕπνος) means sleep or unconsciousness.

Popular as Galvani's name was, in contrast to many other electrical pioneers he never got a unit named after him. Units such as volt (V; A. Volta), ohm (Ω; G. S. Ohm), ampere (A; A.-M. Ampère), and coulomb (C; C. A. Coulomb) are known to most people, who may also think of watt (W; J. Watt) as an electrical unit although it is not—the watt is a unit of any kind of power, electrical or not. There is an instrument called a galvanometer—so named by Ampère—but no galvani unit.

2.2 From ion to electron

An *ion* is an electrically charged atom or molecule, examples being Na^+, H_3O^+, and CH_3COO^-. The name, together with associated names such as anion, cation, and electrode, was suggested by Michael Faraday in 1834, but in a sense very different from the one met in the more recent literature. Among several other discoveries, Faraday is famous for his important experiments on *electrolysis*—yet another of his neologisms—which led to what is currently known as Faraday's two laws.

---

[11] Whewell (1847), vol. 2, p. 534.
[12] http://victorian-studies.net/Bronte-Villette-1.html

According to the first of these eponymous laws, the mass of the element deposited at an electrode is proportional to the charge passing the electrolyte. To measure the charge, Faraday constructed an instrument which he termed a *volta-electrometer* (and later a *voltameter*) but would today be called a coulombmeter. According to the second of his laws, the deposited mass is proportional to the ratio of its atomic weight and valence, the so-called equivalent weight. However, this was not at all how Faraday interpreted his experiments when he reported them in 1834. He thought that the decomposition was caused by the current and not that the ions pre-existed in the electrolyte. Nor did he think about the ions as charged atoms or molecules, such as we do. Although he considered the possibility of discrete atomic or molecular charges, he did not really believe in it and preferred an agnostic attitude with regard to the real existence of atoms. In his own words from the 1834 article in which he introduced the electrochemical laws:

> If we adopt the atomic theory or phraseology, then the atoms of bodies which are equivalent to each other in their ordinary chemical action, have equal quantities of electricity naturally associated with them. But I must confess I am jealous of the term *atom*; for though it is very easy to talk of atoms, it is very difficult to form a clear idea of their nature, especially when compound bodies are under consideration.[13]

Recognizing that words often carry theoretical overtones, Faraday felt a need for new theory-neutral terms that merely described the observed effects in his experiments in electrochemistry.[14] At the time this inevitably meant words derived from Greek or possibly Latin.

Without any formal education himself, Faraday first sought the advice of Whitlock Nicholl, a physician and clergyman who was his personal friend, and it was with his help that he coined *electrolyte* and related terms such as 'electrolytic' and 'electrolytically' in a paper dated 31 December 1833. He also used Nicholl's terms *eisode* and *exode* although he was not quite satisfied with them. Faraday next turned to the erudite polymath William Whewell with whom he exchanged several letters on possible names to be used in the field of electrochemistry. Whewell was an omniscient wordsmith, whose neologisms are known from several areas of science, not to mention the term 'scientist' which he suggested at about the same time (Section 1.4). Late in life he coined the word *astigmatism* for the common eye problem that causes vision to be blurred.

In a letter to Whewell of 24 April 1834, Faraday wrote about the new electrochemical terms he had come up with so far:

> A body decomposed by the passage of the Electric current I call an 'electrolyte', and instead of saying that water is *electro-chemically decomposed* I say it is *electrolyzed*. ... What we have called the poles of the battery I call the *electrodes*. ... [The] evolved substances I call *zetodes*, which are therefore the direct constituents of electrolytes. ... Now can you help me out to two good names not depending upon the idea of a

[13] Faraday (1839), p. 256.
[14] Richeson (1946). Williams (1965), pp.257–69. Ross (1961), reprinted in Ross (1991), pp. 126–72.

current *in one direction only* or upon positive or negative, and to which I may add the prefixes zet or zeto so as to express the class to which any particular *zetode* may belong.[15]

Whewell responded immediately. In a letter of 25 April he commented on how to construct new scientific words. It is, he wrote, 'an additional advantage when they humour philologists so far as to avoid gross incongruities of language. I was well satisfied with most of the terms that you mention; and shall be glad and gratified to assist in freeing them from false assumptions and implications, as well as from philological monstrosities'. To such monstrosities he counted the term zetode, which he at first suggested to replace with *stechion*. Whewell continued: 'I have considered the two terms you want to substitute for *eisode* and *exode*, and upon the whole I am disposed to recommend instead of them *anode* and *cathode*; these ... are good genuine Greek words, and not compounds coined for the purpose'.

As a result of Whewell's intervention the words ion, anion, and cation entered the scientific vocabulary. Whereas we think of the terms anion and cation as based on ion as the primary word with the prefixes *cat-* and *an-* appended to it, to Whewell and Faraday 'anion' and 'cation' came first. They thought of 'ion' as a derived word and one which they did not like very much. Whewell suggested the terms in a letter of 5 May 1834:

If you take *anode* and *cathode*, I would propose for the two elements resulting from *electrolysis* the terms *anode* and *cation* which are neuter participles signifying *that which goes up* and *that which goes down*; and for the two together you might use the term *ions* instead of *zetodes* or *stechions*. The word is not a substantive in Greek but it may easily be so taken, and I am persuaded that the brevity and simplicity of the terms you will thus have will in a fortnight procure their universal acceptation. The *anion* is that which *goes* to the *anode*, the *cation* is that which *goes* to the *cathode*. The *th* in the latter word arises from the aspirate in *hodos* (way), and therefore is not to be introduced in cases where the second term has not an aspirate as *ion* has not.

Although the Whewell–Faraday terms anion, cation, and ion eventually became adopted by all chemists and physicists, for several decades they were not widely used. John Tyndall, Faraday's colleague and first biographer, wrote in 1868 that anode and cathode 'though frequently used, have not enjoyed the same currency as the others', namely electrode, electrolyte, and electrolysis. Moreover, 'The terms *Anion* and *Cation*, which he applied to the constituents of the decomposed electrolyte, and the term *Ion*, which included both anion and cation, are still less frequently employed.'[16]

In his letter of 5 May, Whewell reflected on the objections that Faraday had raised or might raise in their further correspondence. When coining a new word, he said, simplicity and analogy to known words should be taken into consideration in addition to philological criteria. He begged his friend to recall that 'even violent philological anomalies are soon got over, if they are used to express important laws,

---

[15] This and the following quotations are from Ross (1961).
[16] Tyndall (1868), p. 55, accessible as Google Book.

as we see in the terms *endosmose* and *exosmose*; and therefore there is little reason for shrinking from objections founded in ignorance against words which are really agreeable to the best analogies'.[17] In a letter of the next day:

> I still think *anode* and *cathode* the best terms beyond comparison for the two electrodes. ... I am afraid of urging the claim of *anion* and *cation* though I should certainly take them if it were my business, that which goes to the *anode* and that which goes to the *cathode* appearing to me to be exactly what you want to say. To talk of the two as *ions* would sound a little harsh at first: it would soon be got over.

After some further discussion Faraday adopted Whewell's terms in his published paper in *Philosophical Transactions*, one of many in the important series of papers with the common title 'Experimental Researches in Electricity'. Here they appeared with meanings different from those we associate with them today. For example:

> A single ion, i.e. one not in combination with another, will have no tendency to pass to either of the electrodes, and will be perfectly indifferent to the passing current, unless it be itself a compound of more elementary *ions*, and so subject to actual decomposition. ... Compound *ions* are not necessarily composed of electrochemical equivalents of simple *ions*. For instance, sulphuric acid, boracic acid, phosphoric acid, are *ions*, but not *electrolytes*, i.e. not composed of electro-chemical equivalents of simple *ions*.

A variety of other names entered in the correspondence between Faraday and Whewell, most of them strange to a modern ear and none of which, although seriously considered, made it to the scientific lexicon. Examples are *electrobeid*, *zetexode*, *dexiode*, *chemostecheon*, *catazetode*, and *skiaode*. Although ion proved greatly successful, at first it was not much used. In a letter of late 1834 Whewell even suggested that Faraday might eliminate ion and keep only anion and cation, but this did not happen, and fortunately so.

Faraday's electrochemical laws initially met with considerable resistance or were simply ignored. One of the reasons was probably the critique coming from Berzelius, the chemical authority of the period. Berzelius objected not only to Faraday's conclusions but also to his electrochemical terminology which he found to be wholly unwarranted:

> Faraday believes, for reasons which I do not consider valid, that his experiments lead to such changed views in the theory of science that our usual scientific nomenclature is inadequate for a correct expression of the ideas to which the results lead; therefore he has introduced others of which I do not think either that they were necessary in any respect, or that they deserve to be followed.[18]

---

[17] The words 'endosmose' and 'exosmose' (now endosmosis and exosmosis) were coined by the French physiologist Henri Dutrochet in 1827 for the process by which water passes through a semipermeable membrane from one side to the other. The shortened word osmosis (osmose) is from 1854 and due to the Scottish chemist Thomas Graham.

[18] Quoted in Ehl and Ihde (1954).

The terms anion and cation may appear confusing because we associate the first with a negative particle (CH_3COO^-) and the second with a positive one (Na^+). And yet we call the negative electrode a cathode and the positive electrode an anode. But to Faraday and Whewell the terminology was consistent as it was introduced solely to indicate the direction in which deposited constituents of the solution moved when subjected to an electric current. It had nothing whatsoever to do with the nature of the constituents themselves. 'The fundamental meaning of the word *ion* should be carefully kept in view', Frederick Soddy wrote seventy years later. 'The term ... is used strictly in the original sense of Faraday—to express a moving particle carrying an electric charge.'[19] Had Soddy consulted Faraday's works he would have known better.

Apart from the successful neologisms related to electrochemistry, Faraday's name also appears in later eponyms such as the Faraday constant and the unit farad. The first term denotes the electric charge per mole of elementary charges and is thus given by $F = e \times N_A$ with N_A for Avogadro's constant and e for the elementary charge. The farad is the SI unit for electrical capacitance, but curiously the term was originally coined as a unit for charge, namely by the two telegraph engineers Latimer Clark and Charles Bright in 1861, at a time when Faraday was still alive.[20]

The notion of electricity as consisting of tiny subatomic particles was popular in mid-nineteenth century Germany, where it was developed into a speculative atomic theory by Wilhelm Weber, Gustav Fechner, Friedrich Zöllner, and others.[21] Atoms as well as ether might be composed of positive and negative unit charges revolving around each other, they suggested. As far as I know, neither Weber nor other physicists in the German action-at-a-distance tradition proposed special names for the hypothetical elementary charges. The same was the case with Hermann von Helmholtz, who in his Faraday Lecture of 1881 argued confidently that the electrolytic laws should be understood in terms of 'atoms of electricity'. He did not use the term ion.

Faraday had hoped that his new words could be kept free of interpretation. However, his ideas of what happened in electrolytic processes, and with these ideas also the meaning of the words, were dramatically revised with Svante Arrhenius' innovative dissociation theory of 1887 published in the first issue of *Zeitschrift für physikalische Chemie*. Arrhenius had first presented an electrolytic dissociation theory in 1883, but in this case not an ionic theory as he still believed that the charged particles were created by the electric current. According to the 1887 version, the ions were always present in the electrolyte. The effect of the current or voltage was not to produce the ions but only to make them move through the solution. Although the Swedish chemist referred to 'ions' several times, his preferred term for the charged molecular particles was 'active molecules'.

As a result of Arrhenius's ionic theory of dissociation, a cornerstone in the new physical chemistry founded by Wilhelm Ostwald, Jacobus van't Hoff, and others, for a number of years ion became a controversial term in some fractions of the chemical community. Conservative British chemists such as Henry Armstrong and Spencer

[19] Soddy (1904), p. 46.
[20] Clark and Bright (1861).
[21] Wise (1981).

Pickering rallied against the school of physical chemistry and its heretical belief in ions in solutions. In a polemical article of 1896, Armstrong coined the word *ionists* for those who accepted Arrhenius's theory and thereby threatened the sound foundation of chemical science.[22] To the ionists he counted not only Arrhenius, Ostwald, and other physical chemists but also his compatriot, the eminent physicist Joseph Larmor, who was another believer in atomic charges. Armstrong quoted from one of Larmor's recent papers: 'The facts of chemical physics point to electrification being distributed in an atomic manner, so that an atom of electricity, *say an electron*, has the same claims to separate and permanent existence as an atom of matter.'[23]

Although the term ion as used by the chemists typically referred to charged atoms or molecules, in the late nineteenth century it could also be used for the elementary charge of which all matter was supposedly constituted. From a historical point of view there is a connection between ion and electron, such as indicated by Larmor in the phrases cited by Armstrong. At the time the 'atom of electricity' had not yet been discovered but it nonetheless had a name proposed by the Irish physicist George Johnstone Stoney in 1891. The name can be traced back to Stoney's introduction in 1874 of the *electrine* as a unit electrical charge based on Faraday's laws of electrolysis.[24] The electrine was part of Stoney's heterodox system of natural units which also comprised length, mass, and time units called *lenghtine, massine,* and *timine.* Perhaps understandably, this nomenclature did not survive.

Seventeen years later Stoney introduced the word *electron* as a different unit charge whether positive or negative, but not as a constituent of the atom.[25] As Stoney was aware, in Greek the suffix *-tron* indicates a means of doing something, a kind of tool, and not a particle. Consequently, his neologism was not made up of *elec-* and *-tron*, but of *electr-* and *-on* (see also Section 3.1). It was Stoney's name of 1891 that Larmor adopted, but with the significant difference that for him it referred to a real if still hypothetical particle and not to a unit of electricity. He first made use of electron in this sense in 1894 and the next year he wrote of atoms 'to be made up of, or to involve a steady configuration of revolving electrons'.[26] For example, the simplest ideal atom would consist of one positive and one negative electron revolving round each other. Larmor was not the only physicist to write about electrons before they were discovered. Thus, in early 1895 the German-British physicist Arthur Schuster sketched a radiation theory in *Nature* based upon the hypothesis of vibrating electrons within the atom. 'In the existence of the "electron" I firmly believe', he stated without referring to either Stoney or Larmor.[27]

The name that Stoney coined in 1891 was not a neologism as the word electron can be found in the earlier literature if with different meanings. Electron sometimes referred to the Greek word for amber and at other occasions to the natural gold–silver alloy generally known as 'electrum', a Latinized form of electron. Electrum was mentioned by Bacon in his work *Sylva Sylvarum* from 1626. The term reappeared in a very

---

[22] See Dolby (1976) for the controversy over ions.
[23] Armstrong (1896). Emphasis added.
[24] O'Hara (1975).
[25] Kragh (2001).
[26] Larmor (1927), p. 415.
[27] Schuster (1895).

different guise in 1934, when it was used for a hypothetical stellar element or what is currently known as positronium (Section 3.1). Of more interest, in 1858 William Carey Richards, a prolific American author and poet, published a book in verse form with the title *Electron*.[28] In this charming work he followed the Electron, a personification of electricity, from the oldest times to the most recent marvel of electrical technology, the transatlantic telegraph cable. Richards' book is not well known, but it is of interest not only because of its title but also because of its popular and poetical account of the history of electricity.

Although the term electron quickly caught on and became firmly entrenched in the scientific literature, for a while a few other names were in use for this first elementary particle in the modern sense. One of the alternative names was Faraday's old 'ion', which was the term originally used by Hendrik Lorentz, who together with his compatriot Pieter Zeeman qualifies as a co-discoverer of the electron. In 1899 he changed to electron after briefly having considered the name 'lightion' (light-ion) to distinguish the new particle from the much heavier ion associated with electrolysis. Also young James Jeans, soon to become a famous physicist and astronomer, called the particles ions. Following up upon Larmor's speculations he suggested in 1901 that the atom was composed solely of positive and negative electrons except that nowhere in his lengthy paper did he refer to electrons but instead to ions. He regarded the atom to be 'a collection of negative and positive ions, the negative ions each carrying a charge of electricity of amount $-e$, and the positive ions each carrying a charge $+e$.'[29]

Lord Kelvin alias William Thomson was another British physicist who associated the new electron with the old ion. In contrast to Jeans but in rough agreement with J. J. Thomson, he conceived the atom as a number of negative elementary charges embedded in a sphere of positive electricity. Kelvin wanted 'to justify the very modern name *electron*', which he did by coining another name for it: 'The older, and at present even more popular, name *ion* given sixty years ago by Faraday, suggests a convenient modification of it, *electrion*, to denote an atom of resinous [negative] electricity. ... Each atom of ponderable matter is an electron of vitreous [positive] electricity.'[30] Kelvin's electrion (electr-ion) and associated atomic model was not a success and died with him. Despite being proposed by a physicist of the highest possible distinction it was scarcely taken seriously, not even in Britain.

The celebrated discovery of the electron at the end of the nineteenth century had its background in experiments with electric discharges in evacuated glass tubes going back to the 1860s. The term 'discharge' as either a noun or a verb is a polyseme as it has multiple meanings with the same historical root. *OED* lists about forty different meanings. For example, a person can be discharged or released from an institution or a post. In medicine the term may refer to a substance emitted from a body (e.g. vaginal discharge). When physicists began experimenting with electricity in the eighteenth century they found that the electrical charge could be discharged or neutralized by letting it pass as a gas at low pressure. The result was the first 'discharge tubes' that produced different kinds of light and attracted much attention in this and the following

[28] Richards (1858), online as Google Book.
[29] Jeans (1901), p. 425.
[30] Thomson (1897).

century. The tubes were typically known by eponyms such as 'Plücker tube' (Julius Plücker, Germany), 'Geissler tube' (Heinrich Geissler, Germany), and 'Crookes tube' (William Crookes, England).

Experiments with discharge tubes proved that a strange form of rays emanated from the negative pole (cathode) of the tube, but the nature of the rays remained a mystery for decades. Were they electromagnetic pulses in the ether and hence neutral, or were they streams of electrified particles? In 1876 the German physicist Eugen Goldstein coined a name for the rays, namely *cathode rays* or, in his mother tongue, *Kathodenstrahlen*. Further work on the mysterious rays led not only to J. J. Thomson's discovery of the electron but also to C. W. Röntgen's discovery of X-rays (Section 4.2). When the cathode rays first appeared in English scientific literature they were often referred to as *kathode rays* with the spelling taken over from German, although 'cathode' was Faraday's English neologism. Similarly, when writing about the positively charged cations British and American scientists often used the German spelling *kation*. In his 1898 biography of Faraday, Silvanus Thompson noted that 'The words *cathode* and *cation* are now more usually spelled *kathode* and *kation*.'[31]

The Germanism (or what was previously called a Saxonism) 'kathode rays' was for many years employed by British scientists as frequently as the English 'cathode rays'. Even long after the rays had been identified as streams of electrons the *k* survived. For example, in 1912 the British chemist William Ramsay, a Nobel laureate of 1904, wrote a paper on his experiments with 'kathode rays', and the following year J. J. Thomson consistently used the spelling in a critical reply to Ramsay's paper.[32] It took until the early 1920s before 'kathode rays' disappeared from papers and books written in English.

When Thomson discovered the electron in his famous series of experiments with cathode rays in 1897, the particle appeared under the name *corpuscle* rather than electron. He thought that near the cathode the gas molecules in the discharge tube would dissociate 'not into the ordinary chemical atoms, but into these primordial atoms, which we shall for brevity call corpuscles.'[33] The Latin-derived term 'corpuscle' (*corpus* = body) was far from new as it had been employed since the seventeenth century in the broad sense of a minute particle of matter.

According to Thomson, his corpuscle differed conceptually from Larmor's electromagnetic electron, but Thomson was largely alone in using the term corpuscle instead of electron. Thus, in the 1906 Nobel lecture of the celebrated discoverer of the electron one looks in vain for the term electron. Thomson's favoured term can be found as late as about 1920, but typically in conjunction with the electron as in phrases such as 'electron or corpuscle'. Many years later the name 'negatron' (as a counterpart to positron) was suggested, but this name too was mostly ignored (Section 3.1). There is little doubt that Stoney's electron is one of the most successful names ever in the history of the physical sciences. A search for the name on Google Scholar gives a total of about 6.3 million results, and Web of Science reports 4.5 million research papers with the name. For the derivatives electronics, a noun, and electronic, an adjective,

---

[31] Thompson (1898), p. 143.
[32] Ramsay (1912). Thomson (1913).
[33] Thomson (1897). Kragh (2001).

the numbers are 7.4 million and 9.0 million in Google Scholar, and 0.8 million and 2.0 million in Web of Science, respectively. According to *OED*, electron is one of the 2,000 most common words in modern written English. Until about 1980, electron appeared more commonly in books than electronic (including electronics) after which the relative frequency of electron declined. By 2019 the frequency of electronic(s) was about twice that of electron.

The Australian-British physicist William Sutherland adopted the word electron at an early date but thought of it as a structural unit in the ether and not, as Thomson did, as a subatomic particle. In a paper of 1899 he suggested that the ether consisted of doublets of positive and negative electrons kept tightly together. For these hypothetical ether particles, he introduced a name that would reappear much later and then with an entirely different meaning: 'In æther showing no electric charge', Sutherland wrote, 'each negative electron is united with a positive electron to form an analogue of a material molecule, which might conveniently be called a *neutron*.'[34] When the neutron was discovered as a constituent of the atomic nucleus in the early 1930s, Sutherland's proposal was long forgotten (Section 3.3).

One of those who eagerly took up research on the electrons was Owen Richardson, at the time a young physicist at the Cavendish Laboratory and later a Nobel Prize winner. From 1906 to 1913 he was professor at Princeton University. Richardson studied the emission of electrons and other charged particles from a heated electrode placed in a vacuum tube. What he termed the *thermionic* effect in a paper of 1909 was of great importance not only for early vacuum research but also for the tube technology related to wireless communication. Richardson introduced the word not only as an adjective but also, by adding an -*s*, as a noun denoting a whole new branch of science. As thermodynamics was the science concerned with thermodynamic phenomena, so *thermionics* was the new science concerned with thermionic phenomena. He justified his term as follows:

> While fully alive to the desirability of restraining as much as possible the ever-increasing growth of our scientific nomenclature, the author has felt for some time that there is real need of a single word to denote the branch of physics which is alluded to in the literature with varying dignity as 'the emission of ions by hot bodies', 'the leak from hot wires', and a number of other phrases. After giving the matter due consideration, the author ventures to suggest that the word 'Thermionic' ... is very suitable for the purpose. It suggests at once the thermal and electrical nature of the phenomena, while the derivation of the word ion also suggests the kinetic qualities of the ions.[35]

Apart from the word thermionic, a contraction of thermo-ionic, Richardson also coined *thermions* for the emitted particles and in particular for the positive ions. Whereas the first term stuck and thermionic studies developed into a flourishing

---

[34] Sutherland (1899), emphasis added. The ether was often spelled aether or sometimes æther. The letter æ (capital Æ) was part of the Old English alphabet but is today only used as a letter in Danish, Norwegian, and Icelandic.

[35] Richardson (1909), pp. 813–14.

research area, the latter was not widely used. It survived for about a decade and then largely went into oblivion. Thermion briefly resurfaced in connection with the naming of the hydrogen nucleus as a 'proton' in 1921 (see Section 3.3). An American reader of *Nature* favouring 'hydron' over 'hydrion'—two of the alternatives to proton—wrote: 'It may be recalled that the late Lord Kelvin used himself, and tried in vain to induce others to use, the term "electrion" instead of "electron". At this late date it seems quite unnecessary to insist on the retention of the extra syllable simply to have the word "ion" retained in the longer term unless for the sake of euphony, as in the "thermion".'[36]

This section has been about the names of electrical charges, the ion and the electron, as they were formed by physicists and chemists in the nineteenth century. I shall end, as a kind of an appendix, with a key word from a very different science, namely genetics. The word in question, *gene*, has nothing in common with ion and electron except that it became equally popular and for long has entered common language. The coining of the word demonstrates, as Faraday's coining of ion did, that scientists of the past generally reflected seriously about the new words they suggested.

The science concerned with genes is called genetics in the same way that the science concerned with electrons is called electronics. However, the term *genetics* preceded that of gene by two years. The first term was informally suggested by the English biologist William Bateson in 1905 and publicly in 1907, whereas *gene* was only coined subsequently, namely in 1909 by the Danish pharmacist and plant physiologist Wilhelm Johanssen. There existed at the time several names for the hypothetical basic unit of heredity. In his 1868 theory of so-called *pangenesis*, Darwin came up with 'gemmules' for the minute organic units, a Latin word meaning small gems or buds. The Dutch botanist and pioneering geneticist Hugo de Vries suggested 'pangenes' after Darwin's theory. However, Johannsen found none of the words to be satisfactory. In a textbook titled *Elemente der exakten Erblichkeitslehre* (Elements of the Exact Theory of Heredity) he wrote:

> The word 'pangene', which was introduced by Darwin, is perhaps used most frequently in place of *Anlagen*. However, the word 'pangene' was not well chosen, as it is a compound word containing the roots *pan* (the neuter form of Πας, all, every) and *gen* (from γί-γ(ε)ν-ομαι, to become). Only the meaning of the latter comes into consideration here ... It seems simplest to use in isolation the last syllable *gen* from Darwin's well-known word, which alone is of interest to us, in order to replace, with it, the poor and ambiguous word *Anlage*. Thus we will say simply 'gene' and 'genes' for 'pangene' and 'pangenes'. The word gene is completely free of any hypothesis ... [and] offers many advantages as it can easily form combinations with other expressions.[37]

Apart from introducing gene in his book, Johannsen also introduced the important words and concepts *phenotype* and *genotype*. Although he believed that the

[36] Patterson (1921).
[37] Johannsen (1909), p. 124. Online as https://www.biodiversitylibrary.org/bibliography/1060. The German *Anlage* means here plan or predisposition.

genes were real and not just hypothetical entities, he did not conceive them as material particles. The terms gene and genes first appeared in English in a paper of July 1909 by the American plant geneticist George Shull. Johannsen's neologism proved very successful, but initially there was some discussion of how to transfer the German 'Gen' to English in both its singular and plural form. Noting that 'scientific literature is now to have a more Teutonic flavor', a reader of *Science* advocated to take over the German word into English as 'gen'.[38] He regretted that this word had been 'displaced by a linguistic monstrosity', namely 'gene'. After all, the English word for the combustible atmospheric gas was oxygen and not 'oxygene', so why not use 'gen' instead of 'gene'? But of course, gene remained and quickly became a popular term. At the end of the twentieth century it was, like the electron, one of the 2,000 most common words in English (*OED*).

Besides, the word has for a long time become used as a rhetorical and metaphorical figure, much in the same way that DNA has become used (Section 5.2), namely as referring to a person's deeply ingrained and perhaps inherited qualities.[39] Women (and men too, supposedly) may say that they have a 'shopping gene' or a 'clothing gene'. Others may regret that their like of alcohol is in their genes. To speak of a gene for intelligence or one for criminal behaviour makes good sense even though it lacks scientific justification. It is a modern version of the long-forgotten phrenological language (Section 1.5).

The word gene and the concept it stands for is related to a more recent neologism, namely the *meme* introduced by the British evolutionary biologist and writer Richard Dawkins in his influential book *The Selfish Gene* from 1976. Dawkins suggested that certain cultural entities such as words, recipes, pictures, and ideas evolve in a manner analogous to the biological genes insofar that they are passed on from generation to generation. They replicate and mutate like genes, but with the use of human culture and behaviour as their medium of propagation. He motivated his choice of the word meme as follows:

> The gene, the DNA molecule, happens to be the replicating entity that prevails on our own planet. ... We need a name for the new replicator, a noun that conveys the idea of a unit of cultural transmission, or a unit of *imitation*. 'Mimeme' comes from a suitable Greek root, but I want a monosyllable that sounds a bit like 'gene'. I hope my classicist friends will forgive me if I abbreviate mimeme to *meme*. If it is any consolation, it could alternatively be thought of as being related to 'memory'. Or to the French word *même*. It should be pronounced to rhyme with 'cream'.[40]

Despite its inspiration from a biological context, the popular noun meme and its adjectival form memetic belong primarily to the human and cultural sciences, and not to the natural sciences.

---

[38] Cook (1912).
[39] Shea (2001).
[40] Dawkins (1989), p. 192.

2.3 Electromagnetism and some applications

The surname of the Danish physicist and chemist Hans Christian Ørsted, the discoverer of electromagnetism, starts with the letter Ø used only in the Danish-Norwegian alphabet. For this reason, it is often spelled or misspelled Oersted, Orsted, Örsted, or even Oerstedt. In 2017, nearly two centuries after his great discovery, he was honoured eponymously when the company DONG (Danish Oil and Natural Gas) changed its name to Ørsted A/S, a Danish-based multinational energy company and currently the world's largest developer of offshore wind energy. Among the numerous large companies and industrial corporations in the twenty-first century, many bear the name of an inventor or engineer who founded the technologies relevant to the success of the company. Examples are the Linde Group (industrial gases; Carl von Linde), Siemens AG (energy technologies; Werner von Siemens), and Bosch (engineering; Robert Bosch). However, less than a handful of companies are named after great scientists of the past. Ørsted is one of the few and Tesla Inc. (Nikola Tesla) founded in 2003 may qualify as another. As far as I know, no major company has ever been named after Galileo, Lavoisier, Maxwell, or Einstein.

Apart from his work in physics and chemistry, Ørsted was a prolific creator of new words in his mother tongue of which he suggested more than two thousand. With regard to chemical and other scientific terminology he wanted to replace the dominant international Greek-French tradition with words rooted in the local group of languages. In 1814 he published, first in Latin and next in German in *Journal für Chemie und Physik*, a detailed proposal of a scientific vocabulary for all Germanic-Scandinavian languages. Among his general rules for good nomenclature were the following:[41]

- Names have to be true, that is, must not fix false concepts of things.
- A name has to denote the thing and not describe it.
- Names have to be fruitful, so many names can be deduced from one.
- In the building of excellent names, one has to pay attention that the composites are not too long or hard to pronounce.
- Foreign words are to be rejected, if they cannot be adapted to the grammatical rules of the mother tongue.

However, Ørsted's ambitious attempt to create a new scientific language was largely ignored and his numerous neologisms left no traces outside his own country. His proposal of criteria was just one out of many such proposals through the course of history. Some thirty years earlier Lavosier had formulated rules for chemical nomenclature (Section 5.3) and in the 1830s John Herschel reconsidered the issue of scientific terminology (Section 1.6). To mention just one more attempt, in 1947 American physicist Duane Roller suggested a series of criteria for the naming of physical concepts which he formulated in five basic requirements. According to him, an acceptable term should be non-ambiguous, meaningful, international, simple, and euphonious.[42]

---

[41] Surman (2016).
[42] Roller (1947).

Convinced for philosophical reasons of the unity of all natural forces, Ørsted had for long thought that electricity and magnetism were intimately connected or even two manifestations of the same basic force. His hypothesis turned into a fact when he, on 15 June 1820, proved by means of a simple experiment that a current in a wire affected a compass needle some distance from the wire. In an address of 1854, young Louis Pasteur reflected on the discovery: 'Oersted, a Danish physicist, held in his hands a piece of copper wire, joined by its extremities to the two poles of a Volta pile. On his table was a magnetized needle on its pivot, and he suddenly saw (by chance you will say, but chance only favours the mind which is prepared) the needle move and take up a position quite different from the one assigned to it by terrestrial magnetism.'[43]

Without using the term *serendipity*, which at the time was practically unknown, Pasteur described what he thought was a 'serendipitous' discovery, that is, an unexpected discovery made somewhat accidentally as it occurred in the search for something else. Although 'serendipity' is not a scientific term, today it is frequently used in meta-scientific discussions on the nature of discovery and has also been adopted by working scientists. The term, which has roots in Persian history and was coined by the English writer Horace Walpole in a letter of 1754, has an interesting history. According to the American author Ralph Keyes, 'Serendipity is a classic example of a word that began life in obscurity, went into an extended hibernation, then aroused from its slumber to become one of the most used, and overused, words in English language.'[44] The word first appeared in print in 1833 but was for a long time ignored and known only to a few literary and humanist scholars. Eventually it migrated to scientific publications, first perhaps in a 1935 paper by the American bacteriologist Milton Roseneau.[45] Most modern scientists are acquainted with the word serendipity, which they typically use in works of a more popular and general nature. The noun *serendipitist* first appears in James Joyce's *Finnegan's Wake* from 1939, the same work in which he introduced 'quark' (Section 3.3).

Ørsted needed a new name for the extraordinary phenomenon and in his original communication written in Latin he called it *conflictus electrici* or 'conflict of electricity' in the English translation. Later the same year he spoke of 'electro-magnetism' (*Elektromagnetismus*) and 'electro-magnetic effects' (*elektromagnetische Wirkungen*) in a paper in *Annals of Philosophy* originally published in German. The new terms were immediately accepted. Humphry Davy used 'electro-magnetic phenomena' in a paper in *Philosophical Transcations* dated 16 November 1820 and in 1821 Faraday published a 'Historical Sketch of Electro-Magnetism' in *Annals of Philosophy*. Ørsted's term was a neologism, but as it so often happens with this category of words one can find it much earlier if with a very different connotation. The Renaissance Jesuit polymath Athanasius Kircher has been called 'the last man who knew everything.'[46] Among his many scholarly works was a massive work on magnetism from 1641 titled *Magnes sive de Arte Magnetica* (The Lodestone, or the Magnetic Art). Lo

---

[43] Vallery-Radot (1901), p. 76. This is the origin of the popular phrase 'chance favours the prepared mind'.

[44] Keyes (2021), p. 192. See the detailed account in Merton and Barber (2004).

[45] Roseneau (1935).

[46] Findlen (2005).

and behold, in the third edition of this work published 1654 we find on p. 451 the word electromagnetism, or the Latin *magnetismo electri*.[47] Needless to say, the appearance of the word does not indicate that Kircher was somehow aware of the phenomenon discovered by Ørsted or that the latter was inspired by the former.

Ørsted also coined the word *thermoelectric* for the phenomenon that shortly earlier had been discovered by Thomas Seebeck in Germany, namely, in modern terms, the conversion of thermal to electrical energy. He found it remarkable that no liquid was involved and for this reason suggested a new term that distinguished it from the ordinary wet form of galvanic electricity. In a note published in *Annales de Chimie et de Physique* in 1821, he wrote:

> From now on, it will no doubt be necessary to distinguish this new class of electric circuits with its own characteristic name, and for this I propose the expression *thermoelectric* or *thermelectric circuit*; at the same time, the galvanic circuit could be designated the *hydroelectric circuit*.[48]

The neologisms were repeated in a slightly later article in the same journal co-authored by Joseph Fourier, the brilliant French mathematician and physicist. While 'thermoelectric' and 'thermoelectricity' or their corresponding hyphenated forms won quick approval, 'hydroelectric' as Ørsted thought of it did not. On the other hand, much later the term became very common as a name for electric power generated by falling or fast-flowing water. This is yet another example of a scientific term invented in one context but widely used only in another context quite different from the former.

Like many other of the pioneers of electricity, Ørsted got his name associated with a unit, although only at a late date and for a limited period of time. In 1900 electrical units had been named after Volta, Ohm, Ampère, Coulomb, Faraday, Joule, and Henry, and magnetic units after Maxwell and Gauss. No one had thought of Ørsted, but in connection with the 1920 centenary of the discovery of electromagnetism Danish physicists and engineers aired the idea. Ten years later the IEC (International Electrotechnical Commission) recommended the name *oersted* (oe) as the unit for magnetic field strength. The recommendation was confirmed by IUPAP (International Union of Pure and Applied Physics) at a meeting in 1932.[49] However, the oersted unit belonged to the CGS system and when this system was replaced by the SI system in 1960, the unit was abandoned. The SI unit for the magnetic field, or more precisely the magnetic flux density, was now defined as a *tesla* (T) in honour of the Serbian-American inventor and engineer Nikola Tesla.

Electrical conductance, the inverse of the resistance, is measured in terms of the SI unit *siemens* (S) named after the German engineer and industrialist Werner von Siemens. An earlier and now obsolete unit is the *mho*, which is an anonym for ohm, that is, ohm spelled backwards. It was suggested by William Thomson at a

---

[47] The 1654 edition is available online: https://archive.org/details/KircherMagnesSive1654/page/n485/mode/2up

[48] Ørsted (1998), p. 462.

[49] Kennelly (1931).

conference in 1883 concerning practical problems in electrical engineering. Here is how Thomson argued for the new and somewhat strange word:

> For the reciprocal of an ohm in the measurement of resisting power . . . it is suggested to take a phonograph and turn it backwards, and see what it will make of the word 'ohm'. I admire the suggestion and I wish some one would take the responsibility of adopting it; we should have *mho* boxes of coils at once in general use. . . . I do not say that *mho* is the word to be used, but I wish it could be accepted. . . . We shall have a word for it when we have the thing, or rather I should say, we shall have the thing when we have the word.[50]

Although proposed by the admired Thomson, possibly the most famous physicist of the period, the mho unit was not a success and never accepted as an official unit for conductance. As an Austrian electrical engineer commented in 1893: "The name "mho" for the unit of electrical conductivity was probably used by some one because it was introduced by Sir William Thomson; should the new unit be introduced, it would be easy to find an appropriate name (thomson)."[51]

In a paper of 1871 to the London Mathematical Society, Maxwell dealt with the classification and terminology of mathematical terms as they appeared in connection with physical quantities primarily in electromagnetism and thermodynamics.[52] Rather than respecting the convention that new words should be based on the classical languages, the words he suggested and defined were borrowed from common English. One of them was *flux*: 'The flux of heat or of electricity cannot be even thought of in any way except as the quantity which flows through a given area in a given time.' Maxwell looked for words that could indicate the physical meaning of abstract mathematical vector symbols. 'I should be greatly obliged to anyone who can give me suggestions on this subject, as I feel that the onomastic power is very faint in me', he said. For the differential operator ∇ applied to a scalar function P he ventured 'with much diffidence' to call it the *slope* of P:

> We require a vector word, which shall indicate both direction and magnitude, and one not already employed in another mathematical sense. I have taken the liberty of extending the ordinary sense of the word *slope* from topography, where only two independent variables are used, to space of three dimensions.

Maxwell next considered the quantity with the symbol $\nabla\sigma$, where σ denotes a vector function and which contains both a scalar and a vector part. For the latter part,

> I propose, but with great diffidence, to call this vector the *Curl* or *Version* of the original vector function. It represents the direction and magnitude of the rotation

---

[50] Thomson (1884).

[51] Sahulka (1893).

[52] Maxwell's paper 'On the Mathematical Classification of Physical Quantities' is reprinted in Maxwell (1965), vol. 2, pp. 257–66. Some of the content of Maxwell's paper relied on his correspondence with Tait, see Knott (1911), pp. 143–5.

of the subject matter carried by the vector σ. I have sought for a word which shall neither, like Rotation, Whirl, or Twirl, connote motion, nor, like Twist, indicate a helical or screw structure which is not of the nature of a vector at all.

None of the words introduced by Maxwell (including 'concentration' and 'convergence') were neologisms in the strict sense, but they were so in the technical sense of mathematical physics. The notation curl $\mathbf{F}$ for a vector field $\mathbf{F}$ is still common among American authors, but elsewhere it has been replaced by the 'cross product' symbol $\nabla \times \mathbf{F}$.

The symbol ∇ for the operator $(\partial/\partial x, \partial/\partial y, \partial/\partial z)$ did not become known as the slope, as suggested by Maxwell, but as 'nabla' or sometimes 'del', where the latter term refers to a rotation of the Greek Δ, delta. The name *nabla* was first used by Tait, who got it from one of his correspondents, the Scottish classical scholar William Robertson Smith. On 10 November 1870, Smith wrote him: 'The name I propose for ∇ is, as you will remember, *Nabla*. ... In Greek the leading form is ναβλᾰ ... As to the thing it is a sort of harp and is said by Hieronymus and other authorities to have had the figure of ∇ (an inverted Δ).'[53] Maxwell had read Smith's letter but refrained from using 'nabla' except in his private correspondence. In one of his letters to Tait he suggested playfully that perhaps ∇ should be called 'atled', namely delta spelled backwards (an anadrome).

While the Danes succeeded to get the oersted unit accepted at the IEC congress in 1930, the Germans were less fortunate with their proposal of a new unit for frequency called *hertz* (Hz) in honour of their long-deceased compatriot Heinrich Hertz, who died thirty-seven years old in 1894. Maxwell's fundamental electromagnetic theory of light as presented in his 1873 classic *Treatise of Electricity and Magnetism* was at first accepted only by a small group of British 'Maxwellians',[54] whereas it percolated more slowly in Germany and elsewhere on the Continent. Hertz, a young associate professor of physics of Jewish descent in Karlsruhe, Germany, was one of the first to comprehend and develop Maxwell's difficult theory. In a series of watershed experiments from 1888 he detected long-wavelength electromagnetic waves in agreement with Maxwell's theory. These 'Hertzian waves', as they were often called, laid the foundation for the later stormy development of wireless telegraphy alias radio telegraphy and later again for public radio and television communication as known today.

In the fin-de-siècle period people used several names to designate the waves discovered by Hertz and their applications in telegraphy. Among the proposed names was 'telegraphy without connecting wires', but the proposal was found to be 'too cumbrous—an awkward mouthful'.[55] The term 'Hertzian telegraphy' was frequently used as a synonym for the more popular 'wireless telegraphy', and from about 1900 'radio telegraphy' entered the language. As John Munro, an author and engineering professor in Bristol, wrote in 1898, possibly coining the word: '"Wireless telegraphy" is not a bad technical term; but if a more scientific name be desirable would

---

[53] Quoted in *OED*, entry 'nabla'.
[54] The term, in analogy with e.g. 'Newtonian' and 'Mendelian', is not only a post-factum historical construction but was on a few occasions used also in the 1890s. Hunt (1991).
[55] Fahie (1899), p. ix.

not Radiotelegraphy or Ray Telegraphy be preferable to "Space Telegraphy" which Dr. Lodge employs?'[56]

The noun 'radio' referring to either an apparatus or a means of wireless telecommunication is of later origin and may first have been used about 1910. Oliver Lodge, a leading Maxwellian and pioneer of wireless communication, was indeed in favour of 'magnetic space telegraphy' or just 'space telegraphy', a term he and some others advocated. With regard to the Hertzian waves, at the 1894 meeting of the British Association he suggested to rename them 'Maxwellian waves', but he failed due to strong opposition from the Austrian physicist Ludwig Boltzmann.[57] Lodge's compatriot Wiliam Preece, chief engineer of the General Post Office, came up with the term 'aetheric telegraphy' in a paper of 1898.[58] Although it did not catch on, for a long period of time the waves were associated with the ubiquitous ether as in 'ether waves' or similar terms. Still today one can listen to talks or music 'over the ether'.

Lodge was concerned about the lack of connection between the various sciences which increasingly tended to speak in separate languages. In a letter to *Nature* of 1893 he compared the current language of physics to that of the biologists. As to the first, he claimed that it 'consists mainly of adaptions of simple English phrases [and] is full of common words, redefined and made definite in connotation'. The language employed by the biologists, on the other hand, was quite different and in his view problematic:

> Its sentences ... are highly dignified and elaborate structures, not wholly different from a once more prevalent German model. Its words, especially its new words, are hendecasyllabic or, at any rate, polysyllabic. They are extremely classical, and as unlike the language of daily life as can be contrived.[59]

Using 'energy' as an example, he warned of the 'misuse by one science of the language of another science'. Moreover, Lodge emphasized that new technical terms should be used by all sciences in the same sense, which was not the case with the length unit *micron* ($1\mu = 10^{-6}$ m), a neologism first used in spectroscopy and microscopy in about 1880: 'Many biologists call it a micromillimetre, which it is not; and though they may mean the same thing, it can only be by an erroneous, because unconventional, use of the prefix micro. All these things are conventions, and once made the convention should be rigorously adhered to.' The micron unit was officially adopted but abolished in 1967 when it was replaced by the SI unit *micrometre*.

To return to the 1930 IEC congress, the German physicists and engineers had good reasons to honour Hertz with a unit. After all, he was an outstanding and visionary physicist whose work paved the way for electromagnetic telecommunication. However, in part due to anti-German sentiments going back to the Great War their proposal was turned down. It took until 1935 before the IEC officially adopted Hz as a unit of frequency, namely the number of oscillations per second and therefore of

[56] J. Munro, *Electrician*, vol. 40, 21 January 1898, p. 428.
[57] Lodge (1931), pp. 162–3.
[58] Preece (1898).
[59] Lodge (1893). Hendecasyllabic = line of verse involving eleven syllables.

dimension s^{-1}. Although electromagnetic waves provided the historical background for the unit, the unit is for all kinds of waves and periodic phenomena. However, by 1935 Germany had become the Third Reich, which made the name hertz (if not the abbreviation Hz) controversial in some circles.[60] Unfortunately, Hertz was a Jew and for this reason not considered a worthy representative for the new Germany based on the ideology of Aryans as a supreme race.

Hitler's Germany was proud of having an international unit named after a German citizen but not after a German Jew. In 1939 ardent Nazi bureaucrats proposed to solve the dilemma by keeping Hz but associating it with the Aryan Hermann von Helmholtz instead of the Jew Heinrich Hertz. Thus, Hz = H(elmholt)z and not Hz = H(ert)z. The problem or pseudo-problem went all the way to the highest circles in the Nazi bureaucracy, even to Hitler himself. Finally, in the summer of 1941 the Minister of Propaganda, Joseph Goebbels, was informed by telegram: 'The Führer has decided that the hertz should be kept for the designation of frequency.'[61] The hertz survived World War II and in 1960, at the same conference where oersted disappeared, it was adopted as the official SI unit for frequency.

Although Alexander Graham Bell's invention of the telephone in 1876 owed little to science, its further development into an international system of long-distance cable telephony relied heavily on applied electromagnetism. The leading telephone company in the early twentieth century was the technologically innovative American Telephone and Telegraph Company (AT&T) with the Bell Telephone Laboratories as its research branch. One of the problems that faced AT&T and other telephone companies was the attenuation of voice transmission over long distances and how to express it in scientific terms. Telephone engineers were in need of a commonly accepted unit for signal intensity whether expressed in electric or acoustic terms. It was in this context that the unit *decibel* came into being, an eponym which in a not easily recognizable way refers to the celebrated Scottish-American inventor of the telephone. Although the decibel is usually believed to be an acoustical unit, it is a general measure of relative intensity originally based on the electromagnetic signals in telephone wires or cables. More precisely, expressed in decibels the difference in intensity levels between two signals of intensity I_1 and I_2 is given by

$$P_1 - P_2 = 10\log_{10}(I_1/I_2).$$

At a conference in 1923 Bell engineers proposed a logarithmically defined 'transmission unit' (TU) to be used world-wide. However, although the British Postal Office supported the proposal, delegates from Germany and many other countries in continental Europe resisted it. The question evolved into a controversy chiefly between American and British engineers on one side and Germans on the other side. Thomas Purves, Chief Engineer at the British Postal Office, suggested in 1925 'bel' as a better name for the TU unit, obviously a reference to A. Graham Bell. He noted that the

[60] Eckert (2010), pp. 153–5. Wolff (2012).
[61] Eckert (2010), p. 155.

spelling 'bell' was ambiguous as it might associate to telephone bells. As an alternative to 'bel' he briefly considered the name 'heaviside' in honour of Oliver Heaviside, the British physicist and pioneer of telephone transmission theory who had recently passed away. However, Purves concluded that the name was not 'sufficiently neat and crisp'.[62]

There existed at the time another system for the attenuation of telephone currents, a system widely used in Europe and called the 'natural' system because it was based on natural logarithms. At a conference in 1928 Purves' suggestion of 'bel' was confirmed and at the same time a new natural unit called the 'neper' was agreed upon, so named after the Latin form (Ioanne Nepero) of the Scottish inventor of natural logarithms John Napier. The compromise satisfied both the Germans and the British, but the cost was the continuation of an unpractical dual system in European telephony. The Americans didn't care, for the AT&T engineers kept to the 'bel' except that they now changed it to 'decibel' (db, later dB).

As an acoustical unit the decibel won international acceptance at the First International Acoustical Conference in 1937, but also here the German–American rivalry was present and as a result, two units of loudness received official recognition. In addition to the decibel, the *phon* (Greek for sound), originally suggested by the German physicist Heinrich Barkhausen in 1927 and since then used by German engineers, was accepted. It is a shortened version of the Greek *phōnē* (φωνή), meaning sound as in phonetics and microphone. According to *Nature*'s review of the conference, 'The "phon" (a name we owe to Barkhausen) was adopted as a unit of the subjective scale of equivalent loudness, while the use of the "decibel" (a name which originated in America) was restricted to the scale of the associated energy of pressure level.'[63] Barkhausen is best known for his discovery in 1919 that the magnetisation of a ferromagnetic material increases in a series of tiny jumps when the material is subjected to a continuously increasing magnetic field. Physicists refer to the phenomenon as the 'Barkhausen effect', and along with this eponymous term goes another called 'Barkhausen noise'.

After World War II, the decibel unit became gradually accepted also outside the United States and eventually, in 1968, the decibel was made the only official unit for transmission measurements. A variety of other names have been proposed, including logit, decilit, decilog, and decomplog, but none of these have won recognition or just been seriously considered except that 'logit' with a different meaning is a well-known term to statisticians.

The definition of the decibel and also the phon can be seen as related to the *Weber–Fechner law* or relation insofar that this psychological law essentially states that the perceived intensity (p) is proportional to the intensity (S) of the logarithm of the physical stimulus. In another formulation, we perceive equal intensity ratios to be equal intensity intervals. With S_0 denoting the minimum value of the stimulus, the law can be stated mathematically as

$$dp = k\,(dS/S) \ \text{ or } \ p = k\log(S/S_0),$$

---

[62] Kragh (1999), p. 165.
[63] [Anom.] (1937).

where k is a constant. With regard to the name of the law eponymously linked to Gustav Fechner and the physician Ernst Heinrich Weber (a brother to the physicist Wilhelm Weber), it is an appellation that first appeared in the 1890s and is today widely conceived to be a misnomer. According to one author, 'There is no such thing as the Weber-Fechner law', but rather two separate laws first formulated by Fechner in his pioneering work *Elemente der Psychophysik* from 1860.[64] It was in this work that Fechner coined and promoted the word 'psychophysics' for a new science that connected mind to matter and subjective experiences to objective measurements.

Weber did contribute with experiments and an earlier law of his own, but one that was different from Fechner's statement that the perceived action depends logarithmically on the stimulus. Whatever the priority and precise meaning of the Weber–Fechner law, something like it can be found not only in acoustics and electric power transmission but also in areas of astronomy, seismology, chemistry, and mineralogy, not to mention physiology.

In 1856, the British astronomer Norman Robert Pogson proposed a scale for stellar magnitudes by relating the difference of the apparent magnitude of two stars (m_1 and m_2) to the ratio of their observed brightness (F_1 and F_2). In a modern formulation, what is known as the *Pogson equation* states that

$$m_1 - m_2 = -2.5\log_{10}(F_1/F_2).$$

The law implies that a star or some other celestial object (planet, galaxy) of magnitude m is 2.512 times as bright as one of magnitude $m + 1$. The brighter an object appears to the human eye, the smaller its magnitude. For example, the apparent magnitude of the Sun is -26.78, that of Venus -4.6, and of Mars $+0.7$, and for the barely visible Andromeda Galaxy, m equals $+3.4$. When Pogson proposed his magnitude–brightness relation, the Weber–Fechner law was still unknown, and yet Pogson's astronomical law has a great deal of similarity to the latter. On the other hand, there was no historical connection between the two and Pogson's formula cannot be considered simply a special case of the psycho-physical law.[65]

The logarithmic scale also appears in an entirely different area of science, namely as a measure of the acidity of solutions. In the early years of the twentieth century chemists were well aware of the importance of the molar concentration of hydrogen ions, $[H^+]$ or $[H_3O^+]$, as a measure of the acidity of a solution. Among other things, the quantity was crucial to stabilize the acidity of buffer solutions. In a study of 1909 the Danish protein chemist S. P. L. Sørensen—S. P. L. for Søren Peter Lauritz—introduced a new and more convenient measure, namely the now standard pH concept defined as

$$pH = -\log_{10}[H^+]$$

or with $[H_3O^+]$ replacing $[H^+]$. The more acidic a solution is, the less its pH value, whereas basic solutions have pH > 7 and pure water pH = 7. Although in most cases

---

[64] Algom (2021).
[65] Burke-Gaffney (1963).

the quantity varies between 0 and 14, for strong acids or bases it can smaller than 0 or larger than 14. The new name pH is a noun, not an acronym, and is included in the *OED* as such. It soon gave rise to related terms such as 'pH scale' and 'pH meter'.

Sørensen, who mostly wrote in French and Danish, originally used the term 'hydrogen ion exponent' and the symbol either p_H or P_H. During the next decade or so a variety of other symbols were proposed, variants such as PH, Ph, pH^+ and P_H^+, until pH became commonly accepted at about 1920.[66] While the H in pH obviously related to hydrogen, Sørensen did not explain the meaning of the 'p', but he may possibly have thought of 'power', the French 'puissance', or the Latin 'pondus'. In the German version of his paper, he wrote: 'I have chosen the name "hydrogen ion exponent" and the written expression P_H. By the hydrogen ion exponent (P_H) of a solution we understand the Briggs logarithm of the reciprocal value of the normality factor of the solution based on the hydrogen ions.'[67] The new term was used not only by chemists and physiologists but eventually it also entered everyday language. The average layperson may not recall the definition of pH he had learned at school, but he or she most likely knows the term.

The term *buffer* in the chemical sense was another of Sørensen's neologisms in his 1909 paper. In the context of acid–base chemistry it seems not to have been employed by earlier scientists. Thus, in a paper of 1911 a British physiologist adopted the pH concept—he expressed the symbol with the typographically inconvenient $P_{\overset{+}{H}}$ notation—and went on to comment on solutions with a stable pH value. He wrote that 'substances having this effect on addition to a solution are spoken of by Sörensen as "buffer"'.[68] While the context was new, the word was not. For example, it was commonly used in railway technology as 'a round plate or cushion (usually supported by a strong spring) fixed in pairs at the front and back of railway carriages or engines, or on the face of a terminal wall of a line of railway' (*OED*). It was possibly this meaning of the term that inspired Sørensen to adopt it (or the French word *tampon*) in chemistry.

The pH scale is roughly analogous to Mohs's qualitative scale of the hardness of minerals named after the German mineralogist and crystallographer Friedrich Mohs, who in 1812 introduced the scale and the associated scratch test.[69] He assigned hardness 1 to the softest minerals (like talc) and hardness 10 to the hardest (diamond). If one mineral could make a scratch on another, say quartz on beryl, the first was assigned a larger hardness or Mohs number. From Mohs 1 to Mohs 9 the scale has an approximately logarithmic relationship to absolute hardness. The eponymous 'Mohs's scale' or 'Mohs scale' first appeared in English in 1857.

Lastly, a logarithmic scale of the same kind as in the decibel, in Sørensen's pH, and in Pogson's equation appears in seismology. The American seismologist Charles Francis Richter announced in 1935 what he called a 'magnitude scale' for the strength of earthquakes and which from about 1950 became commonly known as the

---

[66] Myers (2010).
[67] Leicester (1968), p. 19. The 'Briggs logarithm' is the decimal or common logaritm with base 10 named after the English mathematician Henry Briggs who introduced it in the early seventeenth century.
[68] Mines (1911), p. 180.
[69] Authier (2013), pp. 349–53.

eponymous *Richter scale*. As he explained in his paper, 'The scale is logarithmic ... A shock of magnitude 1 has ten times the recorded amplitude of the standard shock, magnitude 2 has 100 times the amplitude, etc.'[70] The idea of using logarithms was actually due to Beno Gutenberg, Richter's mentor and a leading German-American seismologist. In an interview in 1978, Richter stated: 'I called it the magnitude scale, and I refrained from attaching my personal name to it for a number of years. And I think it was Professor [Perry] Byerly who started referring to it as the Richter Scale in public.' Moreover, he recognized the similarity of the earthquake scale to that of earlier logarithmic scales in science:

> It became evident that it could be used in a manner to set up a definite scale for which again there was some parallel precedent in the astronomical use of the stellar magnitude scale, which is where I got the word, and also of course in the decibel scale for sound intensities, which is logarithmic. There is also a scale used in—oh, dear, what is it called? [The pH scale]. It is working with soils and expressing their acidity, which is expressed in a logarithm scale again.[71]

Thus, a word in seismology was borrowed from an older one in astronomy. For astronomy's terminological debt to seismology, so to speak, see Section 6.2.

2.4 The many faces of electricity

The Dutch theoretical physicist Hendrik Antoon Lorentz was a great expert on Maxwell's electromagnetic theory which he not only reformulated but also extended and turned into a theory of electrons. Whereas there was no room for electrons or other discrete charges in Maxwell's original field theory, these occupied a central position in the 'Maxwell–Lorentz equations' dating from the 1890s. In 1895 he introduced the non-Maxwellian 'Lorentz force' expressing how a test charge e moving at speed v is affected by a combined magnetic and electric field (E, B). The Lorentz law in SI units states the force to be given by

$$F = e\left(E + v \times B\right).$$

The field quantities in Maxwell's equations can be formulated also in terms of a scalar potential φ and a vector potential A. There are different ways to connect the two potentials, corresponding to different so-called *gauge* conditions. The most popular of the gauges is what until recently carried the name 'Lorentz gauge', which in technical terms is the constraint given by

[70] Richter (1935).
[71] Interview of 1978, Californian Institute of Technology, https://oralhistories.library.caltech.edu/17/1/OH_Richter_C.pdf

$$\nabla \cdot A + \frac{1}{c^2}\frac{\partial \varphi}{\partial t} = 0,$$

where c denotes the speed of light. Google Scholar tells us that the term Lorentz gauge appears in 14,500 scientific articles, with the number for the two past decades being 8,500. However, as we shall see, today it is increasingly recognized that the eponym is unjustified from a historical point of view.

Physicists in the late nineteenth century did not speak of gauge conditions or other terms related to the word 'gauge'. This word has been used in English since about 1500, the noun traditionally denoting 'a fixed or standard measure or scale of measurement, the measure to which a thing must conform' (*OED*). As a technical term it only entered physics in the 1920s and at first in connection with an extended theory of relativity suggested by Hermann Weyl, a German mathematician.[72] Weyl had introduced the German *Eichinvarianz* based on the verb *eichen* meaning to scale or calibrate. Eddington, in his *Space, Time and Gravitation* from 1920, may have been responsible for translating *Eich-* to gauge. From the 1930s onwards words including gauge (gauge invariance, gauge symmetry, gauge theories, etc.) became very popular in theoretical physics. In the category 'particles and fields' alone, the term gauge appears in more than 78,000 scientific papers (Web of Science).

Lorentz is today best known for his non-Einsteinian relativity theory based on electromagnetism and the ether. All physics students are familiar with eponyms such as 'Lorentz transformations', 'Lorentz invariance', and 'Lorentz contraction', terms which as well or better might have been associated with Einstein's name. The contraction of a rod moving through the ether at great speed was proposed by the Irish physicist George FitzGerald in 1889 and is for this reason sometimes referred to as the 'FitzGerald–Lorentz contraction', a combination which can be found as early as 1904—before Einstein's theory—and is still occasionally used. Lorentz's theory led to many of the same predictions as Einstein's, including that the mass of a charged particle should increase with its speed according to $m(v) = m(1 - v^2/c^2)^{-1/2}$. For a decade or so many physicists discussed the theory of relativity as if it was Lorentz's and not Einstein's.

Although the term 'relativity theory' or 'theory of relativity' is firmly associated with theoretical physics and with Einstein's theory in particular, we should not be surprised to find the term earlier in a non-physics context. The prominent American philosopher John Dewey wrote as a young man a paper in the *Journal of Speculative Philosophy* in which he used the phrase thrice if obviously in a different sense than that of the later physicists. For example, 'according to the relativity theory, we must know the relation of our feelings to an object'.[73] While the noun 'relativist' typically referred to a person who believed in or advocated relativism as a philosophical doctrine, from about 1920 it was also used for researchers investigating Einstein's theory of relativity.

Incidentally, 'relativity theory' was not Einstein's first choice of name. It was coined by Max Planck as *Relativtheorie* (relative theory) at a meeting in Stuttgart in 1906, but

---

[72] Jackson and Okun (2001).
[73] Dewey (1883).

only to describe the Lorentz–Einstein equations for an electron in motion. Later at the same meeting the German physicist Alfred Bucherer used the term in the form *Relativitätstheorie* (relativity theory).[74] Einstein, on the other hand, preferred 'relativity principle', which he kept to for several years. The first time he used relativity theory in a title (*Die Relativitäts-Theorie*) was in an article of 1911 published in a Swiss journal.[75] Whether called theory or principle, in some corners 'relativity' was considered a problematic term because it might suggest a scientific legitimation of ethical relativism. Einstein agreed that relativity theory was in some respects a misnomer, but having accustomed himself to the term in the late 1910s he agreed that it would cause more confusion than clarity to adopt a new name.[76]

According to the Russian physicist Yakov Frenkel in a popular talk given in 1931, the theory of relativity 'could be called, with better right perhaps, the theory of the absolute'. With this shift of name 'it would not appeal so much to the present sophisticated generation, and there would be less talk about it'.[77] In the public arena the theory was often boiled down to the silly phrase that 'everything in relative', which was definitely not what Einstein had in mind. This phrase or the equivalent 'all is relative' has been known since antiquity. The influential German philosopher Georg W. F. Hegel used it in one of his major works, the *Wissenschaft der Logik* (Science of Logic) published between 1812 and 1816. 'In Being everything is *immediate*, in Essence everything is *relative*', he stated.[78] The cliché only entered common language with Einstein's theory, undoubtedly inspired by it but almost never referring to it or Einstein's name. 'To the average person this is a huge amount of money, but to Bill Gates it's nothing—it's all relative, isn't it?' This kind of innocent phrase makes perfect sense also to people who have never heard of Einstein and his theory.

For more than a century it has been customary, for both historical and practical reasons, to distinguish between Einstein's 'special' and 'general' theory of relativity, the first dating from 1905 and the latter from 1915. However, this common usage may give the wrong impression of two different relativity theories, whereas in reality there is only one, namely the general theory. To speak of the 'special theory' before 1915 is anachronistic since the term only became meaningful in relation to the general theory of relativity. Einstein was the first to introduce the phrase 'special theory of relativity', which he did in the beginning of a seminal paper of 1916 in which he gave a full exposition of his new general relativity theory, the *allgemeine Relativitätstheorie*. He wrote as follows:

> The theory which is presented in the following pages conceivably constitutes the farthest-reaching generalization of a theory which, today, is generally called the 'theory of relativity'; I will call the latter one—in order to distinguish it from the first named—the 'special theory of relativity', which I assume to be known.[79]

[74] Einstein (1989), p. 254.
[75] Fölsing (1997), p. 209.
[76] Sheldon (1986).
[77] Frenkel (1931).
[78] https://www.marxists.org/reference/archive/hegel/works/sl/slbeing.htm#SL111n_1, §111.
[79] Einstein (1916), p. 769.

To return to Lorentz, when he delivered his Nobel lecture in 1902 he not only referred to Stoney as the originator of 'electron', in connection with his own early work on a refractivity–density formula for transparent substances he also said that the equation has 'often been referred to as the formula of Lorenz and Lorentz'.[80] The general idea that the refractivity index n of a transparent body is related to its density d is old and can be found in the revised edition of Newton's *Opticks* from 1718. One hundred and sixty years later, young Lorentz derived on the basis of electromagnetic theory a definite relationship between the two quantities which for our purpose can be written as

$$\frac{n^2 - 1}{n^2 + 2} = \text{constant} \times d.$$

Unknown to Lorentz in Leiden, nine years earlier the engineer-trained physicist Ludvig Lorenz in Copenhagen had derived the very same formula albeit by a different non-electromagnetic method. This is what Lorentz referred to in his Nobel lecture. Typically named the 'Lorentz–Lorenz formula' it soon became accepted as an important law in optics and electromagnetism.[81] It proved a very useful tool also in physical and structural chemistry, where it helped chemists to gain information on the constitution of molecules.

Although priority to the law clearly belonged to Lorenz, and the order Lorenz–Lorentz was not uncommon in the earliest literature, with the rising fame of the Dutch physicist the order of the double eponym was soon reversed. In many cases Lorenz, who died in 1891, was simply left out. In their authoritative and highly influential *Principles of Optics*, Max Born and his co-author Emil Wolf wrote: 'By a remarkable coincidence, the relation was discovered independently and practically at the same time [sic] by two physicists of almost identical names, Lorentz and Lorenz, and is accordingly called the Lorentz–Lorenz formula.'[82] While Google Scholar gives 110 results for Lorenz–Lorentz and 91 for the reverse order in the period 1880–1929, in the period 1970–2019 the results are 2,980 (Lorenz–Lorentz) and 13,500 (Lorentz–Lorenz).

According to what has been called the 'zeroth theorem of the history of science', the discovery of a law or phenomenon is often attributed to and named after a scientist who was arguably not the first to make the discovery.[83] The zeroth law is essentially the same as Stigler's law of eponymy stating that no scientific discovery is named after its original discoverer. Although the latter version is evidently an exaggeration, as mentioned in section 1.2 there are numerous examples in the history of science of such misattributions. The Lorentz–Lorenz law is one of them and the Lorentz gauge is another. Moreover, in these and some other cases the alleged theorem or law is related to the so-called Matthew effect introduced by Robert Merton in 1968: eminent and

[80] https://www.nobelprize.org/prizes/physics/1902/lorentz/lecture/
[81] See details in Kragh (2018a).
[82] Born and Wolf (1970), p. 87.
[83] Jackson (2008).

famous scientists will often get more credit than a comparatively unknown scientist, even if their work is similar.[84]

With regard to the gauge linking the electromagnetic scalar and vector potentials, Lorenz introduced it as early as 1867 and today it is frequently named the Lorenz gauge.[85] Again citing Google Scholar data, until the year 2000 the gauge was associated with Lorentz's name in 97% of all articles, whereas the percentage dropped to 62% during the two decades 2003–22. That is, more than one third of the modern papers refer to the more historically correct Lorenz gauge.

2.5 Superconductivity and other super-names

At about 1910 the sciences of electricity and magnetism were highly developed. Physicists generally believed that the propagation of electric currents in metals was satisfactorily understood and that no new results of a fundamental nature would emerge from further studies. It therefore came as a complete surprise when the Dutch physicist Heike Kamerlingh Onnes, the head of a large low-temperature laboratory in Leiden, announced in 1911 that at temperature 4.2 K or −269°C the resistance of a mercury wire changed abruptly to zero. The new state needed a name and in early 1913 he coined the term 'supra-conductivity' for it and 'supra-conductor' as the noun for a metal exhibiting the property. The names stayed for a while but were in most cases changed to *superconductivity* and *superconductor*.

Now there was a name for the puzzling phenomenon, but an understanding of what the name covered was notably lacking and would remain so for many years. Einstein was one of a dozen eminent theorists who in vain tried to make sense of the superconducting state. In a paper of 1922, he wrote: 'Given our ignorance of quantum mechanics of composite systems we are far away from being able to convert these vague ideas into a theory. We can only rely on experiment.'[86] This was one of the first times that 'quantum mechanics' appeared in the scientific literature, three years before Heisenberg and others founded the theory that we now associate with the term (Section 4.3).

The word superconductivity was one of the first in the physical sciences with the Latin prefix *super-* (beyond, above), but it was not the first and far from the last. When a solution is made to contain more of the dissolved substance than a saturated solution, it is brought in the form of an unstable *supersaturated* solution, a term used by chemists since the late eighteenth century. During the following century engineers learned to *superheat* steam to above the boiling point of water and chemists to *supercool* a liquid to below its freezing point without turning it into a solid.

With superconductivity the *super-* prefix became increasingly popular and soon attached to a variety of other phenomena, theorems, objects, and theories. *Superfluid*

---

[84] Merton (1973), pp. 439–59. The effect is named after Matthew 25:29: 'For to every one who has will more be given, and he will have abundance; but from him who has not, even what he has will be taken away.'

[85] Jackson and Okun (2001).

[86] English translation by Bjoern Schmekel in Einstein, 'Theoretical Remarks on the Superconductivity of Metals'. Available at https://arxiv.org/abs/physics/0510251.

(and by implication 'superfluidity') for extremely cold helium followed on 1 January 1938, when the Russian physicist Peter Kapitsa explained that 'the helium below the λ-point enters a special state which might be called "superfluid"'.[87] As the electric resistance of certain metals vanishes at about 4.2 K, so the viscosity of fluid helium vanishes at the lambda (λ) point of approximately 2.2 K. In between the appearances of the terms superconductivity and superfluidity, the prefix had made its entry into both astronomy and acoustics.

In a joint paper of 1934 the astrophysicists Walter Baade and Fritz Zwicky, one a German and the other a Swiss-American, introduced the concept and name of a *supernova* and also of a *neutron star*. As they explained at the beginning of their paper: 'The extensive investigations of extragalactic systems during recent years have brought to light the remarkable fact that there exist two well-defined types of new stars or novae which might be distinguished as *common novae* and *super-novae*.'[88] Moreover, they suggested that one such supernova had been observed in our Milky Way, namely the *nova stella* famously discovered by Tycho Brahe in November 1572. While Baade and Zwicky promoted and effectively introduced the word, they were not the first to use it in print. The Swedish astronomer Knut Lundmark had since the early 1920s suggested the idea of two kinds of novae and in early 1933 he wrote about 'super-nova' in a publication from Lund University, Sweden. Nonetheless, it is quite possible or even likely that the word was a neologism coined by Baade and Zwicky and that Lundmark had previously heard about it from discussions and seminars with them.[89]

After 'supernova' and the abbreviation SN had become firmly rooted in the astronomical literature, in a monograph on variable stars the eminent British-American astronomer Cecilia Payne-Gaposchkin expressed reservation with regard to the 'somewhat illogical name supernovae'. In a footnote she explained as well as qualified:

'Nova' (itself an inaccurate term) implies newness, and the supernovae do not excel the novae on this point. The alternative name, 'upper class novae', used by Lundmark, scarcely improves matters by introducing an anthropomorphic parallel; and as the name supernova is no more illogical than a number of others that are in current use, and is, moreover, firmly entrenched, we shall use it in what follows.[90]

A colleague of Payne-Gaposchkin's commented: 'In using the word "supernova" we err linguistically only as much as we do in our common usage of the word "novelty" in the sense of unusual; in unusualness a supernova does surpass the ordinary nova.'[91]

The Latin root word *son* means 'sound'. From a historical point of view, the adjective *supersonic* has two different meanings. First, in acoustics it often refers to sound waves at a frequency above normal human hearing, which is approximately 20 kHz. In this sense the term was widely used in the early twentieth century, but after World War II it was gradually replaced by the synonym *ultrasonic*. Second, supersonic may

[87] Kapitsa (1938).
[88] Baade and Zwicky (1934), p. 259.
[89] Osterbrock (2001).
[90] Payne-Gaposchkin and Gaposchkin (1938), p. 269.
[91] Hoffleit (1939).

also denote objects that move at a speed greater than sound in air (343 m/s = 1,236 km/hour). It began to be used in this sense in the early 1930s and soon became associated with the new applied science called aeronautics. Today it is customary to use the *Mach scale* named after the prominent Viennese philosopher-physicist Ernst Mach who lived 1838–1916. The Mach number is the ratio of the object's speed and the speed of sound, hence Mach 1 = speed of sound, Mach 2 = twice the speed of sound, and so on. Should an object move with the speed of light—which it can only do if it is a photon—it moves at about Mach 900,000.

The eponym was not known to Mach but coined in 1929 by the Swiss physicist and engineer Jakob Ackeret, a specialist in fluid dynamics. According to Ackeret, 'The well-known physicist Ernst Mach has recognized the significance of this ratio with particular clarity … [and] thus it appears to be very justifiable to call the *v/a* [*a* = speed of sound] the Mach number.'[92] The name quickly became a success and was widely used especially after 1947, when the first aircraft pilot, the American Charles Yeager onboard a Bell X-1 rocket plane, broke the sound barrier. It has long been a part of the language of the general public.

The Mach number is not the only eponym named after Ernst Mach. While many laypeople know this number, only few will have heard of *Mach's principle*, a term associated with cosmology and general relativity theory and with no relevance at all for aeronautics. The nature and content of this famous principle—famous among physicists, cosmologists, and philosophers, that is—have given rise to an extensive scholarly literature and are still being debated. In one of its nutshell versions, the space-time metric is determined by the mass of the universe; and in another, the inertia and mass of a body are determined by all other bodies in the universe. Although Mach never stated his idea in these terms or called it a principle, in his important work *The Science of Mechanics* first published in 1883 he came close. His principle or law served as an important guidance for Einstein in his creation of the general theory of relativity and it was actually him who coined the term in a paper of 1918. Referring to his own formulation, he wrote: 'I have chosen the term "Mach's principle" because this principle is a generalization of Mach's claim that inertia has to be reduced upon interaction of the bodies.'[93]

For more than half a century, another Mach eponym frequently turned up in a completely different context, namely in Marxist-Leninist political ideology. In his influential philosophical works Mach advocated 'sensationalism', the view that the observed world consists only of our sensations. Science, he claimed, was a method of ordering sensations in the most economical way and therefore unable to make pronouncements about ultimate reality such as the existence of atoms (Mach, who died in 1916, never accepted atoms as really existing objects). The reason why Mach's philosophy of science became political dynamite was that Vladimir Lenin considered it to be antithetical to dialectical materialism, the one and only correct understanding of Marxism. In his work *Materialism and Empirio-Criticism* from 1909, he vigorously attacked Mach and his Russian followers, the 'Machists' or 'Machians'. For example:

[92] Rott (1985).
[93] Einstein (1918).

The philosophy of the scientist Mach is to science what the kiss of the Christian Judas was to Christ. Mach likewise betrays science into the hands of fideism by virtually deserting to the camp of philosophical idealism. Mach's renunciation of natural-scientific materialism is a reactionary phenomenon in every respect.[94]

When Marxism-Leninism later became the state philosophy of the Soviet Union, *Machism*, *Machian*, and *Machist* became words loaded with strongly negative connotations. It was a most serious matter if a scientist or philosopher in the Stalin era was accused of being a Machist.

After this digression on Mach-related terms, let's return to acoustics and supersonic sound frequencies. The prefixes *super-* and *ultra-* are roughly synonymous with the Greek *hyper-* meaning 'above' or 'beyond'. When the Indian physicist Raghavendra Rao studied sound waves at very high frequencies he thought they deserved a name of their own, not only supersonic or ultrasonic waves but *hypersonic* waves. In a paper in *Nature* of 1937 he introduced the term for frequencies above 1000 megahertz.[95] The *hyper-* prefix has long won acceptance in names belonging to biology and medicine, such as hyperventilation, hypersensitivity, and hyperthyroidism (overactive thyroid). 'He is very hyper', the mother may say about her hyperactive child. More recently the term *hyperauthorship* has been added to the list of hyper- words, a name that refers to scientific articles with a very large number of nominal authors, not only in the hundreds but in some cases even in the thousands.[96]

Although hyperauthorship is common in experimental high-energy physics, there are only few words in the physical sciences with *hyper-* as a prefix. They do exist, though, one example being the group of elementary particles known as *hyperons*. At about 1950 physicists specializing in cosmic rays had discovered a number of new fundamental particles, the nature of which puzzled them. The French physicist Louis Leprince-Ringuet and his colleagues coined in 1953 the word hyperon for those with a mass greater than the nucleon but less than the deuteron.[97] The name was adopted and is still in use, now defined as a baryon composed of a particular mix of quarks.

The history of mathematics provides other and very different examples of the *hyper-* prefix. The conic section called a *hyperbola*, which appeared in English from about 1660, derives from Greek with the literal meaning 'a throwing beyond or above', but it also has the connotations 'extravagant' and 'exaggeration' as in *hyperbole*. Indeed, the adjective *hyperbolic* is a homonym as it can refer to both a rhetorical use ('a hyperbolic speech') and a geometrical property ('a hyperbolic curve'). The mathematical and philosophical literature in the second half of the nineteenth century included discussions of *hyperspace*, a term which typically referred to non-Euclidean geometry and also to speculations of a fourth space dimension. The American mathematician George Halsted wrote a *Bibliography of Hyper-Space and Non-Euclidean Geometry* in 1878 and twenty years later his compatriot, the esteemed astronomer

---

[94] https://www.marxists.org/archive/lenin/works/1908/mec/. 'Lenin' was an adopted name, a pseudonym. His real surname was Ulyanov.

[95] Rao (1937).

[96] Cronin (2001). The most extreme case of hyperauthorship so far is an experimental paper in high-energy physics listing no less than 5,154 nominal authors. See Aad et al. (2015).

[97] Kragh (2023).

Simon Newcomb, published an article with the title 'The Philosophy of Hyper Space'. Today 'hyperspace' is as popular as ever, but now in science fiction stories and movies such as the Star Wars series, where a passage into hyperspace requires the starship to move faster than the speed of light.

In the Victorian period the *hyper-* prefix was often associated with what mainstream scientists considered to be disreputable science or even pseudoscience. Thus, in the late nineteenth century the Frenchman François Jollivet-Castelat invented what he called *hyperchemistry*, a sort of modernized alchemy including element transmutation. Jollivet-Castelat was a co-founder of the Association Alchimique de France which published a review with the title *L'Hyperchimie*.[98] Extravagant indeed.

The *hyper-* prefix is not or only rarely used in chemical nomenclature where, on the other hand, the apparently similar *hypo-* can be found. However, the two prefixes have quite different connotations. *Hypo* in Greek (ὑπό) means under or beneath and is thus an antonym of *hyper* (ὑπέρ). It is known from medical nomenclature in terms such as hypochondria, hypotension (low blood pressure), and hypothermic (low body temperature), and in chemistry it was earlier used to denote compounds in which an element is in its lowest oxidation state. While sodium chlorate is $NaClO_3$, the salt $NaClO$ was called sodium hypochlorite. The prefix *hypo-* was introduced in 1819 by the French chemist Pierre-Louis Dulong to indicate that 'hypophosphorous acid' (H_3PO_2, now called phosphinic acid) contains less oxygen than phosphorous acid (H_3PO_4). When sodium hyposulphite ($Na_2S_2O_3$) was later used large scale as a fixing agent in the photographic industry it became known as just *hypo*.[99] The current name for $Na_2S_2O_3$ is sodium thiosulphate (*thio* = sulphur), a name first suggested in 1869.

The Latin prefix *ultra-*, which as mentioned has the same meaning as *hyper-* (namely 'beyond'), deserves a little more attention. The expensive deep-blue colour pigment *ultramarine* much valued by the Renaissance painters was called so because it was imported from mines in far-away Afghanistan 'beyond the sea'. Of more scientific interest is the term *ultraviolet* rays or radiation coined by John Herschel in 1840. These invisible rays were discovered in 1801 by Johann Ritter, a German chemist and philosopher in the Romantic *Naturphilosophie* tradition. Experimenting with silver chloride exposed to sunlight, Ritter noticed a darkening even beyond the visible violet part of the prismatic spectrum and inferred that the action was due to 'chemical rays'. Much later, in a detailed examination of the colours of the solar spectrum, Herschel suggested a more appropriate name for Ritter's chemical rays:

As orange, indigo, and violet, vegetable tints, are used for these prismatic hues, I may be allowed to express by the epithet lavender the rays which produces the tint in question, rather for the purpose of abbreviating the uncouth appellation of *ultraviolet*, and deciding the ambiguity attaching to the term *chemical* rays (which exist in all regions of the spectrum) than for that of laying undue stress on the observed fact.[100]

---

[98] Kauffman (1985).
[99] Ross (1991), chapter on 'Herschel and Hypo', pp. 194–202.
[100] Herschel (1840), p. 20.

The lengthy title of Herschel's paper in *Philosophical Transactions* referred to another neologism, namely 'photographic', which he had introduced two years earlier as a substitute for 'photogenic' used by his friend Henry Fox Talbot, one of the inventors of the art of photography.[101] Both of Herschel's terms quickly won recognition. As a noun or an adjective *ultra* has more recently entered common language in the sense of people with extremist views or doing extremist things, fanatics of some kind, as in words such as 'ultraconservative' and 'ultramarathon' (running a distance between 50 and 300 km). The plural form 'ultras' usually refers to groups of fanatical supporters of a football team.

Forty years before John Herschel introduced ultraviolet in the language of optics, his father, the famous astronomer William Herschel, announced the discovery of invisible rays at the other limit of the solar spectrum. They came to be known as *infrared* rays, a term not used by Herschel senior in his papers of 1800, where he wrote of 'radiant heat', 'invisible rays of the Sun', 'calorific rays', and the synonymous 'heat-making rays'. As Ritter's chemical rays eventually came to be known as ultraviolet, so Herschel's heat rays might be called *ultrared*, which was indeed a word that turned up in the 1860s. It may first have been used in a paper of 1863 written by the American amateur astronomer and spectroscopist Lewis Rutherfurd.[102] For a couple of decades 'ultrared' (usually in the form ultra-red) was widely used, but at the turn of the century it lost out to infrared (infra-red), the Latin *infra-* meaning below. This neologism seems to have originated in papers from about 1880 by another American, the chemist, photographer, and author John William Draper.[103] Still in the first two decades of the twentieth century ultrared and ultra-red appeared significantly in scientific papers, but after World War II the shift to infrared was complete (Web of Science).

When the Greek word *meta* is used as a prefix it may refer to change or transformation as in the words *metaphor*, *metastasis*, and *metamorphosis*. But most often it refers to something which is above or beyond what follows after the prefix. The term *metaphysics* is usually credited to Aristotle although he did not use it but spoke of 'first philosophy'. Whatever the origin of the word, it does not signify a new branch of physics like astrophysics and geophysics but is a philosophical term relating to the study of the ultimate causes and nature of things. Physicists and other scientists of a positivist inclination often use the adjective 'metaphysical' in a derogatory sense, not only as different from science but contrary to it.

The corresponding term 'metachemical' is rare, but none other than Dmitrii Mendeleev, the lionized founder of the periodic system, used it when he spoke of the 'metachemical electron' that he thought threatened the foundation of traditional chemistry:

> In recent years there have been much talk about the division of atoms into more minute electrons, and it seems to me that such ideas are not so much metaphysical

---

[101] For details, see Batchen (1993).

[102] Rutherfurd (1863).

[103] Draper (1880). In the 1840s Draper thought to have discovered in the solar spectrum what he called *tithonic rays* and suggested a number of related protologisms including 'tithonoscope', 'tithonometer', 'tithonography', and 'diatithonescence'. See Hentschel (2002).

as metachemical, proceeding from the absence of any definite notions upon the chemism of ether.[104]

The now obsolete word 'chemism' means here chemical action or activity. At about the same time that Mendeleev objected to the metachemical electron, William Ramsey, the celebrated discoverer of argon and helium, thought that he had discovered a new inert gas accompanying argon. In an address to the Royal Society of 16 June 1898, Ramsay and his collaborator Morris Travers concluded: 'Inasmuch as this gas differs markedly from argon in its spectrum … it must be regarded as a distinct elementary substance, and we therefore propose for it the name "metargon".'[105] For a year or two, the hypothesis of the new chemical element *metargon* (meta-argon) was discussed in scientific publications after which it disappeared for good as it was realized to be a mistake.

Metachemistry is occasionally used as a term for the philosophy of chemistry, more specifically for the generalized form of chemistry proposed by the French philosopher Gaston Bachelard in his book *The Philosophy of No* published in 1940.[106] Likewise, *metageometry* was a term suggested by the German-American philosopher and science author John Stallo for the non-Euclidean geometries much discussed in the late part of the nineteenth century.[107] Other words used by Stallo and some other authors were *pangeometry* and *transcendental geometry*. When the Russian mathematician Nikolai Lobachevsky in 1855 published a book in French on his new so-called hyperbolic geometry, he chose the title *Pangéometrie*.

A physical system, say an atom or molecule, can be *metastable* if it exists in an intermediate energy state higher than its ground state. In a paper of 1920 the German physicists James Franck and Fritz Reiche introduced the name and notion in an attempt to explain the helium spectrum, which at the time puzzled quantum theorists. The idea of metastable states turned out to be very fruitful and led, among other things, to an understanding of the line spectrum of the aurora borealis, a problem which had defied explanation for half a century.[108]

Finally, after World War II super-something proliferated in the physical sciences. *Supergravity*, a quantum field theory of elementary particles incorporating the principles of general relativity, was proposed in the late 1970s in conjunction with new theories of *supersymmetry*, a property often known under the acronym SUSY. These theories go beyond the standard model by predicting a number of new hypothetical *superpartners* to the known elementary particles. Thus, the electron has a superpartner called a *selectron* and quarks have *squark* partners. Instead of *s*- as prefix, others have *-ino* as suffix such as *neutralino* and *gravitino*. So far, none of these exotic particles have been detected. In roughly the same period string theory, originally a theory

---

[104] Mendeleev (1904), p. 17.
[105] Ramsey and Travers (1898).
[106] Nordmann (2006). The American poet and philosopher Ralph Waldo Emerson used the term 'metachemistry' in the 1840s for his own unorthodox version of a transcendental chemistry. See Browne (1928).
[107] Stallo (1960), p. 259.
[108] Kragh (2009b).

of strong interactions only, turned into the supersymmetric and much publicized *superstring* theory.

Moreover, with the hypothesis of faster-than-light particles or so-called tachyons (Section 3.5), the term *superluminal* became commonly used by physicists. Surprisingly, the first to use the term in English may have been a philosopher and not a physicist. According to the *OED*, it first appeared on p. 236 in Karl Popper's influential *The Logic of Discovery* published in 1959. However, this work was Popper's translation of the older German original *Logik der Forschung* from 1935, where Popper referred to *Überlichtgeschwindigkeit* (faster-than-light velocity), a term used in German physics literature prior to the advent of Einstein's theory of relativity.[109] If we are to believe the Web of Science, superluminal only turned up in a physics paper in 1969, in all likelihood independent of Popper's earlier use of it. Actually, a paper in *Physical Review* from the previous year referred to superluminal, but in this case to sound waves moving faster than light in matter of extreme density. Tachyons travel faster than light propagating in empty space.

On the experimental side, American physicists suggested in 1983 to build a huge and wildly ambitious accelerator project to test these and other theories. Alas, it never became a reality because the politicians deemed the estimated expenses to be too super. It was not for nothing that the planned accelerator facility was called the *Superconducting Super Collider* or SSC. In Japan, physicists studying solar neutrinos upgraded the so-called Kamiokande experiment to the *Super Kamiokande* detector facility.

To mention an example from nuclear chemistry, the heaviest artificially produced transuranic elements, namely those with atomic number greater than 103, became known as *superheavy elements*. This term or the variant 'superheavy nuclei' largely owes its origin to the American physicist John Wheeler, who in the mid-1950s suggested that very heavy radioactive atomic nuclei might have a lifetime long enough to make them detectable. The first scientific papers with 'superheavy' or 'super-heavy' in their titles appeared in the 1960s.[110] According to the Web of Science, by the fall of 2022 the cumulative number of such papers has grown to 2,910 with 335 of them using the hyphenated spelling super-heavy and 450 the spelling super heavy (see also Section 5.5).

2.6 Two household words: plasma; transistor

Most people will undoubtedly associate the term *plasma* with blood, so as will physiologists and medical doctors. Physicists, on the other hand, will think of something very different: a hot gaseous state consisting mainly of atomic nuclei, ions, and free electrons. Whichever of the two present meanings, the term is derived from a Greek word (πλάσμα) which translates as something like 'a substance which can be formed or moulded'. As regards the first and older meaning of plasma, it was introduced in 1836 by Carl Heinrich Schultz, a professor of medicine in Berlin. According to the

---

[109] Popper (1935), p. 172.
[110] Kragh (2018b).

American Journal of the Medical Sciences of the same year: 'The living blood consists of two parts. The first we denominate *plasma*, the second *cruor*. The cruor consists of globules which are suspended in the plasma.'[111] Three years later the word plasma was reinvented, possibly independently, by the prominent Czech physiologist and pathologist Johannes Purkinje, after whom the neurons called Purkinje cells are named. He called the substance found inside the blood cells 'protoplasma' and for the part of the fluid blood left after the suspended cells had been removed he suggested the name plasma. A decade later the word had been adopted in the world of medicine or physiology and composite words like 'cytoplasm' and 'nucleoplasm' soon followed.

The American physical chemist Irving Langmuir, an employee at General Electric's research laboratory, was awarded the 1932 Nobel Prize for his pioneering contributions to surface chemistry. Four years earlier, while investigating electrical discharges in dilute gases, he found a region in the discharge tube which seemed to consist of as many electrons as positive ions and would therefore be neutral. Without any further comments, he stated in a paper on 1928: 'We shall use the name *plasma* to describe this region containing balanced charges of ions and electrons.'[112] Much later, Langmuir's assistant Harold Mott-Smith recalled how the term plasma was imported to physics from the medical sciences. The researchers at General Electric 'realized that the credit of a discovery goes not to the man who makes it, but to the man who names it', he said. The members of the group discussed various names and then

one day Langmuir came in triumphantly and said he had it. He pointed out that the 'equilibrium' part of the discharge acted as a sort of sub-stratum carrying particles of special kinds, like high-velocity electrons from thermionic filaments, molecules and ions of gas impurities. This reminds him of the way blood plasma carries around red and white corpuscles and germs. So he proposed to call our 'uniform discharge' a 'plasma'. Of course we all agreed.[113]

For a while the name caused confusion because it was already widely used in the medical world. 'For a long time we were pestered by requests from medical journals for reprints of our articles', Mott-Smith recalled.

Although Langmuir and his team found the name apt, it was not an immediate success in the physics literature where it only appeared sparingly over the next two decades. From 1929 to 1949 plasma entered the title of 3,018 scientific papers, but almost all of them were in the biomedical sciences. Only about 3% were in journals devoted to the physical and chemical sciences. Plasma still meant blood plasma, but not for long. During the subsequent two decades the number of scientific papers referring to plasma increased tenfold to 30,903. Almost one-third of the papers now belonged to physics, chemistry, engineering, or materials science (Web of Science). As of late 2023 the total number of papers amounts to 1,888,000 with about 789,000, or 42%, belonging to physics and its allied sciences.

---

[111] *American Journal of the Medical Sciences* **18** (1836): 446.
[112] Langmuir (1928). Kragh (2014a).
[113] Mott-Smith (1971).

With the explosive growth of plasma physics in the 1950s new names followed. Plasma oscillations were studied on the basis of quantum mechanics, which in 1956 led the Princeton physicist David Pines to coin a name for the corresponding so-called quasiparticles: 'We introduce the term "plasmon" to describe the quantum of elementary excitation associated with this high-frequency collective motion.'[114] The name quickly caught on. Three years later a Division of Plasma Physics was established under the American Physical Society. In 2001 the term 'plasmonics' turned up as a new research field dealing with surface plasmons in metals and the applications of 'plasmonic' phenomena in nanotechnology and elsewhere.

In the 1960s plasma physics also entered astrophysics and related cosmic problems, giving rise to what was known as *plasma cosmology*. This alternative view of the universe was essentially founded by the Swedish physicist and Nobel laureate Hannes Alfvén, according to whom the universe consisted of equal amounts of ordinary matter and antimatter separated by electromagnetic fields. For a space plasma consisting of both kinds of matter Alfvén coined the word *ambiplasma* based on the Latin prefix *ambi-* meaning 'both ways or sides' (as in ambivalent, ambiguous, ambidextrous, and ambient). Although this theory of plasma cosmology has long been discarded (Section 3.1), plasmas continue to be of focal interest to astrophysicists and cosmologists. It is estimated that more than 99% of all ordinary so-called baryonic matter in the universe is in the plasma state.

Plasma is not the only term that physicists have borrowed from the biologists. Another noteworthy case is the *fission* of a uranium nucleus or some other very heavy atomic nucleus.[115] When Lise Meitner and Robert Frisch at the end of 1938 realized that the bombardment of uranium with neutrons caused a splitting of the nucleus into two, they did not have a name for the astonishing process. 'Fission' was suggested by the American biophysicist William Arnold, who at the time worked at Bohr's institute in Copenhagen with radioactive tracers in biology. Frisch worked there too. Arnold knew that bacteria split into two as they grow and that biologists called it fission. Why not use the same word in nuclear physics? As Arnold recalled many years later:

> I was in the laboratory the day that Dr O. Frisch made the experiment which showed that Uranium atoms split into two parts and released a large amount of energy when hit by neutrons. He said to me, 'you work in a microbiological lab. What do you call the process in which one bacteria divides into two?' I answered, 'binary fission'. He wanted to know if you could use the word fission alone and I said you could.[116]

The biologists were not the only scientists to use the term fission years before it was adopted by the nuclear physicists. During the 1920s and 1930s the same term occurred frequently in the chemical research literature, if of course in a different sense than that used by either biologists or nuclear physicists. Thus, the chemists sometimes characterized the splitting of a bromine molecule into two atoms ($Br_2 \rightarrow Br + Br$) as a fission process. The term fission *could* have migrated from chemistry to physics,

---

[114] Pines (1956), p. 184.
[115] Kragh (2014a).
[116] Arnold (1991).

but it did not. Fission in the meaning of nuclear fission first appeared in print in a paper that Meitner and Frisch published in *Nature* on 11 February 1939. It became an instant success. For the decade 1939–48 Web of Science lists 285 papers with fission in the title, of which about 35 are in the biological sciences and about 145 in nuclear physics; most of the remaining articles are in chemistry, with fission used in a different sense than in nuclear fission. Today, the name fission is usually understood in the latter sense, both by scientists and non-scientists.

By the spring of 1939, when fission had entered the vocabulary of nuclear physics, it was followed by another term, *chain reaction*, referring to the multiplication of neutrons and fission events in a lump of enriched uranium. Both terms appeared in the title of a paper published in *Nature* on 13 May 1939.[117] One might believe that chain reaction was one of those 'old words of nontechnical meaning to which nuclear physics has given new technical meanings', such as stated in a linguistic journal.[118] However, that was not the case. Chain reaction was not new in 1939, but neither was it old and nontechnical, given that it first turned up in chemical texts on reaction kinetics in the 1920s.

The reaction mechanism for the synthesis of hydrogen chloride, $H_2 + Cl_2 \rightarrow 2HCl$, had previously been studied by the German chemist Max Bodenstein, who in 1913 explained it by distinguishing between what he called primary and secondary photochemical processes. The latter kind of process was what today is called a chain reaction, but neither Bodenstein nor others used the term, which was only coined in a 1921 dissertation by Jens A. Christiansen, a young Danish chemist working at Bohr's institute in Copenhagen.[119] Christiansen used the Danish word *kædereaktion* (kæde = chain) and the following year he introduced the German version *Kettenreaktion* in a paper in *Zeitschrift für physikalische Chemie*. Moreover, he argued that chain reactions were not limited to photochemical processes. After the term had been used by German-speaking chemists, it was adopted as chain reaction by their English-speaking colleagues by about 1925. It took only a few years before the term became common and even indispensable in areas of chemistry and biochemistry.

Strictly speaking, Christiansen's chain reaction was not a neologism as it had been used a few times before 1921 in the general sense of a series of events where each event is caused by and follows on from the previous one. Thus, the eminent German-American biologist and embryologist Jacques Loeb spoke of *Kettenreaktionen* in a work of 1907 dealing with so-called artificial parthenogenesis, an asexual form of reproduction. However, this pre-1921 usage was rare and probably unknown to chemists and most other scientists. It more had the character of a protologism than a proper neologism.

The first to use chain reaction in connection with neutron-induced nuclear processes was the Hungarian-American physicist Leo Szilard, who in a British patent application of 1934 referred to a 'chain reaction apparatus', explaining that 'In the chain reaction to be described below, energy is liberated in the form of heat.'[120]

---

[117] Adler and Halban (1939).
[118] Lockard (1950).
[119] Laidler (1993), pp. 256–7.
[120] Feld and Szilard (1972), p. 643. See also Alex Wellerstein, 'Szilard's Chain Reaction: Visionary or Crank?', available at https://blog.nuclearsecrecy.com/2014/05/16/szilards-chain-reaction/

However, Szilard's proposed mechanism, namely neutron bombardment of beryllium, was unrelated to the still unknown uranium fission. When nuclear physicists adopted Christiansen's word in early 1939, it was commonly known in the world of science and the natural choice also for the fission process in uranium.

Moreover, after World War II and the atomic bomb the term became used also in the general and often metaphorical meaning of a series of rapidly occurring events with each event precipitating the next. Figuratively, it came to appear in sentences such as 'Increased oil prices could trigger a chain reaction in economy' and 'The climate change may cause a chain reaction leading to the end of life.' It should be noted that chain reaction is not a very good metaphor for the nuclear processes in uranium or plutonium. In an ordinary chain the pull on a link is merely transmitted to the next link. The number of links remains the same and the pull is not multiplied in strength.

Now to a different subject and a name originating in the research that took place at the Bell Telephone Laboratories in the years following World War II. Earlier work at Bell had resulted in the transmission unit called decibel (Section 2.3) and now the result was the invention of the *transistor*, arguably the most important invention of the twentieth century. The problem that a small team of Bell physicists attacked was to construct an amplifier based on semiconductors such as germanium or silicon and not on vacuum tubes. This work led William Shockley, Walter Brattain, and John Bardeen to the first point-contact transistor publicly announced in the *New York Times* on 1 July 1948. In late December the previous year the transistor effect had been observed in a germanium crystal, except that the term 'transistor' had not yet been coined. The scientists and managers of Bell Laboratories needed a name, a problem they took very seriously.

In a memorandum on terminology dated 28 May 1948 several generic names were proposed and discussed, namely (1) semiconductor triode, (2) surface state triode, (3) crystal triode, (4) solid triode, (5) iotatron, and (6) transistor.[121] With regard to the fifth of the proposed words, the first part of it is evidently the Greek letter *iota* which since the seventeenth century had also been used figuratively as something extremely small or sometimes unimportant ('nothing she said made an iota of difference'). The termination -*tron* was already in use for a variety of instruments and in the ears of Bell scientists therefore a natural one.[122] According to the memorandum, iotatron 'conveys the sense of a minute element . . . however, in view of the many vacuum or gas filled devices such as thyratrons, dynatrons, transitrons etc., it lacks the distinguishing property which would differentiate it from such devices'.

The sixth and eventually successful proposal received the following comment in the memorandum: 'This is an abbreviated combination of the words "transconductance" or "transfer", and "varistor". The device logically belongs to the varistor family, and has the transconductance or transfer impedance of a device having gain, so that this combination is descriptive.' Since there was no clear favourite a ballot was put up

[121] Riordan and Hoddeson (1997), pp. 158–9. https://web.archive.org/web/20080528164454/http://users.arczip.com/rmcgarra2/namememo.gif.
[122] Munns (2017).

with the result that transistor won. The name was due to John Pierce, another Bell scientist. In an interview many years later, Pierce recalled:

> The way I provided the name, was to think of what the device did. And at that time, it was supposed to be the dual of the vacuum tube ... [which] had transconductance, so the transistor would have 'transresistance'. And the name should fit in with the names of other devices, such as varistor and thermistor. And ... I suggested the name 'transistor'.[123]

Pierce's baptizing of the device was confirmed by Brattain in a 1975 interview:[124]

> When the question was asked, 'what should we name it?' Bardeen and I were told, 'You did it, you name it'. We were aware that [Lee] de Forest had called the three-electrode vacuum tube an audion, and that this name did not survive. We also knew that whatever we named it might not take. We knew that a two-syllable name would be better than three or more syllables. J. A. Becker had originated names for other semiconductor devices, namely, *varistor* for the rectifier and *thermistor* for the temperature-sensitive device. We had many suggestions, some ending in *-itron*, which we did not like. Bardeen and I were about at the end of our rope when one day J. R. Pierce walked by my office ... He mentioned the most important property of the vacuum tube, 'transconductance', thought a minute about what the dual of this parameter would be and said, 'transresistance', and then said 'transistor', and I said, 'Pierce, that is it!'

Transistor it was, and transistor it is. In the same interview, Brattain recalled that when he and his colleagues went to a semiconductor conference in England in 1950, the English scientists did not use the term transistor. 'We asked them if by chance they did not think the name *transistor* was appropriate. Their answer was that they thought we had copyrighted the name and were therefore very proper in not using it. We told them that this was not so.' In the early 1950s, the term transistor was extended and modified to the verb 'transistorize', the adjective 'transistorized', and the noun 'transistorization'. These words all referred to the rapid conversion of older electronic systems to new ones based on transistor technology. The first fully transistorized computers were operational in the mid-1950s.

In his further research on semiconductors and transistors, Shockley worried about a strange kind of imperfection that reduced the lifetime of the charge carriers. The imperfection needed a word, he thought, and in 1952 he came up with a most dramatic one: 'For purposes of convenience the word *deathnium* has been introduced to identify the imperfections that catalyse the recombination and thus limit the lifetime.'[125] The suffix *-ium* in Shockley's neologism might indicate that he thought of it as a new chemical element, but apparently he did not. The mysterious substance was eventually identified as trace impurities of copper atoms. In contrast to his colleagues

---

[123] https://www.pbs.org/transistor/album1/pierce/naming.html
[124] Bernstein (1984), pp. 94–5.
[125] Shockley (1952). Riordan and Hoddeson (1997), p. 219.

in semiconductor physics, Shockley took his *deathnium* very seriously and used the term repeatedly, for instance in his Nobel lecture of 1956. By the late 1950s *deathnium* had quietly disappeared never to return. It suffered an early death.

In his paper of 1952, Shockley also introduced some other new words in semi-conductor terminology, namely *source*, *drain*, *grid*, and *gate* for unipolar transistors. The names were chosen after considerable thought on how scientific words should be formed: 'The choice of these names has been based partly on an attempt to find names which describe functions and partly on the value of the names from a phonetic and abbreviational point of view. It should be noted that none of the new subscripts are the same as those encountered in bipolar transistors. Furthermore, it may be noted that the names selected are all monosyllabic.'

About a decade before transistor entered the vocabulary, another very success-ful science-based product was named *nylon*. There are some similarities between the invention and naming stories of the two products, both of which originated in the research laboratory of a private corporation. Working in the laboratory of the large chemical company DuPont, in 1935 Wallace Carothers found a way to pro-duce synthetic fibres from polyamide polymers. Executives at DuPont realized that the new fibres, which the chemists referred to as 'polyamide 66', could be used for women's stockings and within a few years the sensational stockings entered the mar-ket. What should the synthetic product be called? While transistor was chosen by Bell Lab physicists as a meaningful name, the chemists at DuPont's laboratory were not responsible for 'nylon', which was a neologism with no intrinsic meaning and chosen by the company executives solely for marketing reasons.

As transistor was the winner of a contest between several names, so was it the case with nylon, which was chosen only after a long list of other names including oddities such as 'duparooh' and 'wacara' had been considered.

> None of them met everyone's approval. One of the committee members ... suggested
> 'norun' but changed it to 'nuron' when he was reminded that stockings of the new
> fiber would run. 'Nuron' sounded medical; a change to 'nulon' presented the dif-
> ficulty of the redundant phrase, 'new nulon'; and a change to 'nilon' offered three
> confusing pronunciations. Nylon, with an unambiguous spelling and pronunciation
> became the name of the new fiber.[126]

Remarkably, given the huge commercial success of nylon, the name was not trade-marked any more than transistor was. Although from a chemical point of view nylon was a generical name for a group of polyamides, to most people it referred to stockings and as such it could be used also in plural, nylons, or as an adjective as in 'nyloned legs'.

---

[126] Hermes (1996), p. 295. See also Jensen (2005).

3
Fundamental particles

During the eighteenth century several natural philosophers speculated that all matter and all forms of ether were composed of a single undifferentiated substance, force, or particle. In 1816 the British physician and chemist William Prout famously suggested that hydrogen was the primordial *protyle*, a word derived from the Greek *proto hyle* (πρώτη ὕλη) meaning primary matter. The suggestion gave rise to the eponym 'Prout's hypothesis' which states that the atomic weight of all elements are multiples of that of hydrogen. While this hypothesis proved influential in the long run, most of the early speculations were fruitless and the names associated with the hypothetical particles are long forgotten. For example, in 1864 the Danish-American chemist and polymath Gustavus Hinrichs suggested *pantogen* to be the ultimate building block of matter. *Pantos* (παντός) in Greek means 'all' or 'entire' and pantogen is thus the stuff that gave birth to everything.

Hinrichs' name was largely ignored and the same was the case with the *dynamids* proposed in 1857 by Ferdinand Redtenbacher, an Austrian-German physicist and engineer. However, in the latter case it was revived by Philipp Lenard in an atomic theory dating from 1903. According to Lenard, who two years later would receive the Nobel Prize for his work on cathode rays, the dynamids were neutral doublets consisting of pairs of positive and negative electrons located in the central part of the otherwise nearly empty atom. He may have got the name 'dynamid' from Redtenbacher's writings, but it is also possible that he coined the name independently.

Still other and now forgotten names turned up after the electron had made its entry. Who have heard of the *atomecule* coined by the prominent American geologist Thomas C. Chamberlin in 1899 as a hypothetical subatomic entity? Given that *-cule* is a diminutive suffix, the word means 'little atom'. He probably got the funny name from H. M. Stanley, a British philosopher who briefly introduced it in an article in *Mind* from 1884. As far as I know, the word only appears twice in the literature, once in Stanley's paper and once in Chamberlin's.[1] Words of this kind, where the author or speaker invents a word on a single occasion never to return to it, are what linguists call *nonces* or sometimes *occasionalisms*. The inventor of a nonce word may be serious about it, but in other cases it is meant merely as an innocent joke.

A nonce is not, strictly speaking, a neologism because it does not survive. Should this happen and the word be accepted by the relevant speech community, it ceases to be a nonce and becomes a neologism. Linguists and lexicographers do not always agree when a word should be labelled a neologism and when a nonce. According to one formulation, a nonce is 'coined for a particular use and unlikely to become a permanent part of the vocabulary'. Another and somewhat different formula states that nonces are 'one-off coinages that when newly minted seemed apparently bound

---

[1] Stanley (1884). Chamberlin (1889).

to enter the general vocabulary, but soon vanished, lost amid the linguistic ephemera.[2] The Stanley–Chamberlin 'atomecule' never seemed bound to enter either the general or the scientific vocabulary.

This chapter looks at the names of elementary particles, not only of those recognized today but also of those proposed in the past which did not stand the test of time.[3] There are more in the last category than in the first. The issue of naming was taken seriously by physicists in the twentieth century and frequently gave rise to disputes in the community, where the involved physicists sometimes appealed to etymological, euphonic, and semantic arguments to advocate a certain name. As the chosen cases illustrate, whereas names are sometimes genuine neologisms, in other cases they are borrowed from earlier history and other sciences but used with a new meaning. In addition, the cases indicate that during the past fifty years or so there has been a marked change in how physicists coin new words and the public adopts them. While in the earlier period the names of particles were as a rule with few exceptions based on ancient Greek and sometimes Latin words, this is no longer so. It goes without saying that the chapter is incomplete as it only deals with a select number of the many particles which have been either hypothesized or discovered since the beginning of the twentieth century.

3.1 The positive electron

The charge of the electrons discovered in the late 1890s (Section 2.2) was assumed to be positive as well as negative, but to the surprise of the physicists only the latter particles turned up experimentally. Claims to the contrary proved untenable and by 1906 the fundamental dissymmetry between the two charges was broadly recognized.[4] Yet, matter is electrically neutral, so some kind of 'positive electron' must necessarily exist. The word appeared frequently in the physics literature from about 1900 well into the 1920s, but from 1910 onwards these positive electrons were generally thought to be associated with the much heavier hydrogen ion H^+, eventually baptized the proton.

Did the true positive electron as a mirror particle of the familiar negative electron exist in nature? Yes, it did, namely in the cosmic rays first identified in the 1910s. When the rays were rediscovered or confirmed by Robert Millikan in 1925, he coined the name 'cosmic rays' which eventually replaced other names such as 'penetrating radiation' and 'high-altitude rays' from the German *Höhenstrahlung*. In the United States, the rays were sometimes referred to with the eponymous 'Millikan rays' suggested in an editorial in the *New York Times* of 23 November 1925:

> The mere discovery of these rays is a triumph of the human mind that should be acclaimed among the capital events of these days. The proposal that they should bear the name of their discoverer is one upon which his brother-scientists should insist.

[2] Mattiello (2017), p. 25.
[3] Parts of the chapter rely on Kragh (2023).
[4] See Kragh (1989) and Dahl (1997), pp. 257–64 for the history of pre-positron positive electrons.

'Millikan rays' ought to find a place in our planetary scientific directory all the more because they would be associated with a man of such fine and modest personality.[5]

Although Millikan refrained from using the eponym, neither did he object to it.

In May 1933 the American physicist Carl Anderson announced that he had detected in cosmic ray experiments a positive electron or what he called a *positron*. In addition to this very successful neologism he proposed for symmetry reasons to call the negative electron a *negatron*, but this name was rarely used. When a word is first fully accepted and widely used in the literature, such as was the case with electron, it is very hard to change it to another word. Unknown to Anderson, the term 'negatron' had previously been used in a different context, namely for an electronic oscillator with negative resistance patented by the British inventor and engineer John Scott-Taggart in 1919.[6] In the early 1930s this device was still known to specialists in wireless technology but not, apparently, to Anderson.

Although Anderson introduced the word positron to the scientific world, he may not have been the one who originally coined it. It seems to have been triggered by a telegram from Watson Davis, the director of *Science Service*, who in February 1933 wired him: 'Why not christen your new particle "positron"?' In reply, Anderson wrote that he and his colleagues had already contemplated the word: 'We have been discussing in the laboratory for some months past the desirability of calling the free positive electron, positron, and then using the similar contraction, negatron, for the free negative electron.'[7]

The name may appear to be a natural contraction of 'positive electron', but not all physicists liked the term and nor did all linguists. Robert Oppenheimer objected that it—in contrast to electron—was a 'terrible word, because it starts with a Latin syllable and ends with a Greek syllable'.[8] In papers from late 1933 and early 1934, Oppenheimer and his co-authors avoided the term and instead wrote of 'positives'. As Ray Monk writes in his biography of the great physicist, 'It would not be until the summer of the following year [1934] that Oppenheimer resigned to the word "positron", which he regarded as a barbaric mixture of Latin (posi-) and Greek (-tron).'[9] Many years later the American linguist Charles Whitmore, a professor of rhetoric at the University of Michigan, referred to the same issue:

It is not clear, at first sight, whether the division should be posi-tron or posit-ron. If the latter (analogous to posit-ive), the r is an insertion, and the preferable form would be positon; but this, although I have occasionally seen it, seems never to have become current, and negatron seems to have won no acceptance at all, electron being too firmly established.[10]

---

[5] Partially quoted in *Science* **62** (1925): 461–2. Menzies and Sloat (1926).
[6] Blake (1926), p. 255. Scott-Taggart (1933), p. 358.
[7] Davis (1933).
[8] Interview of 9 August 1971 with Wendell Furry, one of Oppenheimer's collaborators. Kragh (2023), p. 86.
[9] Monk (2012), p. 208.
[10] Whitmore (1955). Also noted in Asimov (1974), p. 138.

Anderson may have shared or come to share some of the reservations with respect to his neologism. Late in life he recalled: 'I do not like the name particularly well, and although I have never discussed it with Professor [Paul] Dirac, my feeling is that he may find it wholly inelegant.'[11] He did not elaborate or explain why he did not like the inelegant word.

The kinds of hybrid Greek–Latin words that Oppenheimer found to be barbaric are known from many other terms in science as well as elsewhere.[12] Thus, in 'television' the prefix *tele-* is Greek while the root *vis* is the Latin *visio* meaning 'sight'. The first practical demonstration of television took place in 1926 and two years later a reader of the journal *Science* objected to the word: 'I am ... horrified at the continued creeping into our language of hybrid words. If we are not very careful one of the most objectionable hybrids so far will become fixed in our vocabulary.... The word I have in view now is "television". Can anything worse be imagined?'[13] Another reader of *Science* agreed. After having noted modern scientists' regrettable lack of knowledge of Greek and Latin, he wrote:

> Mongrel or hybrid words constitute a linguistic crime. Generally speaking, a [scientific] word should be all Latin or all Greek. . . . There is no palliation for the linguistic crimes of 'hypersensitive' and 'hypertension' when we can say 'supersensitive' and 'supertension', or 'television' when we might have invented something like 'teleopsis'.[14]

Far from considering hybrids to be linguistic crimes, an American chemist defended them, again in the columns of *Science*, as signs of a living and progressive language. Responding to earlier objections, he wrote: 'Our language contains far too many words like "morphometer" or "psephometer" formed with undue consideration of a dead language and too little consideration of a living language and a progressive people. The English language would be in better shape if some people knew less Greek and Latin.'[15]

There are numerous other examples of hybrid words in science and medicine. One of them is *odoriphore*, a term which means 'odour bearer' and in biochemistry is used for a group of atoms which confer a particular smell on a chemical compound. The first half is Latin, the second Greek. The same meaning is covered in the all-Greek word *osmophore*. Both terms have found their way into the physiological literature. The rather frightening anatomical term *postzygapophysis* consist of one Latin prefix (*post*) and two Greek words (*zyg*, yoke; *apo*, away) which have been added to the Greek *physis*. For more on hybrid words in science, see Section 4.3.

To return to the positron, the rapid rise of this word at the expense of positive electron can be illustrated by the research papers carrying one of the two terms

---

[11] Anderson and Anderson (1983).
[12] Flood (1960), p. xiii.
[13] Wiley (1928). The abbreviation TV for television, a word which can be found as early as 1900, dates from 1948.
[14] Lay (1930). Since about 1880 entomologists have known a genus of flies called *teleopsis*, where the Greek word 'opsis' means 'seeing' or 'view'. It is from this word that 'optics' is derived.
[15] Fraps (1931).

in their titles. According to the Web of Science, during 1933–6 papers with either positron(s) or positive electron(s) are distributed as follows:

	1933	1934	1935	1936
positron(s)	8	17	21	9
positive electron(s)	28	18	10	0

After that time no more research papers were published in the latter category. As far as the other of Anderson's neologisms is concerned, there appeared only a single paper in the 1930s with negatron(s) in its title. As Watson Davis noted with regard to this name: 'Since the electron for forty-odd years has been called by its old name it seems unlikely that scientists will take kindly to the new one. "Positron", since its coining, has been firmly written into the literature and promises to stick.'[16] Although a rarity, negatron has not completely disappeared from the physics literature. As late as 2001, a specialist in cosmic rays wrote: 'Contrary to common usage where neg-atively charged electrons, i.e., negatrons, are commonly referred to as electrons and positively charged electrons as positrons, we will strictly distinguish . . . between the two kinds of particles to avoid confusion. We shall use the term electrons only for both kinds of particles, negatrons and positrons, combined.'[17]

With a single exception no alternative names were proposed for Anderson's positron. The exception was a young British astrophysicist who would later become an important if controversial figure in the new fields of history and philosophy of sci-ence. Herbert Dingle objected to what he (like Oppenheimer) thought was an 'ugly' word because of its hybrid character. As a possible substitute he suggested a word derived from Sophocles' tragedy *Elektra* from about 410 BC. 'I venture to propose the name "oreston" for the newcomer. The word is euphonious, pure Greek, and since, in one of the most beautiful Greek stories, Orestes and Elektra were brother and sis-ter, it implies an appropriate relation between the two particles.'[18] Dingle's fanciful suggestion, clearly a nonce word, was politely ignored.

Another word, which from a modern point of view is synonymous with positron, is the *antielectron* famously predicted and so named by Paul Dirac in 1931 as a conse-quence of his relativistic quantum theory of electrons. In his paper of that year he also introduced the term *antiproton* for the equally hypothetical negative proton. *Antineu-tron* followed four years later. This was not quite the first time that the *anti-* prefix turned up in physics, for as early as 1898 the German-born British physicist Arthur Schuster speculated light-heartedly that there might exist regions of the universe filled with 'antimatter' consisting of 'antiatoms'.[19] The letter to *Nature* in which he published the suggestion carried the title 'Potential Matter—a Holiday Dream', thus indicating that he did not take his antiatoms seriously. The word was a nonce and meant to be

---

[16] Davis (1936).
[17] Grieder (2001), p. 760.
[18] Dingle (1934). Wildman (1933).
[19] Schuster (1898).

nothing but, and yet, many years later, antiatoms and antimatter were taken seriously and routinely called so by physicists.

With the discovery of the antiproton in 1955 a few physicists speculated that there might be as much matter as antimatter in the universe. Or perhaps there existed a separate universe made up almost entirely of antimatter. In 1956 the American nuclear physicist Maurice Goldhaber came up with the protologism *anticosmos*, which, however, remained a protologism.[20] A few years later Hannes Alfvén developed the speculation into his theory of plasma cosmology (Section 2.6). He commented on the terminology associated with the new theory:

> Strictly speaking, the term 'antimatter' is a misnomer, since antimatter is every bit as much matter as its 'ordinary' counterpart.... Yet antimatter has become the standard term, and we have to accept it. What we can do, however, is to coin a new word for 'ordinary' matter: we shall call it *koinomatter*, after the Greek word 'koinos', meaning common or well known.[21]

According to Alfvén, antiatoms of all kinds existed and 'antihydrogen, antioxygen, and anticarbon may combine into complex organic compounds'. With the demise of Alfvén's alternative cosmology his words 'ambiplasma' and 'koinomatter' disappeared from the dictionary of science. However, antimatter lived on and has now appeared in more than 3,000 scientific papers. The term was not only used by physicists but also entered science-fiction literature and novels such as Dan Brown's most popular *Angels and Demons* from 2000 turned into a blockbuster movie nine years later. Brown's novel is essentially a story of one gram of antimatter produced at CERN and the disaster that will follow if it annihilates with ordinary matter.

Physicists were not alone in coining scientific terms with the *anti-* prefix. They were preceded by early immunologists who came up with currently well-known terms such as *antibody*, *antitoxin*, and *antigen* in addition to the much older *antiseptic*. The Nobel laureate Paul Ehrlich introduced 'antibody' (German: Antikörper) in 1891, but it took a while before the term won acceptance. It first appeared in the title of an article in 1904, whereas today the total number of such articles exceeds 176,000. Another German Nobel laureate in physiology, Emil von Behring, was responsible for 'antitoxin' in 1890, and 'antigen' was coined by the Hungarian bacteriologist László Detre (alias Ladislas Deutsch) in 1903.[22] When physicists later began speaking of antiparticles and antimatter, most likely they were unaware of the immunologists' usage of anti- terminology. In a discussion of Alfvén's theory a Soviet astrophysicist referred to the possibility of 'the formation of cosmic bodies and antibodies'. What he had in mind was celestial bodies made of antimatter and not the antibodies belonging to immunology.[23]

Whereas immunologists and medical doctors had no problem with antibody, according to Arthur Quiller-Couch, an esteemed writer and professor of English

---

[20] Goldhaber (1956). The term 'anti-universe' has been used in some later cosmological theories.
[21] Alfvén (1966), p. 34.
[22] Lindenmann (1984).
[23] Vlasov (1965).

literature, the term should never have been coined. Having first come across 'this abominable term' when perusing a textbook on pathology, he wrote:

> Now I do not doubt the creature thus named to be a poisonous little wretch. Those who know him may even agree that no word is too bad for him. . . . I say that for our own self-respect, whilst we retain any sense of intellectual pedigree, 'antibody' is no word to throw at a friendly bacillus. Is it consonant with the high dignity of science to make her talk like a cheap showman advertising a 'picture-drome'? It is, in fact, a barbarism, and a mongrel at that. The man who uses it debases the currency of learning.[24]

Today the antielectron and the positron are just two different names for the same elementary particle, but in the early and mid-1930s, when Dirac's theory of antiparticles was still somewhat controversial, this was not the case. Anderson's discovery of the positron was wholly unrelated to Dirac's theory and many physicists in the period took care to distinguish between the two particles and their names. Likewise, the hypothetical negative protons discussed in the 1930s and 1940s were in most cases not conceived as antiprotons. Whereas positrons became popular and even useful, few of the early texts referred to antielectrons. More recently, positrons have turned up in PET scanning technologies, PET being an acronym for positron-electron tomography. Devices of this kind go back to the 1960s.

If readers come across the term 'positon' in physics papers ca. 1938–50, they may think it is a misprint of 'positron', but most likely it is not. Nor is 'negaton' a failed spelling of 'negatron', the name that Anderson unsuccessfully suggested for the negative electron. In fact, for a decade or so the two words without *r* appeared fairly frequently in the physics literature. It seems that these words were suggested and promoted by Niels Bohr, who first used them in an article in the *Encyclopædia Britannica Book of the Year* dated 14 January 1938. Three years later, in a long letter to Robert Millikan, he discussed 'the question of nomenclature of the elementary particles' and in this connection he advocated the two words as supplements to the generic term electron:

> The whole question of nomenclature is very delicate since, on one hand, due regard must of course be taken to the experimental pioneer work and, on the other hand, it is of importance in the future literature that all names are in harmony with philological harmony. . . . [T]he word electron may be reserved for the comprehension of light electric particles and the words negaton and positon well fit for a description in which it is essential to emphasize the charge of the special particles concerned. In all these questions, however, I am of course as eager as everybody else to conform with the general custom.[25]

---

[24] Quiller-Couch (1923), p. 29.
[25] Letter of 9 April 1941, quoted in Kragh (2023).

Probably unknown to Bohr, the negaton–positon terminology had been suggested a decade earlier by Henry Hubbard, a member of the U.S. National Bureau of Standards. From 1924 onwards, Hubbard designed a series of periodic charts of the elements, starting with Rutherford's neutron (proton–electron) in the early versions. In notes to his revised chart of 1928, he wrote: 'The positive atom of electricity, the "proton" is 1845 times heavier than the negative. (Since "electron" is ambiguous, the term "positon" is suggested as more explicit, and "negaton" would avoid ambiguity as the corresponding name for the negative electron).'[26]

The term positon never became popular in English-language papers and books and negaton even less so. On the other hand, the first word was and still is routinely used in French publications simply because positon was accepted as the French word for positron, which was considered an Anglicism not suited for the French tongue. According to *Le Grand Robert de la Langue Française*, both terms are acceptable but positon is the preferred one as far as pronunciation and orthography is concerned.

The terms suggested by Bohr found favour among several physicists in continental Europe but not among American physicists who saw no reason to replace positron with positon or characterize the negative electron as a negaton. At an international but predominantly European conference held in Cracow in October 1947, a commission under the International Union of Pure and Applied Physics (IUPAP) considered the nomenclature of elementary particles. Bohr was not present, but the leading French physicist Pierre Auger proposed the same names that Bohr found appropriate, namely positon and negaton. Although the commission expressed sympathy to the proposal, it refrained from explicitly recommending it. In a non-committal formulation, the commission 'looks with favour upon the terms *positon* and *negaton* as means to distinguish between the two signs of the [electronic] charge'.[27]

In a short note of 1945 the American physicist Arthur Ruark coined the word *positronium* for a hypothetical short-lived atomic system of one electron and one positron orbiting around their common centre of mass. Four years later the first evidence for the exotic atom was reported, and today positronium chemistry and physics has grown into an established branch of science. Although the name dates from 1945, the concept goes back to 1934 when the Croatian (then Yugoslavian) physicist and astronomer Stjepan Mohorovičić suggested the existence of what he called 'electrum' and conceived as a chemical element residing in the stellar regions.[28]

If Mohorovičić's name appears familiar to geologists it is probably because of the 'Mohorovičić discontinuity' between the Earth's crust and mantle discovered in 1909 by Stjepan's father, the geophysicist and seismologist Andrija Mohorovičić. The eponym first appeared in the geological literature in 1936, but it is difficult to pronounce and spell correctly, which may have been the reason why it is usually called the 'Moho discontinuity' or just 'Moho'. The shift from Mohorovičić discontinuity to

---

[26] Quoted from *OED*, entry 'negaton'. The 'positon' as a name for the positive electron had been proposed in 1923 by the American engineer and unorthodox scientist Arvid Reuterdahl, who claimed that the positon and the electron were special cases of the more fundamental 'energon'. Kragh (1990).

[27] Futscher (1949).

[28] Ruark (1945). Kragh (1990).

the contracted Moho discontinuity seems to have started in the 1960s. Modern geologists also refer to it as the 'M-discontinuity'. The eponym honouring the Croatian seismologist is not easily recognized in Moho and not at all in M-discontinuity.

As far as the word positronium is concerned, according to modern physics nomenclature it should be divided as positr-onium and not positron-ium. Physicists use the suffix -*onium* to denote a bound state of a particle and its antiparticle of which positronium e^+e^- is one example. Another example is the quark–antiquark system which since 1975 has been known as charmonium (see Section 3.5) and yet another is the *protonium* made up of a proton and an antiproton, $p\bar{p}$. Confusingly, chemists use -*onium* in a different sense, namely to denote a class of cations of which the best known is the ammonium ion NH_4^+. Chemists as well as physicists speak of *muonium*, but whereas in chemical terminology it is an electron revolving around a positive muon (μ^+e^-), in physical terminology the same word denotes a hypothetical muon–antimuon system ($\mu^+\mu^-$). Positronium and the chemical muonium have certain properties in common with ordinary elements and have officially been assigned the chemical symbols Ps and Mu, respectively.

3.2 Meson or mesotron?

At the time when Anderson was awarded the 1936 Nobel Prize for the discovery of the positron, he and his co-worker Seth Neddermeyer had obtained preliminary evidence for yet another elementary particle in the cosmic rays. When they announced the discovery the following year they called it a 'heavy electron' or 'penetrating particle'. A new, shorter, and more appropriate name was proposed in a note to *Nature* published 12 November 1938: 'We should like to suggest . . . the word "mesotron" (intermediate particle) as a name for the new particles. It appears quite likely that the appropriateness of this name will not be lost, whatever new facts concerning these particles will be learned in the future.'[29] The two American physicists knew that the particle found in the cosmic rays puzzled the theorists and therefore wanted to base the name on a firm empirical fact, namely that it was heavier than the electron and lighter than the proton.

In fact, 'mesotron' was not the original choice of Anderson and Neddermeyer but due to their boss Millikan. Anderson recalled that at first the note to *Nature* referred to 'mesoton' without *r*, but then Millikan intervened:

He immediately reacted unfavorably and said the name should be *mesotron*. He said to consider the terms *electron* and *neutron*. I said to consider *proton*. Neddermeyer and I sent off the *r* in a cable to *Nature*. Fortunately or not, the *r* arrived in time, and the article appeared containing the word *mesotron*. Neither Neddermeyer nor I liked the word, nor did anyone else that I know of.[30]

---

[29] Anderson and Neddermeyer (1938).
[30] Anderson and Anderson (1983), p. 118.

Although Anderson and Neddermeyer never referred to 'mesoton' in print, this term was used by several authors in the late 1930s and early 1940s. For a period of time, it seems to have been the favoured word in France, possibly for the same linguistic reasons that made 'positon' more popular than 'positron'. Millikan, however, was committed to 'mesotron', a term he promoted and advocated to Bohr early on. 'I am writing to you to express the hope that the name for this particle which Anderson has suggested and which seems to be the most appropriate, namely "Mesotron" . . . will be generally adopted.'[31]

Although the Caltech name was widely used for several years, eventually it lost out to the abbreviated form *meson* first suggested by the Indian physicist Homi Bhabha in early 1939. A few weeks earlier he told Bohr about his discussions with colleagues in Cambridge, urging Bohr to support the renaming: 'I have sounded various people on this point in Cambridge, including Dirac, & they all agreed that meson is better than mesotron. [Maurice] Pryce and I are, therefore, going to see if we cannot get either [Ralph] Fowler or [Charles Galton] Darwin or both to write a letter to Nature on this subject, and we should therefore very much value your views.'[32] Bohr agreed that Bhabhas's meson was preferable to the Anderson–Millikan mesotron.

As Bhabha argued in his article in *Nature*, the *tr* in mesotron was redundant, 'since it does not belong to the Greek root "meso" for middle; the "tr" in neutron and electron belong, of course, to the roots "neutr" and "electra"'. He consequently suggested to call the new particle a meson instead of a mesotron. In England, Darwin commented in public on the philological aspects of the names. As he pointed out, the termination -*tron* was in wide use for various instruments, including the cyclotron and dynatron, and it would only cause confusion to use it also for particles. Sure, one had to take into consideration the words electron and neutron, such as Millikan had done, 'but the division of these as elec-tron and neu-tron is one which the most illiterate would scarcely make.'[33] Darwin apparently counted Millikan among the illiterates. He suggested that -*on* taken by itself might be the most natural suffix for a particle, meaning a renaming of positron and electron to *poson* and *negon*, respectively. Darwin's proposal was ignored.

For about a decade both terms were used for the Anderson–Neddermeyer particle, generally with Americans preferring 'mesotron' and Europeans 'meson'. In June 1939 a large conference on cosmic rays took place in Chicago with Arthur Compton as its principal organizer. In a preface to the proceedings he referred to the lack of agreement regarding nomenclature: 'In the original papers and discussion no less than six different names were used. A vote indicated about equal choice between *meson* and *mesotron* with no considerable support for *mesoton, barytron, Yukon* or *heavy electron*.'[34] Continuing his campaign for mesotron Millikan addressed Bhabha on the terminological issue:

[31] Millikan (1939).
[32] Bhabha to Bohr, 17 December 1938, quoted in Kragh (2023).
[33] Darwin (1939). Munns (2017).
[34] Compton (1939).

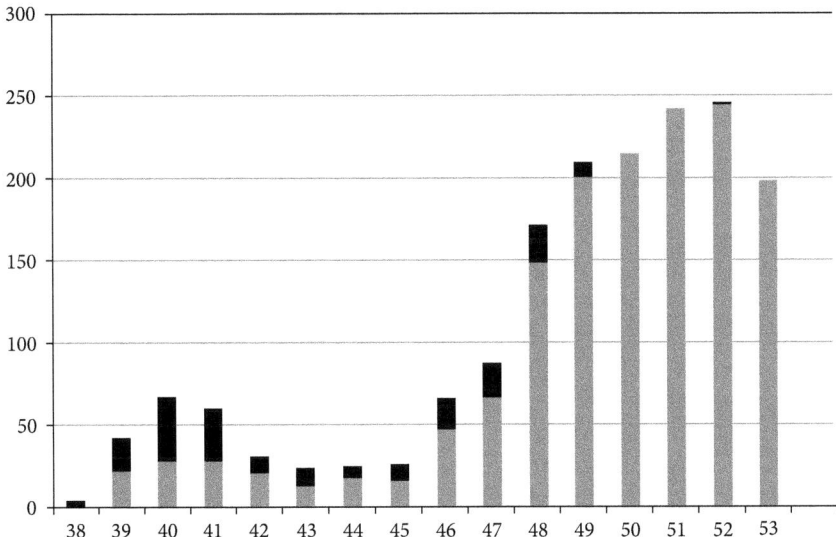

Figure 3.1 Number of research articles 1938–53 with 'meson(s)' (grey) and 'mesotron(s)' (black) in the title. Data from Web of Science.

I have been looking up recently the current usage of the terms 'mesotron' and 'meson', and I estimate that there are about five times as many of us who are preferring the original name mesotron to those who have changed over. Would it not be desirable for us to get together on the name mesotron, which is most generally used and which the majority of us think in view of the fact that we already have electrons and positrons is in every particular the logical one to use for another particle?[35]

However, the anti-meson campaign met with little success. Despite the opposition of Millikan and other American physicists, the use of meson began to increase and that of mesotron to decrease. To give an indication of the trend, in the period 1938–43 the situation was even with a total of 116 papers with mesotron in the title and 112 with meson. During the next six-year period 1944–9 the corresponding numbers were 89 and 495 (Figure 3.1). After 1949, only a single research paper was published with mesotron in the title. As another example of the complete victory of meson over mesotron, consider the 1948 Solvay physics conference devoted to elementary particles. Whereas meson(s) appeared more than 800 times in the proceedings, mesotron(s) was conspicuously absent.

The British-American physicist William Swann shared Millikan's strong dislike of meson. As Millikan told a Californian colleague in a letter of 1946, Swann 'feels very vigorously about it that the use of "meson" is a very unfortunate one, not only because it violates all historical and etymological properties but is also so close in name to a word [maison] that has come in French to be used as a word for a house of ill fame

---

[35] Quoted in Kragh (2023).

[a brothel], that he will not tolerate its use at all.'[36] Indeed, Swann stuck to mesotron as late as 1954, in the last research paper ever to include the term in its title.

As alluded to by Compton in 1939, at the time there were half a dozen or more names in play for what eventually became the meson. Some of them were mentioned casually in papers just once or twice, or they appeared only informally. One of those names was *dynatron*, which was an unfortunate term given that it was in wide use for an electronic tube device invented by Albert Hull in 1918. In Section 3.1 we saw an analogous example of how a name used in wireless technology, namely negatron, was also proposed in elementary particle physics. Another of the ephemeral meson names of the late 1930s was the eponymous *yukon*, a word referring to the Japanese physicist Hideki Yukawa, who in 1935 had predicted on theoretical grounds a nuclear particle with a mass of approximately 200 times that of the electron.

For nearly a decade it was generally but erroneously assumed that the hypothetical Yukawa particle was the same as the observed Anderson–Neddermeyer particle, alias the mesotron or meson. With the recognition in about 1947 that there are at least two kinds of mesons, the question of terminology needed to be reconsidered. For a couple of years, the generally adopted terms became μ-meson and π-meson (mu-meson, pi-meson), where the first name related to the Anderson–Neddermeyer particle and the second to the strongly interacting Yukawa meson. In the early 1950s physicists began changing to the abbreviated and currently used names *muon* and *pion*. These words seem to have been introduced by Enrico Fermi, first orally in 1950 and then in print the following year.[37] The number of papers with one of the two names in the title gives an indication of how fast they were adopted: 1950–2: 41 (3) and 1953–5: 239 (16), where the numbers in parentheses are those with muon in the title. The term meson is still in use in modern particle physics, but now for a class of strongly interacting particles composed of a pair of a quark and an antiquark. While the pion is a meson, the muon is not or not any longer.

From a linguistic point of view, the shortened forms muon and pion are examples of what is called *middle clipping*, which is the removal of a syllable or word element from the middle of a word.[38] Thus muon = mu(mes)on and pion = pi(mes)on. Other physics examples are the instrument ammeter = am(pere)meter, the unit Hz = H(ert)z, and Eddington's wavicle = wav(e part)icle referred to in Section 4.3. Words can also be shortened by front or back clipping, where either the first or the last word element is removed, as in phone = (tele)phone, lab = lab(oratory), and math = math(ematics). Finally, both front and back clipping can be used, such as is the case with flu = (in)flu(enza).

3.3 Nuclear particles

Apart from the ordinary hydrogen isotope H-1 or ^{1}H, all atomic nuclei are made up of tightly bound *protons* and *neutrons*, which are also known collectively as

[36] Quoted in Kragh (2023).
[37] Fermi (1951).
[38] Caso (1980).

nucleons. Although these are elementary particles, they are composed of even more fundamental particles called *quarks*. How and when did these and related names come into being?

The words 'proton' and 'neutron' are usually credited to Rutherford in 1920 although he did not refer to them in print that year. Moreover, in both cases they can be found many years earlier in the scientific literature if in somewhat different meanings. As far as proton is concerned, the word appeared in *Philosophical Magazine* of 1908, in a speculative paper written by two amateur physicists, Alfred and Augustus Jessup. The neutron, conceived as an ether particle made up of a positive and a negative unit charge, can be found even earlier, namely in 1899 and in the same widely read physics journal. 'In æther showing no electric charge each negative electron is united with a positive electron to form an analogue of a material molecule, which might conveniently be called a neutron.'[39] The author was William Sutherland, an Australian physicist, whose suggestion was taken over by the eminent German chemist Walther Nernst in his textbook *Theoretische Chemie* first published in 1903.

There is only a *t* to distinguish 'neutron' from 'neuron', but of course the two terms are entirely different in both substance and etymology. Nonetheless, it may not be out of place to digress momentarily on the neutron-minus-*t* word. *Neuron* (νεῦρον) and its plural form *neura* (νεῦρα) are classical Greek words which were used by some of the ancients in relation to the anatomy of the nervous system. However, this usage seems to have been forgotten and was only reintroduced in 1891 by the German anatomist Heinrich Wilhelm Waldeyer, who used neuron as an alternative for 'nerve unit'. Two years earlier he had coined another successful term, namely *chromosome* based on the Greek names for colour (*chrōma*) and body (*sōma*).[40] Waldeyer's neuron terminology soon won general recognition except there was some discussion concerning the parts of the nervous system to which it should apply. Moreover, in French the term was adopted with an *e* in the end (neurone), a spelling form which was also used by many British and American specialists.[41] The present trend in English medical language is to use the spelling neuron only.

As far as the proton–neutron case is concerned, it is one of many cases in which there are little-known precedents to a new scientific word. According to the standard definition of *neologism*, it is 'a relatively recent or isolated term, word, or phrase that may be in the process of entering common use, but that has not been fully accepted into mainstream language.'[42] Thus, it does not need to be absolutely original or freshly coined. If it is, it is better called a *protologism*, that is, a brand new word which has not yet been communicated by others or is known only to a very limited group of people. Neither Jessups' proton nor Sutherland's neutron was well known, if known at all, in the physics community and in all likelihood Rutherford was unaware of them. They were protologisms, whereas Rutherford's identical words of 1920–1 qualify as neologisms.

---

[39] Sutherland (1899).
[40] On Waldeyer and the term chromosome, see Zacharias (2020).
[41] Metha et al. (2020).
[42] Malmkjaer (2006), p. 601.

In 1921, Rutherford explained how he had come to the word proton:

> The question for suitable name for this unit was discussed at an informal meeting of a number of members of Section A of the British Association at Cardiff this year [1920]. . . . Finally, the name 'proton' met with general approval, particularly as it suggests the original term 'protyle' given by Prout in his well-known hypothesis . . . The need of a special name for the nuclear unit of mass 1 was drawn attention to by Sir Oliver Lodge at the Sectional meeting, and the writer then suggested the name 'proton'.[43]

The new word was not an immediate success and initially some alternatives were suggested. Given that the proton was identical to the hydrogen ion H^+, why not choose a name that reflected the close relation to the lightest of all elements? Lodge suggested 'hylon', whereas Rutherford's former collaborator Frederick Soddy preferred the more direct contraction 'hydrion', which for a period was used by some chemical authors. Other scientists found it unnecessary with a new name for the 'positive electron'. In his 1924 Faraday Lecture, Millikan addressed the issue of terminology:

> When used without a prefix or qualifying adjective, the word electron may indeed signify . . . both the generic thing, the unit charge, and also the negative member of the species, precisely as the word man in English denotes both the genus homo and the male of mankind. There is no gain in replacing 'positive electron' by 'proton', and there is a very distinct loss, logically, etymologically, and historically.[44]

The neutron is a different story, for Rutherford thought of it as a proton–electron composite (p^+e^-) and not as an elementary particle. At first he conceived the neutron—'an atom of mass 1 which has zero nucleus charge'—as if it were an atom of an unusual chemical element: 'Its presence would probably be difficult to detect by the spectroscope, and it may be impossible to contain it in a sealed vessel'.[45] Given that the neutron had no external electron it was a strange element indeed, and yet the German chemist Andreas von Antropoff, a professor at the University of Bonn, decided in 1926 to place it in his own version of the periodic table. As a name for this element of atomic number zero ($Z = 0$) he proposed 'neutronium'.[46]

When James Chadwick discovered the 'real' neutron in 1932, at first he took over both Rutherford's name and conception of it. Misled by imprecise mass measurements he was convinced that the neutron was formed by a close alliance of a proton and an electron. It took more than a year before the neutron became a proper elementary particle. From a linguistic point of view, neutron is a hybrid of the same category as positron. It consists of the Latin form *neutr-* with the Greek ending *-on*. As pointed out by Charles Whitmore in 1955, should the name be in pure Greek, it

[43] Note by Rutherford in Masson (1921).
[44] Millikan (1924).
[45] Rutherford (1920).
[46] Fontani, Costa, and Orna (2015), p. 444.

would have the longer and more awkward name *oudeteron*.[47] The Greek word *oudeis* (οὐδείς) means 'not even' or 'not anything'.

The rise of experimental nuclear physics in the 1930s was closely connected with new machines that could accelerate charged particles to high energies. The most important and best known of these 'accelerators' was the *cyclotron* invented by Ernest Lawrence in the early part of the decade. The frequently used *cyclo*- prefix comes from the Greek *kyklos* (κύκλος), meaning circle or ring. An encyclopaedia provides 'all round education'. The name cyclotron had been informally used early on by Lawrence and his group in Berkeley, but it was only officially introduced in a footnote to a paper from September 1935 written by Lawrence and two collaborators and then as just an alternative to a more cumbersome name:

> Since we shall have many occasions in the future to refer to this apparatus, we feel that it should have a name. The term 'magnetic resonance accelerator' is suggested. . . . The word 'cyclotron', of obvious derivation, has come to be used as a sort of laboratory slang for the magnetic device.[48]

The US patent granted Lawrence on 20 February 1934 was not for the cyclotron, but for an 'apparatus for the acceleration of ions'. In the post-World War II period the cyclotron was followed by a family of new, more expensive and ever more powerful accelerators, most of them with the characteristic -*tron* suffix (synchrotron, Bevatron, Cosmotron, Tevatron, etc.). The term synchrotron is formed by blending two words and omitting parts of them: synchrotron = synchro(nize) + (cyclo)tron. Another example of word formation of this kind is permafrost = perma(nent) + frost, which will be dealt with in Section 4.1.

The family of high-energy accelerators became known as 'atom smashers' in the popular press, a term that entered the vocabulary at about 1950. Another early but less known atom smasher was the *betatron* constructed in 1940 by Donald Kerst as a machine accelerating electrons up to energy of 2.3 MeV.[49] Kerst initially referred to the machine as an 'induction accelerator' and it was only after an internal naming contest that betatron' was chosen, a name that refers to the high-speed electrons emitted by beta decay. Other proposals included 'inductron' and 'rheotron'. The first syllable in the latter term is derived from the Greek *rheos* meaning stream, flow, or current. *Rheology* is the branch of physics dealing with the flow of liquids and other states of matter. A *rheostat* is a variable resistor used to control the electric current through a circuit.

Although accelerator may be generally associated with high-energy physics, the term has or has had a number of other connotations, some of them scientific and others not. Chemists have used the word to designate a substance that causes a chemical process to run faster, but now they usually call such a substance an 'accelerant'. In contrast to catalysts, accelerants may be consumed in the process in which they take part (see also Section 5.2).

---

[47] Whitmore (1955).
[48] Lawrence, McMillan, and Thornton (1935).
[49] Livingston (1966), pp. 186–201.

The close similarity between the proton and the neutron, apart from their electric charge, was first expressed terminologically by the Dutch physicist Frederik Belinfante in 1939. His common name for the two nuclear particles was *nuclon*, but this word was only used by a handful of physicists and by 1942 it had passed into oblivion. The Danish theorist Christian Møller at first liked the name but soon had doubts if it was correct from a philological point of view. He consulted his colleague Léon Rosenfeld on the matter. 'Is "nucle" really the root, meaning that "us" [in nucleus] and "on" are endings?' he asked. Rosenfeld, who knew the classical languages, answered at length:

> Physicists have used the innocent *Greek* ending '-on' in the sense of 'elementary particle' because the first isolated elementary particles happened to have the purely Greek names 'electron' and 'positron'. ... Now the word 'neutr-on' has been formed with this Greek ending and the *Latin* root 'neutr-'. Moreover, 'posit-on' and 'negat-on' which really should be understood as abbreviations of the more correct words 'positive-on' and 'negative-on'. Again, 'meson' and 'deuteron' are purely Greek words. For 'nuclear particle' one has the choice between the Greek word 'karyon' (which is not very attractive) and the word 'nucle-on' formed like 'neutron' with the Latin root 'nucle-'. You will see that the root is 'nucle-' and not 'nucl-' from e.g. the English adjective 'nuclear' (nor 'nuclar'!).[50]

Stimulated by Rosenfeld's comments, in a brief letter to *Physical Review* published in early February 1941, Møller suggested to replace Belinfante's 'nuclon' with *nucleon* as the common name for protons and neutrons. While the first was a protologism, the latter was a neologism that quickly caught on and within a few years had entered the physicists' standard vocabulary.

The term 'nucleonics' followed after 'nucleon', but contrary to what one would expect without the former being derived from the latter or having any connections to it. Nucleonics was coined as a term analogous to electronics and like this word it mostly referred to applied and engineering physics. The discipline of nucleonics grew out of the Manhattan Project, the name being coined in a 1944 memorandum written by Zay Jeffries, a General Electric engineer. According to the memorandum, 'following the lead of "electronics", we propose that the word "nucleonics" shall refer to both science and industry in the nuclear field'.[51]

Nucleons are strongly interacting particles or *hadrons*, a name introduced in 1962 by the Russian physicist Lev Okun, who coined it from a Greek word ἁδρός meaning large or massive. Moreover, nucleons are composed of three quarks, the proton of two 'up quarks' and one 'down quark' and the neutron of two down quarks and one up quark. The latter terminology was a result of the accumulating evidence in the early 1960s that hadrons and antihadrons consist of more fundamental entities with fractional electric charges $\pm 1/3\,e$ and $\pm 2/3\,e$. For these still hypothetical sub-hadronic particles, the young physicist George Zweig suggested in 1964 the name *aces*, but the name failed to catch on and was soon forgotten.

---

[50] Letter of 7 December 1940, quoted in Kragh (2023).
[51] Johnson (2012), p. 127.

Physicists adopted another name, catchier and more literary, proposed at about the same time by Murray Gell-Mann, who in 1969 would receive the Nobel Prize for his work on elementary particles.[52] Gell-Mann famously found inspiration for his neologism *quark* in a passage in James Joyce's notoriously difficult novel *Finnegan's Wake* from 1939. The word only appears once in the novel and with no indication whatsoever of its meaning:

> Three quarks for Muster Mark!
> Sure he has not got much of a bark
> And sure any he has it's all beside the mark.

Gell-Mann may have been fascinated not only by the queer name 'quark', but also that Joyce referred to three of them. According to the recollection of the physicist Stephen Wolfram, the word originated in a discussion between Gell-Mann and Richard Feynman. One day 'Murray came to him [Feynman] very excited and said he'd found the word "quark" in James Joyce. In telling this to me, Feynman then went into a long diatribe about how Murray always seemed to think the names for things were so important. "Having a name for something doesn't tell you a damned thing", Feynman said.'[53] Perhaps not, but quark became very popular, not only in the physics community but also in the broader public. Google Scholar gives more than one million results for the word and according to *OED* it occurs about twice per million words in modern English. The word quark may be nonsense, but it is definitely not a nonce.

Although Feynman eventually converted into what he called a 'quarkerian', he had his own ideas of the constituents of hadronic particles.[54] He suggested that they were made up of hard point-like particles that in certain respects differed from the quarks. For the hypothetical constituent he coined the word *parton*, a name simply based on 'part'. Compared to Gell-Mann's quark the neologism was unimaginative and inelegant, but it was important enough to make its way into the *OED*.

Quarks are not alone in the tiny atomic nuclei. They have to be bound together, which is done by exchange particles called *gluons*, a name which was introduced in the late 1960s and obviously is based on the term glue combined with the standard termination -on. At the time the gluons were hypothetical objects, but a decade later they were indirectly observed in experiments. In a comment on the names proposed by American nuclear and particle physicists, the historian John Heilbron commented: 'The playful names coined by high-energy physicists have been criticized as inelegant, non-ancient, capricious and misleading. No doubt it is unlucky that quark means garbage in German, but gluon is an inspired put-on: it looks Greek, means nothing in German, puns in English and satisfies Bacon's requirement that a word express a clear and distinct idea.'[55]

To go back in time to the days of early nuclear physics, people were in need for a unit that would express the probability that a certain nuclear reaction would occur,

---

[52] Johnson (1999), pp. 214, 224–5. Gell-Mann (1995), p. 180.

[53] https://blogs.scientificamerican.com/observations/remembering-murray-gell-mann/

[54] Feynman used the term 'quarkerian' in a 1973 interview by Charles Weiner. https://www.aip.org/history-programs/niels-bohr-library/oral-histories/5020-5

[55] Heilbron (2002).

what was and still is called its 'cross section'. The unit represented by the cross-sectional area of a typical atomic nucleus, with a radius of about 5×10^{-13} cm, became known as a *barn* equal to the microscopic area 10^{-24} cm^2. The curious origin of this unit with its undignified name is related to the secret Manhattan Project in the early 1940s, when two physicists working at Purdue University in Pennsylvania, Marshall Holloway and Charles Baker, came up with the name in December 1942. At first they considered 'oppenheimer' and 'bethe' in line with the tradition of naming a unit after some great and relevant (and normally deceased) scientist. But they found both names to be inappropriate and then suggested a non-eponym which could also serve the purpose of a code word, which was desirable at the time:

> The rural background of one of the authors led to . . . the 'barn'. This immediately seemed good and further it was pointed out that a cross section of 10^{-24} cm^2 for nuclear processes was really as big as a barn. Such was the origin of the 'barn'.[56]

The suggestion appeared in a secret Los Alamos report declassified in 1948 and was officially adopted two years later at a meeting in Paris by a new Joint Commission on Standards, Units and Constants of Radioactivity. However, the commission only did so because there was no other option. In a report it was stated that the commission 'felt that there is not any new unit involved which would demand an international definition, but that the expression "barn" . . . should be accepted since it is in common usage in the United States'.[57] In 1960 IUPAP recommended the unit. As Holloway and Baker realized, the barn (symbol b) is too big a unit for most nuclear and particle reactions, which is the reason why physicists often used subunits such as nanobarn (nb = 10^{-9} b) and picobarn (pb = 10^{-12} b).

The prominent British biologist and popular science writer Lancelot Hogben later decried the barn as 'the nadir of verbal vulgarity in natural science'.[58] Whether because the term was considered vulgar or for some other reason, in 1976 the Council of Ministers of the European Economic Community (EEC) proposed the abolition of the barn and that the area unit should instead be written as 100 fm^2. The length unit fm = fermi = 10^{-15} m was named after the famous Italian-American nuclear physicist Enrico Fermi and first introduced in a paper from 1956: 'We shall measure all distances in terms of 10^{-15} m as a unit and shall call this unit the fermi'.[59] Confusingly, the term fermi was also used as an area unit to replace the barn (1 fermi = 1 barn = 10^{-24} cm^2).

The eponymous fermi length unit is now obsolete as it has been replaced by the more systematic 'femtometre' (femtometer in American spelling) conveniently with the same symbol fm. The prefix *femto-* might be thought to be of Latin or Greek origin like *milli-* and *nano-* but is actually derived from the Danish and Norwegian 'femten' meaning fifteen. It was officially adopted by the General Conference on Weights and

---

[56] Holloway and Baker (1972). Glasstone (1958), p. 293. 'Big as a barn' is a common American colloquialism.
[57] Paneth (1950).
[58] Hogben (1970), p. 146.
[59] Hofstadter (1956).

Measures in 1964 and at the same meeting the prefix *atto-* (10^{-18}) was adopted. This too is of Danish-Norwegian origin: atten = eighteen.

3.4 Photons and neutrinos

In 1905 Einstein introduced the revolutionary hypothesis that light consists of small parcels of energy given by the relation $E = h\nu = hc/\lambda$, where ν is the frequency, λ the wavelength, h Planck's constant, and c the speed of light. He spoke of 'light quanta' (*Lichtquanten*), a term rarely used any longer. Einstein's light quantum was in the late 1920s renamed the *photon*, one of the most popular and widely used terms not only in physics but also in many other contexts. It is commonly stated in textbooks and articles that what Einstein called a light quantum was in 1926 christened a photon by the American physical chemist Gilbert N. Lewis. However, this is a gross mistake, for although Lewis did indeed write of photons in 1926, he used the name for an unconventional conception of light that had nothing in common with Einstein's view.[60] Moreover, a few scientists had used the term before Lewis promoted it in connection with his own unorthodox theory of the nature of light.

Leonard Troland, an American physicist and psychologist, was much interested in photometric measurements of light impinging on the human eye, and it was in this context that he invented the term 'photon' in 1916, a decade before Lewis. As Stoney's electron was not a particle but a unit of electric charge, so Troland's photon was not a particle but, in his case, a unit for the illumination of the retina. Although Troland and a few other American writers continued to use photon until about 1923, the psycho-physiological unit remained little known. The same was the case with roughly similar works in Europe, where the Irish physicist John Joly in 1921 proposed a photon theory of light stimulation. In France, René Wurmser adopted the same word in a 1924 paper on photosynthesis published in *Annales de Physiologie*. These early works were even less noticed than Troland's and they were unknown to Lewis when he independently reintroduced the name in 1926. Incidentally, the word *photosynthesis* was relatively new as it had been coined by an American botanist only in 1893 as a replacement of the broader term 'assimilation' used until then.[61]

Given that photon was originally conceived as a unit of light intensity associated with photochemical and biological processes, it is ironic that Einstein laid name to a unit of a roughly similar kind. Based on thermodynamics and quantum theory, in a paper of 1912 he had formulated a photochemical law stating that for each light quantum (or photon) absorbed by a chemical system only one molecule is activated. This second law of photochemistry, as it is often called, had been enunciated from different arguments four years earlier by the German physicist Johannes Stark. It is therefore known as the Stark–Einstein photochemical law although Stark's name is often left out. To honour Einstein for his contribution, in about 1930 some chemists began referring to the unit 'einstein' for the energy of one mole—Avogadro's number—of

---

[60] For more details on the naming history of the photon, see Kragh (2014a) and Hentschel (2018), pp. 27–38.

[61] Gest (2002).

photons regardless of their frequency. 'The name one light equivalent or one Einstein, symbol E, is proposed for 6.06×10^{23} quanta.'[62] The einstein unit with symbol E was mostly used in photobiology, but it never became popular and is today obsolete.

Far from being an imponderable light quantum, according to Lewis the photon was a carrier of light which he characterized as 'a new kind of atom'. In his article in *Nature*, he took 'the liberty of proposing for this hypothetical new atom, which is not light but plays an essential part in every process of radiation, the name *photon*.'[63] This was obviously an entirely different meaning of the term than the one associated with Einstein's light quantum. What happened next is puzzling, namely that a large part of the physics community adopted Lewis' word without caring or even knowing about his theory of light. In contrast to the name photon, this theory was a complete failure such as indicated by the lack of references to Lewis' *Nature* article. For a period of more than forty years it was only cited twice, in both cases in 1927 and by Lewis himself.

The rapid assimilation of Lewis' name—or the Troland–Lewis name—is shown by a count of titles of research papers including photon(s) and light quantum (quanta):

	1926–35	1936–45	1946–55
light quantum (quanta)	20	0	5
photon(s)	19	29	243

For example, the participants in the famous fifth Solvay conference held in October 1927 used photon and not light quantum in their published reports appearing in the conference proceedings titled *Électrons et Photons*. The name became disseminated to the young generation of physicists through the early textbooks in quantum mechanics. While the first of these, George Birtwistle's *The New Quantum Mechanics* published in 1928 and with preface dated 1 October 1927, stuck to light quanta, Paul Dirac's highly influential *Principles of Quantum Mechanics* of 1930 kept consistently to photon.

Like other popular science words, photon entered into several other contexts and derived names. The adjective 'photonic' appeared as early as 1929 (*OED*). More importantly, as electron gave rise to 'electronics', photon gave rise to the noun *photonics*, a term that emerged in the late 1960s. The French physicist Pierre Aigrain is often credited as the father of photonics if not strictly speaking of the word. This new branch of applied physics is about harnessing photons, as electronics is about harnessing electrons and nucleonics is about harnessing atomic nuclei. According to a paper of 1975, 'photonics is the name given to the field of science covering systems in which photons are the principal carriers of information.'[64]

As mentioned in Section 3.3, by 1930 the word neutron existed as a term for Rutherford's proton–electron composite which shortly later metamorphosed into

---

[62] Bodenstein and Wagner (1929), p. 456.
[63] Lewis (1926).
[64] Krasnodebski (2018).

Chadwick's elementary particle of the same name. Confusingly, in this period there was a third neutron candidate of a very different kind, namely a hypothetical particle that soon came to be known as the *neutrino*. In an open letter to Lise Meitner and other physicists of 4 December 1930, Wolfgang Pauli suggested that 'there could exist in the nuclei electrically neutral particles that I wish to call neutrons, which have spin ½ and obey the exclusion principle'.[65] Given that neutron was already in use—although Pauli may not have been aware of it—the name was unfortunate and initially caused some confusion. In a lecture given in Pasadena in June 1931, Pauli still referred to his particle as a neutron residing in the nucleus. Oppenheimer, who was among the participants listening to Pauli, suggested the name 'magnetic neutron' which he and his co-author John F. Carlson used in a paper of 1932.

The present name for the magnetic neutron was coined by Fermi, who in 1933, and informally possibly earlier, invented the term *neutrino*, an Italian diminutive of neutron. The Italian suffix *-ino* means 'little' or 'cute', while *-one* is the suffix for something big or tall. Fermi's name was a genuine neologism. Moreover, it was more than just a simple renaming of Pauli's particle, for Fermi conceived the neutrino to be a massless particle that did not belong to the nucleus but was created together with an electron in the very moment when the nucleus decayed. Francis Perrin, a French physicist, may have been the first to use the term in a scientific publication, in December 1933, and four months later it appeared in the title of a paper written by Hans Bethe. Having first been coined and accepted, the name and its associated concept remained a stable part of the physicists' vocabulary. Although it was later recognized that there are three different kinds of neutrinos, no alternative name has ever been proposed for the elusive neutral particle eventually discovered in experiments of 1956.

The neutrino and the electron belong to the same family of elementary particles. According to the present standard model, massive particles are either quarks or *leptons*, where those in the latter group comprise three charged particles, the electron, the muon, and the much heavier tau (τ) particle, sometimes called a tauon. To each of these correspond a type of neutrino. The term 'lepton' was originally proposed by Christian Møller—the physicist also responsible for nucleon—in 1947, but in the more restricted sense of only electrons and their associated neutrinos. Møller believed that somehow these two particles belonged together and he looked for a common name for them. He once again asked his friend Rosenfeld for assistance in what he felt was a serious philological question. In a letter of 12 October 1946, he explained:

I have presented the question before a philologist, who after long considerations proposed a name derived from the Greek leptos = fine, delicate. I could think of using the word *Lepton*. It would be appropriate both with regard to content and form—and it sounds good in both English, French, and German, and it cannot be confused with other words referring to elementary particles. People I have talked with so far about the proposal believe that it is appropriate. What's your opinion?[66]

---

[65] Brown (1978) with a full translation of Pauli's letter.
[66] Quoted in Kragh (2023).

Rosenfeld liked the word, which first appeared in print in a paper by Møller from 1947. Over the next few years it and the corresponding adjective 'leptonic' began to be used in the physics literature.

The discovery in the mid-1970s of the tau (τ) particle showed that Møller's name was not quite as appropriate as he had thought. With a mass almost twice that of a proton ($m_\tau = 3484m_e$) the particle is anything but light, and yet it is classified as a lepton. Evidence for the tau was first found in electron–positron experiments by a group led by the American physicist Martin Perl, who in 1995 would be awarded a Nobel Prize for the discovery. Perl and his group first referred to the new particle with the letter U (for unknown) and after further evidence had shown it to be real, they considered a lower case Greek letter in analogy with the letter μ for the muon. They settled on τ because the particle was the third charged lepton to be found, after the electron and the muon, and triton (τριτον) means third in Greek. Aware of the large mass of the tau, Perl concluded that a particle's elementarity has nothing to do with the size of its mass. 'Indeed, we emphasise this by referring to the τ by the oxymoron "heavy lepton".'[67]

Are there more elementary particles with the termination -ino than the neutrino? Yes, there are, but none of them exist or at least, none have been detected so far. In 1946 the French physicist Jean Thibaud hypothesized the existence of tiny electron-like particles with mass and charge much smaller than those of the electrons. He appropriately called them *electrinos*, but neither the name nor the concept survived for more than a few years. The electrino does not exist.

Many years later the -ino suffix returned, this time in connection with particles predicted by supersymmetric and similar theories (see Section 3.5). Perhaps they belong to the same depressing category as the nonexistent electrino, but this it too early to say. One of those hypothetical particles is the *photino*, a name which should not be read as a contraction of photon and neutrino but as a very little (and perhaps cute?) photon. More precisely, it is a superpartner of the photon (see Sections 2.5 and 3.5). Although the supersymmetric photino dates from the late 1970s, the name can be found earlier, namely in a footnote to a paper that Roger Penrose, Nobel laureate and famous mathematical physicist, wrote in 1968 on his so-called twistor theory. Penrose also introduced the 'anti-photino'.[68]

3.5 Wimps and other exotic words

The adoption of the term barn as a unit for cross section in nuclear reactions is an early example of how unconventional and sometimes whimsical words entered the nomenclature of physics in the post-World War II era. As noted by the linguist Arthur Caso, there was in physics and some other branches of science a trend toward improvisation and an almost complete break with the established tradition of

---

[67] Perl (1978). In nuclear physics, 'triton' is sometimes used for the nucleus of the H-3 isotope in analogy with 'deuteron' for the H-2 isotope. Triton is also the name of one of Neptune's moons.

[68] Penrose (1968).

forming new words preferably from Greek roots.[69] While scientists in the past more or less consciously strove to create a terminology removed from familiar associations, many of their modern colleagues thought that scientific words should operate within the framework of ordinary speech. Elaborating on Caso's insight, another linguist wrote:

> While a great deal of scientific vocabulary is still formed in the traditional manner from Latin and Greek roots and affixes, more of the words of modern science manipulate common elements in ways which do not conform to the same linguistic requirements expected in the past by both the scientist and the layman. . . . New meanings are now more freely created by composing them from known words through the use of conjoining, abbreviating, and metaphoric strategies.[70]

The linguists were not alone in noting the new trend in terminology. As observed by one physicist, 'In the latter half of this century . . . the language of physics is undergoing a shift from its classical and honorific traditions to a metaphoric mode.'[71]

In 1948, the year in which barn first appeared in public, American physicists Parry Moon and Domina Spencer proposed a systematic reform of physical terminology that included a large number of new words based on Greek. As they optimistically stated, 'Experience has shown that many scientific words taken from ancient Greek are accepted by the entire scientific world, regardless of language.'[72] Moon and Spencer argued that a great deal of the traditional words in physics were ambiguous or in other ways unsuitable and therefore ought to be replaced by new words derived from Greek. Here are a few of their proposed names:

force = *kratos*;	power = *dynamos*;	charge = *elektros*
field = *zenos*;	current = *hermos*;	flux = *phantos*

As one might expect, this attempt to reform the language of physics and turn back to the purity of Greek terminology was not welcomed by the physics community. None of the about two dozen of Moon–Spencer protologisms made it to become a neologism actually used in the scientific literature.

Attempts to apply quantum mechanics to Einstein's theory of gravitation—what broadly speaking is known as quantum gravity—began in the early 1930s. The first elementary particle associated with quantum gravity was the hypothetical *graviton* conceived as a quantized unit of gravitational energy. Today the graviton has entered more than 5,600 research papers in theoretical physics. To find its origin we have to consult an obscure paper from 1934 in the Soviet ideological journal *Pod Znamenem Marksizma* (Under the Banner of Marxism) in which Dmitri Blokhintsev and F. Galperin speculated about a possible connection between gravitation and the new neutrino:

[69] Caso (1980).
[70] Raad (1989).
[71] Stahl (1987).
[72] Moon and Spencer (1948a).

One has to surmise that the gravitational energy, just like the electromagnetic energy, are radiated by energy *quanta* rather than by energy waves—in the former case by quanta of electromagnetic energy (light quanta, photons), in the latter—by the quanta of gravitational energy (*gravitons*). . . . Just like light quanta, gravitons should have mass only when moving with the velocity of light. They are not electrically charged.[73]

Presumably unaware of the Russian paper, from about 1940 a few Western physicists began talking about 'gravitons', a natural term for the hypothetical gravitational quanta. It only became commonly used in the 1960s.

Møller's *nucleon* and *lepton*, and also Okun's *hadron*, were still in the classical tradition, such as was the *graviton* of Blokhintsev and Galperin. On the other hand, Feynman's *parton* and the contemporaneous *gluon* only paid lip service to the tradition by the suffix *-on*. Gell-Mann's *quark* belonged to the same modern category as *barn*, only it was more imaginative. When physicists discovered that there are many kinds or 'flavours' of quarks and had to find names for them, they suggested exotic and whimsical words, or what linguist Hanna Pulaczewska has called 'fantasy-names', with no connection at all to the classical languages.[74] Another and quite appropriate term for this kind of modern or postmodern names is 'anything-goes names'.[75] Although there clearly is a trend towards this category of names and away from the Greek–Latin naming tradition, scientists still occasionally propose names in the latter tradition. A noteworthy example from astrophysics is the term *asteroseismology* dating from about 1983 (Section 6.2). Another modern example is the *ekpyrotic* model of the universe so named in 2001 (Section 6.5).

The physicists described some quarks as 'strange' and some as 'charmed', and others again they characterized with terms such as 'beauty' and 'truth' (later to be replaced with the more prosaic 'top' and 'bottom'). Quarks could also have different 'colour charges' that related to their flavour and which superficially mirrored the three electrical charge states positive, negative, and neutral. The technical term 'colour' was suggested by Gell-Mann, who called the three quantum states blue, red, and white. He might as well have called them x, y, and z. Needless to say, the colour and flavour used to describe quarks are purely metaphoric. The field theory of strong interactions describing the different states of quarks is known as 'quantum chromodynamics', usually abbreviated to its acronym QCD in analogy with quantum electrodynamics = QED. The term 'chromo' is from Greek *chrōma* meaning colour and alludes to the quark colours. It appears in several other scientific names such as in the metallic element *chromium*, in *chromosome* ('coloured body'), and in *chromatography* ('colour writing').

---

[73] Quoted from the translation in Gorelik and Frenkel (1994), p. 96. *OED* refers to a speculative non-relativistic theory of gravitation proposed by the Indian amateur physicist Shah Sulaiman in 1935. However, Sulaiman's classical gravitons were entirely different from those associated with quantum gravity.

[74] Pulaczewska (1999), pp. 217–24.

[75] Roelli (2021), p. 497. 'Anything goes' may be a reference to the anarchistic theory of science proposed by the philosopher Paul Feyerabend in his book *Against Method*. As there are no rules in scientific reasoning (according to Feyerabend), so there are or should be no rules for word formation in science.

In the terminology of Caso, the charm, colours, and flavours of quarks are examples of expansion of meaning, namely that a new denotation with no relation to previous denotations is given to an established word. When asked of how he thought of the concept and word 'charm', theoretical physicist Sheldon Glashow (to whom it is due) alluded to its meaning as an amulet. 'Charm, in the sense of a magical device to avert evil', he said, referring to its crucial role in the new quark theory of elementary particles.[76] Without charm, the theory would collapse. According to the Merriam-Webster dictionary, one of the meanings of the noun 'charm' is 'something worn about the person to ward off evil and ensure good fortune'.

Another of the fantasy words of high-energy physics was the *charmonium* introduced in 1975 by Thomas Appelquist and David Politzer to denote particles consisting of a pair of charmed quark and antiquark. As recalled by Politzer in his 2004 Nobel lecture, it was only with some difficulty that the word was accepted:

> Appelquist and I hurriedly dashed off a short version of our work to *Physical Review Letters*, where it was immediately and unequivocally rejected by senior editor . . . It was against that journal's policy to let authors engage in the coining of frivolous, new terminology. In the case at hand, our friend and colleague Alvaro De Rújula, on hearing of our work, had coined the term 'charmonium', which in a single word was able to transmit the central new idea of the paper to any serious particle physics reader.[77]

After some discussion a compromise was reached, namely that charmonium could be used in the text but not in the title. Even weirder than barn, quark, and charmonium was the fantasy-name *boojum* proposed by David Mermin, a distinguished American solid-state physicist and science writer. In studies of the superfluidity of the rare helium isotope He-3, Mermin and other physicists had found a peculiar vortex-like flow pattern whose behaviour reminded him of a passage in Lewis Carroll's poem *The Hunting of the Snark*:

> In the midst of the word he was trying to say
> In the midst of his laughter and glee,
> He had softly and suddenly vanished away—
> For the Snark was a Boojum, you see.

Mermin introduced the 'boojum' in a conference proceeding of 1977 and then, after lengthy discussions with the editor, also in a paper with two co-authors published in the high-ranking *Physical Review Letters*.[78] Among the references in that paper was one to Webster's *New International Dictionary*, where Carroll's term appeared. Believe it or not, today 'boojum' is an accepted technical term in physics which has appeared in approximately 300 research papers.

---

[76] Calder (1977), p. 99.
[77] https://www.nobelprize.org/uploads/2018/06/politzer-lecture.pdf
[78] See Mermin (1981) for a detailed and charming account for how 'boojum' eventually became an accepted physics term.

The name *tachyon* dating from 1970 is in the classical tradition insofar that it is derived from a Greek word (*takhys*, ταχύς) meaning swift or fast. The same word had been used as a prefix more than a century earlier, as in the surveying instrument called a 'tachymeter' invented to measure distances and level surfaces. In 1867 the German physician and naturalist Hermann Lebert coined the word 'tachycardia' for unusually rapid heartbeat (*kardia* = heart). A tachyon is a hypothetical particle moving faster than the speed of light, something Einstein originally but erroneously thought was ruled out by his theory of relativity.

For a period of about two decades, tachyons were much discussed and searched for, but none have been found and in all likelihood never will be found. As early as the 1920s the Russian physicist Lev Strum pointed out that special relativity theory does not forbid superluminal signals or particles if only they remain superluminal. However, his insight was ignored and it took until the 1960s before the possibility of 'meta particles', as they were first called, attracted attention. After the American physicist Gerald Feinberg in 1967 developed a relativistic quantum theory for the superluminal particles, the subject became popular among physicists, philosophers, and the lay public. 'In anticipation of the possible discovery of faster-than-light particles, I named them tachyons, from the Greek word *tachys*, meaning swift'.[79] Ordinary particles moving slower than light were called *bradyons* (slow) and those moving just at the speed of light, such as photons, were sometimes referred to as *luxons*. The latter word is a hybrid as its root *lux* is the Latin word for light, the equivalent of the Greek *phot*.

Dark matter first appears in the astronomical literature in the French form 'matière obscure' in an article written by Henri Poincaré in 1906. In 1930 the Swedish astronomer Knut Lundmark used the same term in German ('dunkle Materie') and three years later it reappeared in a paper, also in German, by Fritz Zwicky who is often but wrongly said to have coined it. Zwicky wrote of 'the surprising result [that it] would then follow that dark matter is present in very much greater density than luminous matter'.[80] He did not italicize or put the term in inverted commas, or otherwise call attention to it. The later so popular term existed but was used very scarcely by astronomers until it was realized in the 1970s that much of the universe is filled with a strange form of nonluminous and therefore invisible matter more abundant than ordinary matter.[81]

Bibliometric data suggest when dark matter turned from a hypothesis to an observationally supported reality. The term appeared only once in the title of a scientific paper during the period 1930–49 and not at all in 1950–69, after which it exploded: during the next two decades about 1,900 papers were written with 'dark matter' in the title (Web of Science).

Although dark matter had been discovered by the late 1970s, its nature remained a complete mystery which has still not found its final solution. The main question was whether it was 'baryonic' or 'non-baryonic', meaning whether it consists of ordinary matter particles or not. The word *baryon* for nucleons and other heavy particles

---

[79] Feinberg (1970).
[80] Zwicky (1933).
[81] Bertoni (2018).

was suggested in 1953 by Abraham Pais, who in the classical tradition derived it from the ancient Greek word *barys* (βαρύς) for 'heavy' or 'weighty'. As early as 1920 an Australian physicist suggested 'baron' for what came to be known as the proton, but whereas baryon was a successful neologism, baron remained a protologism.[82] The root appears in many other words such as the element barium, the instrument barometer, and the isobars known from meteorology. Curiously, when astronomers and astrophysicists discuss non-baryonic matter in the universe, they include the electrons residing in all ordinary matter. In the vocabulary of particle physics, the light electrons are non-baryonic, but in the vocabulary of astrophysics they are baryonic.

If dark matter were baryonic it might consist of what astronomers called MACHOs, an acronym for 'massive compact halo objects'. On the other hand, it might also be non-baryonic, made up of hypothetical particles predicted by theory and with no place in the standard model. Such particles were collectively known as WIMPs or 'weakly interacting massive particles', a terminology introduced in the early 1980s. The English noun wimp typically refers to a feeble person, a coward or weakling, a connotation that the physicists may have had in mind when they suggested the acronym. Incidentally, the acronym WIMP is also used in computer language, where it stands for 'Windows icon mouse (or menu) pointer'. More recently, WIMPs have been followed by other hypothetical dark matter candidates such as FIMPs (feebly interacting massive particles) and SIMPs (strongly interacting massive particles).

Among the chief candidates for dark matter is the *axion*, a very light neutral elementary particle independently theorized by Frank Wilczek and Steven Weinberg in 1978. So far, some 8,500 scientific papers have been written on the still hypothetical axion. The name suggested by Wilczek may seem to be a classical hybrid composed of the Latin *axle* and the Greek termination *on*, but the origin of the term was more mundane. As Wilczek recalled:

> A supermarket display of brightly colored boxes of a laundry detergent named Axion had caught my eye. It occurred to me that 'axion' sounded like the name of a particle and really ought to be one. So when I noticed a new particle that 'cleaned up' a problem with an 'axial' current, I saw my chance. (I soon learned that Steven Weinberg had also noticed this particle, independently. He had been calling it the 'Higglet'. He graciously, and I think wisely, agreed to abandon that name.) Thus began a saga whose conclusion remains to be written.[83]

The study of strong interactions in the 1970s gave rise to new words related to mixed states of quarks and gluons. For these states or fields two American particle physicists introduced, after having consulted classics scholars, what 'we call meiktons (pronounced "make-ton"), from the classical Greek for a mixed thing—the terms

---

[82] Pais (1953). Kragh (2023).
[83] Wilczek (2016), who also revealed the origin of the name in his Nobel lecture of 2004. The Axion dishwashing liquid was introduced by Colgate-Palmolive in 1968 and is still sold under that name in Asia and Latin America.

hermaphrodite or hybrid have also been used'.[84] Strange terms such as *meiktons* and *hermaphrodite mesons* appeared in the physics literature in the 1980s but eventually they were replaced with the less exotic but polysemic term *hybrids*.

Other playful words were considered when physicists developed the first theories of what came to be known as 'grand unified theory', or GUT for short. In 1974 Howard Georgi and Sheldon Glashow proposed a theory of this kind which predicted new and extremely heavy particles that would cause transitions between quarks and leptons and thus allow for the proton being unstable. At one stage Glashow—who a few years later became a Nobel laureate—contemplated calling the hypothetical particles either 'ponderons' or 'vector basket-balls'.[85] None of the words made it into print, though. The first was a classical hybrid word insofar that *pondus* is Latin for weight and -*on* is a Greek termination. On the other hand, the second word clearly belongs to the non-classical whimsical category. This was also the case with the *anyon* particle introduced in the early 1980s (Section 4.4).

Most WIMP particles are predicted by theories of supersymmetry or supergravity (Section 2.5). They are weakly interacting in the sense that they interact only by the weak and gravitational forces, and in this respect they are similar to the well-known neutrinos, which may be why most of them end on -*ino*. One of the candidates for exotic dark matter is the *neutralino*, a stable combination of three superpartners. In addition to the *photino* mentioned in Section 3.4, the zoo of particles predicted by supersymmetric theories includes *gluino, axino, gaugino, gravitino, sneutrino, zino, smuon*, and *higgsino*—and there are more of them.

The latter particle, the higgsino, is the superpartner of the *Higgs boson*, which famously was discovered in CERN experiments of 2012 and based on theoretical work from the mid-1960s by Peter Higgs in England, Robert Brout and François Englert in Belgium, and half a dozen other theorists. In contrast to super-particles, the Higgs (as it is often called in shorthand) is a crucial part of the standard model. The discovery promptly resulted in a Nobel Prize shared between Higgs and Englert (Brout had passed away in 2011).

Whereas there is a long tradition in physics for naming units after outstanding scientists, when it comes to particles eponyms are practically absent. Prior to the Higgs, the only examples were *fermion, boson, pomeron*, and *skyrmion*. However, the latter two are so-called quasiparticles (Section 4.4) and fermion and boson refer to a group of particles with either half-integral or integral spin and not to an individual elementary particle. While fermion is named after Enrico Fermi, the name boson honours the Indian physicist Satyendra Bose. Originally quantum physicists spoke of Fermi–Dirac statistics and Bose–Einstein statistics, but in a lecture in Paris from 1945 only published three years later Dirac suggested the new names which were quickly adopted. 'It is useful to have a new word to describe particles which obey these statistics', Dirac said. 'I will give the name *fermions* to those particles which obey Fermi statistics and for which we can never find two particles in the same state.'

[84] Chanowitz and Sharpe (1983).
[85] Crease and Mann (1996), p. 402.

Bose had found a new type of statistical mechanics in 1924 and therefore 'I shall call the particles obeying these statistics *bosons*.'[86]

The particle predicted by Higgs, Brout, Englert, and others was a massive boson which became known as either the Higgs particle or more frequently the Higgs boson. The latter name was well established already in the 1970s when it and its associated term 'Higgs mechanism' for the generation of mass entered a large number of scientific papers. The number of papers with either of the terms in the title grew as follows (Web of Science):

	1970–9	1980–9	1990–9	2000–9
Higgs boson	22	200	566	756
Higgs mechanism	21	28	27	46

But why Higgs rather than associating the particle with some other name? Eponymies in science are invariably connected with priority attributions and in the case of the Higgs boson it was far from evident that Higgs' name and his name only should enter the eponym. As detailed in a study by Arianna Borrelli, both before and after the discovery there were proposals of several other and longer names that better represented the actual history leading up to the discovery of the Higgs boson. At a symposium in 2013, Steven Weinberg commented that the predicted particles 'came to be called Higgs particles, though they could as well have been called Brout-Englert-Guralnik-Hagen-Higgs-Kibble particles'.[87] Of course, by that time nobody thought seriously to rename the Higgs boson, which by then had been adopted by the public as the 'God particle' and become a valuable brand to CERN and the high-energy physics community in general. The unfortunate name or nickname 'God particle' aroused public attention with a book from 1993 written by Leon Lederman, a leading particle physicist and recipient of the 1988 Nobel Prize.[88] It was chosen for marketing reasons and nothing but.

---

[86] The lecture was only published 1948. See Dirac (1995), p. 1253.
[87] Quoted in Borrelli (2015), p. 46.
[88] Lederman (1993).

4
More physics names

Among the many scientific advances in the nineteenth century the laws of thermodynamics stand out as possibly the most fundamental. The now household word 'energy' acquired a new meaning and by the end of the century it was all over in the natural sciences as well as beyond the domain of science. With energy followed other terms associated with thermodynamics such as energetics, entropy, and enthalpy. As usual, there were also words invented in the period that were soon forgotten and are unlikely ever to be resurrected. Section 4.2 starts with the sensational discovery in 1895 of 'a new kind of rays' or what soon became known as either X-rays or Röntgen rays. The new rays of the fin-de-siècle period gave rise to a series of words some of which, such as 'radioactive', entered common language whereas others were used only by the scientists and some have just disappeared.

What is generally recognized as perhaps the most important fundamental theory ever, quantum mechanics, dates from 1925 but was preceded by the so-called old quantum theory by a quarter of a century. As detailed in Section 4.3, quantum physics, whether in the old or new version, required its own language and terminology. The mathematics of quantum mechanics is clear and operational, but ordinary language is ill-suited to describe the weird quantum world. Reflecting the origin of quantum mechanics in German-speaking countries, some of the key words were originally in German and used internationally either in this language or in combination with English words. They are science examples of hybrid words. Because of the public and cultural visibility of quantum mechanics, some of the terms associated with this theory have been adopted in the common language with new meanings. The best known of these terms or phrases is probably 'quantum jump'.

Section 4.4 is devoted to the research area called solid-state physics and its transition to the modern discipline which since about 1980 has carried the name condensed matter physics. In relation to these branches of physics the section considers various technical terms with 'quasi-' as a prefix and also the accelerating use of acronyms in both physics and other areas of science.

4.1 Sciences of heat and energy

Given the massive role that *energy* plays in all areas of modern life, it is hard to believe that the concept of energy, in the scientific sense of the term, is less than two centuries old. During the Napoleonic era, energy in this sense was still unknown. The concept had to be invented, and the invention was intimately related to the fundamental insight dating from the 1840s that there is something in nature which is always conserved, which cannot be created and can never perish. According to a recent cultural study of the term energy, it originated in a cultural context from which it migrated to the field of science and then, eventually, migrated back to culture with a new meaning. The word is all over, but it should be kept in mind that it, like so many other words, has a long and complex history. According to Peter Hjertholm, a historian of ideas and culture:

> When we use the term [energy] in our modern scholarly discourses and in our daily conversations, we often do so by way of analogies from the science of energy, often without much conscious thought.... The problem with this language practice is that our scientific conception of the term predisposes us to think that energy has *always* been a scientific concept or part of scientific discourses.[1]

The Greek word *energeia* was originally coined by Aristotle and is composed of *en* (in) and *ergon* (work). *Energeia* was a technical term in Aristotelian philosophy commonly translated into English as 'actuality' or 'activity'. However, Aristotle's 'activity' differed very much from what we associate with both this term and the term energy. Thus, the Greek philosopher described notions such as happiness and contemplation as forms of *energeia*. Aristotle also invented another term that would later be revived, namely *entelecheia* (entelechy) formed by the Greek words *enteles* (complete), *telos* (end, purpose), and *echein* (to have). It can be translated as something which makes real or actual what is otherwise merely potential, such as the entity that turns inert matter into a living body.

'Energie is the operation, efflux or activity of any being; as the light of the Sunne is the energie of the Sunne, and every phantasm of the soul is the energie of the soul.'[2] Thus wrote the English Platonist philosopher Henry More in 1642, apparently using the word energy (or 'energie') in the modern sense in the middle of the sentence but only there. For More, energy was a metaphysical concept. As documented by Hjertholm, by 1800 the term energy had long been part of English cultural and religious language in meanings roughly similar to Aristotle's *energeia*. Theologians spoke of 'divine energy' in the seventeenth century and in 1775 Samuel Johnson argued in a letter that when the 'powers of nature have attained their intended energy, they can be no more advanced'.[3] 'Energy' and 'Energetick' also appeared several times in Johnson's dictionary published twenty years earlier.

Blount, in the 1707 edition of his dictionary, included energy as well as 'energetical', explaining that 'Energetical Bodies are Bodies which are eminently active, and very efficacious in producing their Operations'. Another of the dictionaries of the early Enlightenment period, the 1720 edition of Phillips' *New World of Words*, added that apart from 'effectual Working, Efficacy, Force', the term energy also had other and more specific meanings: 'In *Rhetorick*, a Figure wherein great Force of Expression is us'd: In a Medicinal Sense, a stirring about, or Operation of the Animal Spirits and Blood.'

To move ahead in time, in a series of lectures given in 1802 the English natural philosopher and polymath Thomas Young transformed the well-known energy into

---

[1] Hjertholm (2023), pp. 9–10.
[2] *OED*. Hjertholm (2023), pp. 111–12.
[3] Hjertholm (2023), p. 87, pp. 252–9.

a scientific term. However, he only used the word in its more restricted meaning corresponding to a body's kinetic energy or what at the time was known as its *vis viva*, either mv^2 or, as was later realized, $\frac{1}{2}mv^2$. 'The product of the mass of a body [m] into the square of its velocity [v] may properly be termed its energy', he said, adding: 'This product has been called the living or ascending force ... and some have considered it as the true measure of the quantity of motion; but although this opinion has been very universally rejected, yet the force thus estimated deserves a distinct denomination.'[4]

The current much broader meaning was only adopted with the acceptance of the fundamental law of energy conservation and then, ironically, with 'force' being used instead of energy. Among the precursors of the law was the leading German chemist Friedrich Mohr, who may today be best remembered for the 'Mohr method' to determine the total chlorine content in water. He, like the pioneers of the law of energy conservation, Julius Robert Mayer in Germany and James Prescott Joule in England, spoke consistently of force (or *Kraft* in German). Likewise, Hermann von Helmholtz's important memoir of 1847, in which energy conservation was first fully formulated, carried the title *Ueber die Erhaltung der Kraft* or in English 'On the Conservation of Force'. Another early and more widely read work was William Grove's *Correlation of Physical Forces* from 1846, published in several editions with no mention of energy in its modern meaning.

The term energy only came in general use at about 1860 and it took a decade or so before it replaced force. The entry on energy in the *Encyclopaedia Britannica* illustrates the changed popularity of the word. All that the 1842 edition had to say was this: 'ENERGY, a term of Greek origin, signifying the power, virtue, or efficacy of a thing. It is also used figuratively, to denote emphasis of speech.' By comparison, the ninth edition of 1879 included a detailed seven-page entry in which the author, William Garnett, described physics to be simply the science of energy.[5] From a modern point of view, but not from the point of view of contemporary scientists, the usage may appear confusing or even inconsistent. After all, energy and force are different entities measured in different units, the first in joule J and the second in newton N, the connection being $J = N \times m$.

The still commonly used term *electromotive force* (emf) of a battery or other electrical source apparatus is a reminiscence from a time when the modern concept of energy had not yet been introduced. Volta used 'electro-motive' for his battery in 1800 (Section 2.1) and the following year he coined 'electromotive force', defining it as the action that caused unlike charges to separate and keep them separated.[6] However, the electromotive force is not a force and also not an energy but the potential energy per charge and thus has the same unit as volt, namely joule per coulomb. Although it is definitely a misnomer, electromotive force has for long been so firmly established that it will probably remain in the scientific vocabulary.

[4] Young (1807), p. 52. The factor $\frac{1}{2}$ in the expression of kinetic energy was a result of the later law of energy conservation.

[5] Smith (1998), p. 2.

[6] Varney and Fisher (1980).

A few British writers referred to the term energy in works from about 1850 and three years later one of them, the Scottish physicist and engineer William Rankine, introduced it formally. His carefully phrased definition was this:

> The term *energy* is used to comprehend every affection of substances which constitutes or is commensurable with a power of producing change in opposition to resistance, and includes ordinary motion and mechanical power, chemical action, heat, light, electricity, magnetism, and all other powers, known or unknown, which are convertible or, commensurable with these.[7]

Rankine distinguished between what he called 'actual energy' (roughly the same as kinetic energy) and 'potential energy', introducing the latter concept in its general sense and not only in the sense of mechanics. He defined the actual or kinetic energy as 'a measurable, transferable and transformable affection of a substance, the presence of which causes the substance to change its state in one or more respects'.

As Rankine recalled in a letter to *Philosophical Magazine* of 1864, he found inspiration in Aristotle's terminology: 'The step which I took in 1853, of applying the distinction between "Actual Energy" and "Potential Energy", not to motion and mechanical power alone, but to all kinds of physical phenomena, was suggested to me, I think, by Aristotle's use of the words δύναμις [dynamis] and ενέργει [energeia].'[8] Helmholtz and other German scientists were slower to adopt the term. It took until the mid-1870s before energy (Energie) appeared with the same frequency as force (Kraft) in German publications.[9]

In a critical review of the sixth edition of Grove's *Correlation of the Physical Forces*, Maxwell called attention to 'the importance of the study and special cultivation of scientific language'. He found the terminology of physics in general and of thermodynamics in particular to be in an unsatisfactory state, such as illustrated by Grove's confusingly frequent use of the term force.

> The fathers of dynamical science found a number of words in common use expressive of action and the results of action, such as force, power, action, impulse, impetus, stress, strain, work, energy, &c. ... But the equivalent words Force, *Vis, Kraft*, came most easily to hand, so that we find them compelled to carry almost all the ideas above mentioned, while the other words which might have borne a portion of the load were long left out of scientific language.[10]

According to Maxwell, the phrase 'physical forces' was misleading as most of Grove's so-called forces (which included light, magnetism, motion, and chemical affinity) were not forces in Newton's well-defined sense. Maxwell further referred to the metaphorical use of the word force: 'It may be a legitimate metaphor to speak of the force of public opinion as being brought to bear on a statesman so as to exert

---

[7] Rankine (1853).
[8] Rankine (1864).
[9] See Caneva (2021), pp. 236–7.
[10] *Nature* **10** (1874): 302–4.

an overpowering pressure upon him, because here we have an action tending to produce motion in a particular direction; but when we speak of "the Queen's Forces", we use the term in a sense as unscientific as when we speak of the Physical Forces.' Rather than using common terms in science he recommended new words or what he called neologies (that is, neologisms). These were important, but as Maxwell stated in an obituary of Faraday, new ideas were not the product of new words, but the other way around. The coining of words came last in the process of science:

> We have, first, the careful observation of selected phenomena, then the examination of the received ideas, and the formation, when necessary, of new ideas; and lastly, the invention of scientific terms adapted for the discussion of the phenomena in the light of the new ideas.[11]

The new science of energy came to be known as *thermodynamics*. The first syllable in this word is from the ancient Greek word *thermos* (θερμός) meaning warm or hot, a word that corresponds to the Latin *calor* found in terms such as *calorie*, *caloric*, and *calorimeter*. Noting that there was no special designation for the science of heat, Whewell coined the term *thermotics*, which he considered better than the *thermology* proposed by Auguste Comte.[12] He argued that his own suggestion agreed with the names of corresponding sciences such as acoustics and optics. Neither Comte's 'thermology' nor Whewell's 'thermotics' won approval.

The terms calorie and calorics reflect the once so influential belief of heat being a kind of imponderable, self-repelling fluid called *caloric*, a hypothesis that was widely adopted in physics and chemistry from the late 1780s to about 1830. For example, in his famous table of chemical elements or 'simple substances' from 1789, Antoine-Laurent Lavoisier included caloric (*calorique*) among the substances that belong 'to all the kingdoms of nature'. The notion of three 'kingdoms' of nature goes back to Linnaeus, who distinguished between stones, plants, and animals corresponding to the sciences mineralogy, botany, and zoology. John Dalton was no less convinced of the reality of caloric, which formed an essential part of his atomic theory published two decades after Lavoisier's *Traité Élémentaire de Chimie*. He described caloric as an 'elastic fluid of great subtility'.

The new term first appeared in the collaborative work *Méthode de Nomenclature Chimique* from 1787 written by Lavoisier, L. B. Guyton de Morveau, C. L. Bertholet, and A. F. Fourcroy (see Section 5.3). It may have been invented by Morveau.[13] When Lavoisier discussed the term in the first chapter of his *Traité*, he carefully motivated it in accordance with the principles of chemical nomenclature that he and his collaborators had stated two years earlier. 'This substance, whatever it is, being the cause of heat, or, in other words, the sensation we call *warmth* being caused by the accumulation of this substance, we cannot, in strict language, distinguish it by the term *heat*; because the same name would then very improperly express both cause and effect.' Referring to *Méthode*, Lavoisier wrote:

[11] Maxwell (1965), II, p. 359.
[12] Whewell (1847), vol. 2, p. 508. Comte (1835), vol. 1, p. 194.
[13] Morris (1972).

In the work published by Mr de Morveau, Mr Berthollet, Mr de Fourcroy, and myself ... we thought it necessary to banish all periphrastic expressions, which both lengthen physical language, and render it more tedious and less distinct, and which even frequently does not convey sufficiently just ideas of the subject intended. Wherefore, we have distinguished the cause of heat, or that exquisitely elastic fluid which produces it, by the term of caloric.[14]

Lavoisier admitted in some of his texts that the subtle caloric fluid might turn out to be hypothetical, and yet there is little doubt that he conceived it as a real if extremely tenuous material quantity. Like other forms of matter, it would obey Newton's law of gravitation. In *Méthode*, Lavoisier coined the word calorimeter (*calorimètre*) for the apparatus that he and Pierre-Simon Laplace had invented in 1783 and initially called just a 'machine'. The new name was explicitly theoretical as it signified an apparatus for measuring caloric. Noting that it was a hybrid Latin–Greek construct, he apologized for the coining which preferably should be in Greek only.[15] But unfortunately the obvious choice—thermometer—was already in use for an instrument measuring temperature and not heat. Although the idea of caloric has long gone, scientists still refer to words such as calorimeter and calorimetry.

Whereas Lavoisier introduced and promoted caloric, he never referred to *calorie* as a unit of heat. This extremely popular term first appeared in print in 1824 and was possibly coined by the French industrial chemist Nicholas Clément. Although the calorie unit was soon adopted by chemists and physicists, the term calorie only became truly popular at the end of the century when used as a nutritional unit for the energy content in food and beverages.[16] Still today, most nutrition facts labels include information of how many calories there are in a product. Whereas in Europe the energy content is given in both calories (kcal) and joules (kJ), in Canada and the United States the labels refer only to calories, an energy unit which has long been obsolete in scientific circles. Many people presumably believe that food and beverages contain calories in the same literal sense that they contain proteins, carbohydrates, vitamins, and sodium. But of course, there is not a single calorie floating around in a bottle of beer or buried in a slice of pizza. The funny term 'empty calories' for food with much sugar but practically no nutrients came into common use in the early 1960s.

Although the caloric theory disappeared almost two centuries ago, there are still vestiges of it in common language where we often speak of heat as a kind of substance. We say about a poorly insulated house that it loses a lot of heat during the winter; and we conceive a radiator as a device that contains heat. From the point of view of physics, this is just wrong as heat, in contrast to energy, is not a thermodynamic function of state. It has been seriously (but probably unrealistically) proposed to rid the physics vocabulary of the noun heat and to do the same with work.[17]

---

[14] Lavoisier (1965), pp. 4–5.
[15] See Roberts (1991) for a detailed account of the history and naming of Lavoisier's calorimeter.
[16] Hargrove (2006).
[17] Romer (2001).

Moreover, in everyday language we often conflate heat with the concept of temperature. 'Let's increase the heat, it's no more than 16 degrees Celsius.' The word *temperature* in its present meaning only entered English language in the mid-seventeenth century and then as a direct result of the invention of the thermometer. It is based on and etymologically related to the Latin *tempare* with multiple meanings including to moderate or restrain oneself.[18] From this term also came 'temper' and 'temperament', referring to a person's attitude or mood. 'He couldn't control his temper.' During the renaissance period the term 'temperature' was known but at the time used as a synonym of 'temperament'. As late as 1720 John Locke wrote in his posthumously published *Elements of Natural Philosophy* that 'Bodies are denominated *hot* and *cold* in proportion to the present *temperament* of that part of our body, to which they are applied; so, that feels hot to one, which seems cold to another.'[19]

Apart from in thermodynamics, *thermos* or *thermo* enters into a variety of other science names. The best known of these is undoubtedly the *thermometer*, an instrument developed in the seventeenth century from its predecessor called a *thermoscope*. Whereas the latter name first appeared in print as 'thermoscopium' in 1617, thermometer—a measurer of heat—dates from 1624 (Section 1.1). The thermoscopes constructed in the early part of the century enabled natural philosophers to examine (*-scope*) changes in sensible heat but was not supplied with a scale. It did not show changes in temperature, a concept which had not yet been defined as separate from heat. Indeed, temperature was only established with the thermometer, namely as the physical quality measured by this instrument. It took more than a century before heat was similarly defined as the quantity measured by the calorimeter invented by Joseph Black and perfected by later scientists.[20]

Other and later thermo-words included the following. Reactions between atomic nuclei at extremely high temperature or energy, such as they occur in the interior of the stars, are called *thermonuclear*; the *thermosphere* is a layer in the Earth's atmosphere at height ca. 500–1,000 km and with temperatures that can rise to about 2,000°C; the *thermostat*, a name invented by the English consulting chemist Andrew Ure in 1830, is a device which regulates the temperature; and *thermophiles* are organisms which thrive at a temperature as high as 100°C or even higher. In Section 2.3 we noted Ørsted's coining of *thermoelectricity*.

Thermos (noun, singular) typically refers to the popular vacuum bottle used to keep coffee or tea hot, but the clever commodity was originally designed to keep liquid gases very cold. The Scottish chemist James Dewar, a pioneer of cryogenics (low-temperature science, see later in this section), was the first to liquefy hydrogen at a temperature that he estimated to be approximately 20 K or −253°C. In 1892 he invented what for a time was called the 'Dewar flask'—or just a 'dewar'—for this and similar purposes. He did not patent the flask, which in the first decade of the twentieth century was transformed into a consumer product and then named a 'Thermos flask'.[21]

[18] https://www.merriam-webster.com/words-at-play/origin-of-temperament-and-temperature
[19] Locke (1758), p. 100. Emphases added.
[20] Oliveira (2018).
[21] Rowlinson (2012), pp. 89–95.

The word thermodynamics only appeared in the 1850s and then in its hyphenated form *thermo-dynamics*. At a meeting of the Royal Society of Edinburgh on 1 May 1854, William Thomson coined the word, which he five years earlier had used as an adjective in 'thermo-dynamic engine'. It was, however, another Scotsman who promoted the term and made it enter the scientific language. In 1857 Rankine wrote an entry in John Nichol's popular *Cyclopaedia of the Physical Sciences* where he not only used the word thermo-dynamics but also and for the first time distinguished between what he called the first and second laws of the new science. Moreover, he coined the word *energetics* for a science even more comprehensive and fundamental than what he understood as thermodynamics:

> The laws of Thermo-dynamics may be regarded as particular cases of more general laws applicable to all such states of matter as constitute ENERGY, or the capacity to perform work, which more general laws form the basis of the SCIENCE OF ENER-GETICS, a science comprehending all special branches, the theories of motion, heat, light, electricity, and other physical phenomena.[22]

The second law of thermodynamics was stated in two different forms in the 1850s, first by Thomson as an inbuilt tendency in nature towards what he consistently called the *dissipation* of energy. Imagine two bodies at different temperatures placed in contact with heat being transferred from the warmer to the colder with work being done; although the energy is conserved, it has dissipated in the sense that the system's capacity to perform work has diminished.

Rudolf Clausius in Germany reached independently the same conclusion but framed it in a significantly different language. According to Clausius, the law was about an impossibility, namely that a self-acting cyclic machine cannot possibly convey heat from a body of lower temperature to another at a higher temperature. Having suggested several formulations, including one from 1862 in terms of what he called *disgregation*, in 1865 Clausius came up with a final version that built on a new concept and a new associated word. 'The entropy of the universe tends to a maximum', he summarized. The word *entropy* with symbol S was a neologism that he coined for what he at first referred to as *Verwandlungsinhalt*, German for content of transformation. Clausius explained his reasons for the new word:

> We might call S the *transformational content* of the body . . . but as I hold it to be better to borrow terms for important magnitudes from the ancient languages, so that they may be adopted unchanged in all modern languages, I propose to call the magnitude S the *entropy* of the body, from the Greek word τροπή, *transformation*. I have intentionally formed the word *entropy* so as to be as similar as possible to the word *energy*; for the two magnitudes to be denoted by these words are so nearly allied in their physical meanings, that a certain similarity in designation appears to be desirable.[23]

---

[22] Cited in Smith (1998), p. 165. In the 1890s the term 'energetics' came to signify a kind of alternative science movement promoted in Germany by Wilhelm Ostwald and Georg Helm in particular. See Deltete (2003).

[23] Clausius (1867), p. 357.

Although entropy eventually became a most successful scientific name, for a long time the concept was considered abstract and difficult, and for this reason it only permeated slowly into physical theory, and even more slowly into neighbouring disciplines such as chemistry, geology, engineering, and biology. The French philosopher Henri Bergson stated in his *L'Évolution Créatrice* (Creative Evolution) from 1907 that the law of entropy increase was 'the most metaphysical of the laws of physics'.[24] While it took many years before the term entropy was generally adopted, Clausius' symbol S was an instant and lasting success. It has been used for entropy ever since Clausius introduced it.

In a letter to his friend Peter Tait of 1867, Maxwell sketched a thought experiment that apparently contradicted the second law of thermodynamics. He imagined the eponymous 'demon' functioning as a valve between a hot and a cold gas but in such a way that 'the hot system has got hotter and the cold colder & yet no work has been done'.[25] Although the famous thought experiment is invariably known as *Maxwell's demon*, the Scottish physicist did not use the term demon either in his letter to Tait or at any other occasion. He called it a 'finite being' and referred to 'the intelligence of a very observant and neat-fingered being'. The more dramatic and popular 'demon' was first introduced by William Thomson in 1874. Five years later he explained that 'The word "demon", which originally in Greek meant a supernatural being, has never been properly used to signify a real or ideal personification of malignity'.[26]

Thomson's more dramatic and imaginative demon immediately caught on, with the result that today more than 5,000 academic works in physics and philosophy have discussed Maxwell's 'demon' and only a handful his 'being'. As an eponym it refers to Maxwell, but the metaphor is due to Thomson. Maxwell insisted that his being was a far more mundane thing than a demon; it was not supernatural and did not need to be endowed with intelligence, but could be just a sophisticated kind of valve. In a letter of 1870 he likened the being—or demon—to 'a mere guiding agent' such as 'a pointsman on a railway with perfectly acting switches who should send the express along one line and the goods along another'.[27] By using the pointsman as a metaphor Maxwell wanted to highlight that the action of the being—and perhaps of the meaning of entropy—was first of all a matter of information. He used this and other metaphors primarily for illustrative purposes, namely to describe the unfamiliar in terms of the familiar.

Clausius did not pay much importance to the concept he had invented and named, and Thomson never used it. On the other hand, in his *Sketch of Thermodynamics* from 1868 Tait adopted 'the excellent word Entropy' while at the same time complaining that Clausius had used it in an unsuitable sense:

> It seems more convenient, however, to treat as the Entropy of a substance the availability of its contained heat, etc., for the production of work, than its unavailability; so that we shall ... use the excellent term Entropy in the opposite sense to that in

[24] Bergson (1908), p. 264.
[25] Letter of 11 December 1867 reproduced in Garber, Brush, and Everitt (1995), pp. 176–8.
[26] Thomson (1879). On Maxwell's demon and other demons in the history of science, see Canales (2020).
[27] Garber, Brush, and Everitt (1995), p. 205.

which Clausius has employed it, viz., so that the *Entropy of the Universe tends to zero*, which is Thomson's theory of dissipation, rather than the unmodified nomenclature of Clausius, according to which the *Entropy tends to a maximum*.[28]

Tait's reversed use of entropy was more than just a misappropriation of a technical term. It marked the beginning of a public controversy between Tait and Clausius concerning the meaning of thermodynamics and the priority of having founded the new science.

In the first editions of his widely read *Theory of Heat*, Maxwell followed Tait's problematic reinterpretation of the word entropy. Only in the fourth edition of 1875 did he admit that the result had caused 'great confusion into the language of thermodynamics'. Still in the early years of the twentieth century it was rare to find entropy mentioned in textbooks on physical chemistry and chemical thermodynamics.[29] Walther Nernst was awarded the 1920 chemistry Nobel Prize in part for his heat theorem or so-called third law of thermodynamics dating from about 1910. Today this law is often stated in terms of entropy—namely, that entropy changes become zero as the temperature approaches absolute zero ($\Delta S \to 0$ for $T \to 0$)—but this was not Nernst's formulation. He disliked entropy and avoided as much as possible to use Clausius' name.[30]

While not embracing entropy, early chemical thermodynamics was supplied with a new and lasting name for the heat content of a system. The term *enthalpy* coined by the Dutch physicist Heike Kamerlingh Onnes in 1909 has for long been familiar to students of chemistry. It may have been derived from a Greek word *enthalpos* (ενθαλπος) with the meaning 'to warm within' or 'to put heat into'. Kamerlingh Onnes did not actually publish the neologism, which he apparently came up with in informal discussions with his colleagues at the Leiden low-temperature laboratory. It first appeared in print in a footnote to a paper by an American research fellow staying at the laboratory: 'This name has been used by Kamerlingh Onnes to indicate the function $(\varepsilon + pv)$.'[31]

One of the reasons why the entropy law became controversial was the consequence that it, if applied to the universe as a whole, implied: that the universe would irreversibly end in a high-entropic 'heat death' (*Wärmetod* in German) with no life and no activity. After all, entropy is a measure of disorganization and lack of structure. The pessimistic scenario caused much alarm in the half-century from about 1865 to 1915, when philosophers, theologians, and scientists discussed the possibility of processes that might halt or counter the disturbing growth in cosmic entropy.[32] Although these discussions did not lead to anything of lasting scientific value, they did result in a number of new words most of which were, however, soon forgotten.

The German physicist Felix Auerbach speculated in the early twentieth century that the continual evolution of life might be secured by an anti-entropic force which

---

[28] Tait (1868), p. 29. Smith (1998), pp. 256–8.
[29] Kragh and Weininger (1996).
[30] For Nernst's dislike of entropy and his reasons for avoiding the term, see Kragh and Weininger (1996).
[31] Howard (2002) on the origin of the term enthalpy. The symbol ε denotes the internal energy and pv the product of pressure and volume.
[32] Details in Kragh (2008).

he called *ectropy* or *extropy*. Another and more easily recognizable antonym to entropy is the *negentropy* (negative entropy) coined by the distinguished French physicist Léon Brillouin in 1950 in the context of information theory. And then there was the old Aristotelian term *entelechy* which was revived by the German embryologist Hans Driesch in about 1910 to signify a vital principle serving the purpose of maintaining the high degree of organization in living organisms.

Despite the troublesome birth of the name, today entropy is one of the most used technical terms not only in the physical sciences but also in mathematics, engineering, biology, and the medical sciences. According to Web of Science, since 1900 there have appeared a total of about 73,000 papers with entropy in the title. Of these, 34,600 are classified as physics and engineering, 13,300 as materials science, 8,400 as mathematics, and 2,400 as astronomy. In 1999, a journal simply called *Entropy* was established as an outlet for research focusing on entropy and information theory. Not all physicists agree that the term entropy deserves to be so popular. According to Arieh Ben-Naim, an Israeli physical chemist, the second law of thermodynamics is essentially probabilistic and concerned with information, something not covered by Clausius' term:

> The term entropy, as originally coined by Clausius, is an unfortunate choice. Moreover, it is also a misleading term both in its meaning in ancient and in contemporary Greek. ... I believe that the time has come to reach the inevitable conclusion that 'entropy' is a misnomer and should be replaced by either missing information or uncertainty. These are more appropriate terms for what is now referred to as 'entropy'.[33]

However, the suggestion to replace entropy with 'missing information' or some other word has been largely ignored.

Like many other scientific terms, entropy and the adjective entropic have become popularized and employed figuratively in areas widely different from those in which the terms originated. The educated layperson will not be puzzled to read about 'cultural entropy' or 'political entropy' where Clausius' term signifies, for example, decline into disorder or tendency towards inactivity. 'A solar spot has burst and expanded its heat into the great pool of entropic peace' wrote Arthur Koestler in his 1946 novel *Thieves in the Night*. Fourteen years later the American novelist Thomas Pynchon wrote a short story called 'Entropy' in which he used the entropy metaphor to describe the decline of the established humanist society. Pynchon recalled: 'Since I wrote this story I have kept trying to understand entropy, but my grasp becomes less the more I read ... the qualities and quantities will not come together to form a unified notion in my head.'[34]

Although James Dewar is today recognized as a pioneer of *cryogenics*, he never used this word or the corresponding adjective *cryogenic*. The names are derived from a Greek word for cold or frost, *kryos* (cryos, κρύος), combined with the common ending for generation or to generate. The Latin equivalent for the Greek *kryos* is *frigus*, a

[33] Ben-Naim (2008), p. xv and p. xvii.
[34] Quoted in Mirowski (1989), p. 62.

term we have in, for example, the 'refrigerator' or 'fridge' maintaining a cold environment. While the modern refrigerator is a machine, the name first appeared around 1800 for the 'ice houses' widely used for food preservation and other purposes. Until recently a woman with no desire for sex was said to be frigid, a term which is now considered offensive.

Cryogenics and cryogenic are both names of a relatively recent date, the adjective being older than the noun, denoting the science dealing with very low temperatures. The first recorded use of cryogenic appears in a paper of late 1894 in which Kamerlingh Onnes (Section 2.5) referred to his new low-temperature physics laboratory in Leiden as his *kryogeen laboratorium*, or in the English version of 1895, 'cryogenic laboratory'.[35] Incidentally, the word 'cryostat' as a device for maintaining very low temperatures—an advanced refrigerator—is another neologism of the Leiden laboratory. It took many decades before cryogenic appeared significantly in scientific papers and at no time was it nearly as popular as 'low temperature'. According to Web of Science, in the three decades between 1960 and 1989 the first term appeared in the title of 2,709 papers, whereas many more papers (about 17,000) referred to the traditional low temperature, with or without a hyphen.

Still in the late 1930s cryogenics was nearly absent from the literature. As a name for a scientific discipline this term only appeared a decade later and then only on a couple of occasions. It began to be more commonly used at about 1960. With the journal *Cryogenics* founded this year by Kurt Mendelsohn, a German-born British specialist in low-temperature physics, the word acquired authority and came to denote the fast-growing discipline and community of scientists and engineers engaged in research related to temperatures below 120 K or −153°C. The terminology related to this kind of research was taken up by an international working group at a meeting in Paris in late 1969. As Nicholas Kurti commented in his report from the meeting, the use of words involving *cryo* was 'the most hotly debated topic in low temperature terminology'. The working group found the term 'cryotemperature' to be 'etymologically so offensive that its use should be discouraged', presumably a reference to the term being a Greek–Latin hybrid. On a more general note:

> The working group ... hoped that if certain terms were agreed upon by a substantial majority of the low temperature community, the rest would follow suit. In particular, much would depend upon the attitude of low temperature periodicals and to what extent they would be willing to adopt a firm editorial policy to encourage the use of generally accepted and universally understood terminology.[36]

The suggestion to avoid the term 'transition temperature' in studies of superconductivity and to replace it with the eponymous 'Onnes temperature' (after Heike Kamerlingh Onnes) was not followed by the low-temperature community. Noting that over the past two decades several terms had been used for the science of producing cold, the group in Paris recommended yet another one:

---

[35] Scurlock (1992), p. 2.
[36] Kurti (1970).

The working group received from several independent quarters pleas for the introduction of a new all-embracing term 'cryology' which in accordance with the generally accepted linguistic usage of '-logy' would cover everything that treats or is concerned with 'frost' i.e. low temperatures, in particular those below 120 K. The working group was strongly in favour of the introduction of the term 'cryology' although it realized that its general acceptance would take a long time. Derivative words such as cryologist and cryological should also be encouraged.[37]

Many of the linguistic suggestions of the working group were accepted by the relevant communities, but the general term *cryology* was not. Scientists and engineers working with very low temperatures—the 'cryogenicists'—simply ignored it, something which often happens with recommendations made by nomenclature commissions. What is more, the cryogenics working group may have thought that their favoured word cryology was a neologism, but if so they were mistaken.

Glaciology is older than and very different from cryogenics, and yet what the two areas have in common is that they both deal with cold matter. In the first issue of *Journal of Glaciology*, a journal founded in 1947, the leading British glaciologist Gerald Seligman wrote a brief terminological paper in which he pointed out that cryology was sometimes used as an alternative name for glaciology. He did not like the alternative.[38] Seligman's paper deserves to be quoted at some length because of its general relevance to the issue of scientific words and names.

Shortly before the war this new word [cryology] for the study of glaciology was coined in Central Europe. The Greek noun κρύος [cryos] means 'cold'. It is not clear why this word should be applied exclusively to cooled water; it could equally well be used for any cooled substance ... An objection has been raised to linking the Latin *glacier* to the Greek λόγος [logos] and one or two highly ingenious alternatives have been offered to satisfy the mind of the protesting purist. But it seems a pity to introduce a new word when we already have one which has been in use for so long. ... In America the word 'cryology' is coming into fashion to describe the study of refrigeration. For this its use is far less illogical and unnecessary. More than one prominent American glaciologist has written to express approval of it in this sense and abhorrence of its use for the scientific study of ice. One of them has also pointed out that in English-speaking countries cry-ology has a slightly ridiculous ring. It was partly in an endeavour to make the word 'glaciology' universal and to combat the unwelcome newcomer, that this Society adopted the former word in its new title. It is to be hoped that 'cryology', so far as the scientific study of ice is concerned, will not be heard of again.

Although the word cryology never replaced glaciology, it later turned up elsewhere in the earth sciences, in permafrost studies and Antarctic research. Seligman's wish

---

[37] Scurlock (1992), p. 4.
[38] Seligman (1947). The society to which the note refers is the International Glaciological Society founded in 1936 with Seligman as the originator.

was not fulfilled. The term only appears rarely, though, but sufficiently often that it is accepted as a word in the *OED*.

In a classified memorandum of 1943 the American geologist and palaeontologist Siemon Muller coined the term *permafrost* for what had traditionally been called 'permanently frozen ground'. The new word became quickly rooted in the scientific vocabulary and has appeared in the title of more than 6,000 research publications. However, initial concerns were raised about the etymological purity of the word. *The Military Engineer*, a journal published by the Society of American Military Engineers, objected that the new term did not live up to accepted linguistic standards:

> During World War II, 'permafrost' came to be used for permanently frozen ground. This word was made by mangling the Latin root *perman* (verb *permanere*, to remain) and combining it with the English word 'frost'. Such combinations are frowned upon by purists. However, the real disadvantage lies in the difficulty of deriving verbal nouns or verbal adjectives from 'permafrost'. ... The use of such expressions as 'permafrost investigations' or 'permafrost research' brings out a further awkwardness. As it can have no adjective, we are forced to use a noun as an adjective. 'Cryopedologic research' or 'investigations in Cryopedology' are not only euphonious and correct expressions, but permit the implication that the research has an engineering as well as a purely scientific objective.[39]

The term *cryopedology* was used a few times, but it never succeeded in challenging permafrost.[40] However, sometimes words that nearly disappear are resurrected many years later. This is what happened to cryopedology, which is recognized today as a subdiscipline of soil science with its own textbooks and series of international conferences. *Pedology*, the second part of the word cryopedology, is considerably older as it was constructed in 1862 from the Latin word element *ped-* with dual meanings. One of them is 'soil, ground, earth' and the other is 'foot', as in pedestrian, pedestal, and pedal (the Greek *ped-* refers to children as in pedagogy). The term pedology remained dormant for about three decades until it was revived by Russian soil scientists and by the late 1920s also adopted in Western Europe and the United States.[41]

There are several other names in science carrying the *cryo-* prefix. For instance, the Greenlandic mineral *cryolite* (ice-stone) got its name in the late eighteenth century because of its visual resemblance to ice. (The native Greenlanders had their own name for it: they called it *orsuksiktæt*, meaning 'the stone looking like a seal's blubber'.[42]) *Cryotherapy* denotes surgical and other medical techniques to cool parts of the body. A *cryotron* is a switching device developed in the 1950s that can turn on and destroy superconductivity. Then there is the *cryosphere*, which is today defined as the part of the Earth's surface dominated by ice and snow. This term was originally suggested in a somewhat different meaning, namely as a special zone of the lithosphere including

---

[39] *The Military Engineer* **40** (1948): 305–8. Available at https://www.jstor.org/stable/44567266
[40] Bryan (1946). Bockheim (2015).
[41] Simonton (1999).
[42] Kragh (1995).

an atmospheric component. The name was proposed in 1923 by the Polish scientist Antoni Dobrowolski, but it became more widely used only in the last part of the century.[43] *The Cryosphere*, a scientific journal established in 2007, is devoted to the study of 'all aspects of frozen water and ground on Earth and on other planetary bodies'.[44] The following year an International Association of Cryospheric Science was founded under IUGG, the International Union of Geodesy and Geophysics.

4.2 Invisible rays

If one insists on a date for the birth of so-called modern physics, 8 November 1895 is a good choice. This was the day—actually the evening—when Wilhelm Conrad Röntgen, a professor of physics at the University of Würzburg, Germany, recognized that he had discovered a new kind of penetrating rays that seemed to defy explanation in terms of known physics. In early January 1896 his communication 'Ueber eine neue Art von Strahlen' (On a New Kind of Rays) appeared in the proceedings of the Physical-Medical Society of Würzburg. The magic rays caused an enormous stir, especially among physicists and medical doctors, but also in the public at large. Rarely has a discovery been received so enthusiastically by scientists and non-scientists alike. In the year of 1896 there appeared no less than 1,044 publications on the new rays, including 49 books.

Röntgen needed a name for the rays, which he introduced in his original paper almost in passing: 'Boards of pine two or three centimetres thick absorb only very little. A piece of sheet aluminium, 15 mm thick, still allowed the X-rays (as I will call the rays, for the sake of brevity) to pass, but greatly reduced the fluorescence. Glass plates of similar thickness behave similarly.'[45] That was all. In suggesting 'X-rays' he presumably alluded to the unknown nature of the rays. 'X' or 'x' often signifies something or someone which is anonymous or unknown, as in 'the mysterious Mr. X', or something which is unspecified as in 'an amount of x dollars'.

Röntgen's German paper quickly appeared in an English translation in *Nature* on 23 January 1896. A month later, J. J. Thomson offered an extensive commentary on 'The discovery of Prof. Röntgen of the rays which bear his name'.[46] The term X-rays was absent from Thomson's account. Thus, there was from an early date two names for the new rays, which is further illustrated by the Nobel Prize awarded to the German physicist in December 1901, the first physics prize ever. The prize was given 'in recognition of the extraordinary services he has rendered by the discovery of the remarkable rays subsequently named after him'.

The eponymous name 'Röntgen rays' was informally suggested at a public meeting of the Würzburg Physical-Medical Society of 23 January, where one the attendants, the Swiss anatomist and medical doctor Albert von Kölliker, proposed to honour Röntgen in this way. Unsurprisingly, the proposal was accepted with applause.

[43] Barry, Jania, and Birkenmejer (2011).
[44] https://www.the-cryosphere.net/about/aims_and_scope.html
[45] Röntgen (1896).
[46] Thomson (1896).

Whether due to von Kölliker's proposal or not, the eponym was quickly adopted by German scientists and physicians. There were a few exceptions, though, such as aired in a book of 1897: 'The term Roentgen [Röntgen] rays, which has been so popular in Germany, honours more the reverence for famous men than for the mother tongue. They are not Röntgen rays, but energy rays, or possibly ether rays, and may be called Röntgen's energy rays.'[47]

Over the next decade *Röntgen rays* and *X-rays* were used indiscriminately in English and American publications if most often with 'Roentgen' instead of 'Röntgen'. While the rays came to be known as X-rays (with or without a hyphen) in the English-speaking world, in many other languages, including German, Dutch, Russian, Swedish, and Danish, they were called and are still called Röntgen rays. In French the name is 'rayons X'. Initially the term Roentgen rays (or Röntgen rays) was used more frequently than X-rays, but the practice was soon reversed. Web of Science gives 74 papers with the eponym in the title for 1900–10 and 337 papers with X-rays. Sometimes the capital X was used, whereas in other cases the x was not capitalized. A major reason for the reversal was simply that English-speaking persons found it difficult to pronounce Röntgen's name. As the *British Medical Journal* commented in 1896 on a suggestion of using 'roentography' as a term for the new method of X-ray photography:

> We should be glad if the discoverer's name could be thus perpetuated in his intellectual offspring; but, unhappily Professor Röntgen has not the good fortune on which Byron congratulated Goethe, of having a name sufficiently euphonious for the articulation of posterity.

Responding to the 'universally admitted difficulty' of finding an acceptable terminology for the art and activity of X-rays, a reader of the journal advocated 'electrography'. As he wrote: 'Out of it the words "electrogram", "electrographic", and "electrograph" readily offer themselves as accurate and expressive. Common agreement has authorized the use of similarly constructed synonyms in the cases of the telegraph, the phonograph, and in photography.'[48] The *British Medical Journal* considered a number of other proposals without endorsing any of them:

> 'Shadowgraphy' is an impossible monster. 'Skiagraphy' has the authority of the Greek dictionary, but with a different meaning. We rather incline towards 'Skiography', which is legitimate in structure, and yet, being new, is open to no question as to validity of title. 'Radiography', and its more thoroughbred equivalent 'actinography', albeit they come trippingly enough off the tongue, are a trifle vague. ... 'Scotography', which has recently been suggested ... expresses the fact that in the new method there is 'no light, but only darkness visible', but it is hardly ... in accordance with the principles of scientific nomenclature to name a thing according to what it is not instead of what it is.[49]

---

[47] Cited in Glaser (1993), p. 229.
[48] Wade (1897).
[49] [Anon.] (1896).

Of these names, 'skiagraphy' and 'actinography' were used in English for X–rays at a few occasions, *skia* meaning shadow and *actino* ray or beam in ancient Greek. John Herschel constructed in 1840 an 'actinograph', an instrument he used for measuring solar radiation. At about the same time Henry Talbot referred to the new method of photography as a 'sciagraphic process'.[50] The prefix *scoto* derives from the Greek *skotos* (σκότος) for darkness. It appears in 'scotopic vision', which refers to the ability to see in darkness or dim light and also in 'scotophyte', a plant that flourishes in the dark. Although 'electrography' and 'electrogram' did not catch on, in other areas of medical research related words became widely used. The most popular of these words derived from electrogram by inserting a term between the two syllables in 'electrocardiogram' abbreviated ECG or EKG (Greek *kardia*, heart). The term was coined by the Dutch physiologist Willem Einthoven in 1893, before Röntgen discovered his rays. In 1924, Einthoven was awarded the Nobel Prize in medicine 'for his discovery of the mechanism of the electrocardiogram'.[51]

Perhaps understandably, Röntgen never used the eponymous word but always referred to X-rays. Indeed, it may seem immodest and unnatural for a scientist to refer to something named after him- or herself. Without having checked, I doubt if Arthur Compton ever referred to the effect named after him as the 'Compton effect' or that Hans Geiger referred to his famous counter as the 'Geiger (or Geiger-Müller) counter'—but perhaps they did.

Eponyms are normally nouns and only rarely used as or modified to verbs. But it does happen, perhaps most notably with the word 'boycott' which is named after the nineteenth-century British landowner Charles Cunningham Boycott. Of the few eponymous verbs referring to scientists, we have already mentioned 'galvanize' (Section 2.1). The only commonly known and widely used word of this kind is 'pasteurize' named after the famous French chemist and bacteriologist Louis Pasteur. This term and the noun 'pasteurization' were first used in the late nineteenth century, at a time when Pasteur was still alive (he died in 1895). Röntgen may not seem suitable for this kind of words, but 'ich röntge' has been and can still be used in German.[52] In English one could similarly, in the early part of the twentieth century, meet phrases such as 'I was roentgened' and a scientist could even be a 'roentgenologist' or what soon became known as a radiologist (*OED*).

Although Röntgen is by far best known for his rays and associated apparatus, he has also left a few other marks in the history of science. For one thing, the *roentgen* unit (symbol R) for ionizing radiation was internationally adopted in 1928 and later redefined several times. However, it is no longer or only rarely used today. In a very different area of science Röntgen's name turns up in the artificially produced element *roentgenium* with symbol Rg and atomic number 111 (Section 5.5). A few atoms of this element were synthesized by a team of nuclear scientists in 1994, but it took another ten years before the element and its proposed name were officially approved by IUPAC, the International Union of Pure and Applied Chemistry. Finally, a crater

[50] Batchen (1993).
[51] Cooper (1986).
[52] Mosskop (1995).

on the Moon of diameter 126 km is named after the German professor and so is a minor planet.

Röntgen's discovery of X-rays almost immediately gave rise to a series of new discoveries of which radioactivity was the most important. On 29 February 1896, Henri Becquerel reported to the French Academy that a fluorescent salt of uranium emitted rays that looked conspicuously like but were not X-rays. In contrast to Röntgen's rays, those found by Becquerel seemed to be emitted spontaneously, without any external agent causing them. Since the phenomenon was also observed in metallic uranium, Becquerel referred to the rays as 'uranium rays' or 'uranic rays'. This was initially the favoured term, but when it was realized that the rays were also emitted by some other elements people began speaking of 'Becquerel rays', a term used as early as 1896.[53] At a time when it could still not be excluded that the Becquerel rays were of photochemical origin, the British physicist Silvanus Thompson suggested an alternative name for them: 'The writer ventures to give to the new phenomenon ... the name of *hyperphosphorescence*.'[54] Thompson's inapt name never entered the English or any other dictionary.

Pierre and Marie Curie famously discovered two new radioelements much more active than uranium. In their two papers of 18 July and 26 December 1898 to *Comptes Rendus*, the journal of the French Academy, they proposed to call one of the elements *polonium* ('from the name of the country of origin of one of us') and the other *radium*.[55] With regard to the latter name it is generally assumed that the Curie couple gave it the name because radium is highly radioactive, but this was not the case. They suggested the name because the element, or rather its salts, radiated with a faint blue light when observed in the dark. Thus, the name derives from the Latin *radius* in the same way that the verb 'radiate' and the noun 'radio' do.

The first of the papers in *Comptes Rendus* introduced in a casual manner the neologism 'radio-active substances', which somewhat strangely only appeared in the title and not in the text. The adjective radio-active was given more prominence in the second paper of 26 December which also, for the first time, referred to the noun *radioactivity*: 'One of us [Marie Curie] has shown that radio-activity appears to be an atomic property, persisting in all the chemical and physical states of the material.' It should be noticed that the two authors did not propose the names in any formal sense but merely introduced them without any comment and as if they were already known by the readers. For the rays emitted by the new radioactive substances they continued to refer to them as Becquerel rays.

Radio-active and radio-activity quickly caught on, both terms normally written with a hyphen that only disappeared for good in about 1920. Rutherford's monograph of 1904 carried the title *Radio-Activity*, whereas his more extensive book of 1906 called *Radioactive Transformations* referred to 'radioactivity'. The terms introduced by the Curies had the advantage over Becquerel rays in that they were more generally applicable and did not refer to the emitted rays only. Radioactive can be

[53] Malley (2011) says on p. 25 that 'Becquerel rays' were first used by the Curies in 1898. However, Thompson used it in his *Philosophical Magazine* paper of 1896.

[54] Thompson (1896), dated 6 June 1896.

[55] English translations in Romer (1970), pp. 69–74.

used as both an adjective ('radioactive decay') and an adverb (an atom can 'decay radioactively'). On the other hand, there is no direct counterpart to Becquerel rays in the form of radioactive rays which strictly speaking is a misnomer although one frequently used. A radioactive body emits rays, but the rays consisting of alpha, beta, or gamma particles are not themselves radioactive. Historians of science have occasionally referred to the pioneers of radioactive research as 'radioactivists', which is a nice term but not one that was used in the past.

While Web of Science only lists 16 publications from 1900 to1910 with Becquerel rays in the title, in the same period there were 246 with radioactivity and 208 with radioactive. About one third of the articles used hyphenated terms. When Becquerel gave his Nobel lecture in 1903, he used the terminology invented by the Curies.

During the early phase of research on radioactivity several new words were suggested most of which are no longer part of the scientific dictionary. One of these apparently ephemeral words was 'metabolon' coined by Rutherford and Soddy in a 1903 paper dealing with the new disintegration theory. 'It seems advisable to possess a special name for these now numerous atom-fragments, or new atoms, which result from the original atom after the ray has been expelled, and which remain in existence only a limited time', they wrote. 'We would therefore suggest the term *metabolon* for this purpose.'[56] The word coined by Rutherford and Soddy is from ancient Greek and means 'changeable' or 'to change'. It is essentially the same word as the much better known 'metabolism' introduced about 1880 for the chemical processes that convert food to energy in the cells of an organism. Metabolon was used only by a minority of the physicists and chemists investigating radioactivity and by 1910 it had largely disappeared. It looked as if it was a protologism and nothing more. However, about eighty years later metabolon was reinvented in a completely different field of research, namely biochemistry, and this time with more success.

Paul Srerer, a prominent Hungarian-born American biochemist, was undoubtedly unaware of the Rutherford–Soddy neologism (or protologism) when he suggested the very same term in 1985. He looked for a term that described various molecular complexes of metabolic enzymes which until then had been designated with half a dozen different names:

> Communication about such complexes might be facilitated if a single word were available for them. Such invented words in biochemistry, often coined long after a phenomenon is described, have served useful purposes since replacing a phrase of many words with a single word allows the thinking about an area to become more focused and productive ... It seemed clear that no simple word could convey in its own structure the concepts I have discussed so ... I propose, therefore, the word 'metabolon' for a 'supramolecular complex of sequential metabolic enzymes and cellular structural elements.'[57]

In contrast to the Rutherford–Soddy metabolon, Srerer's version has won wide acceptance in scientific articles and also appears regularly in textbooks on biochemistry.

---

[56] Rutherford and Soddy (1903). Soddy (1904), p. 123.
[57] Srerer (1985).

'Radiochemistry' and 'radiochemist' were other, more successful neologisms that followed the discovery of radioactivity. By 1910 these were well-established words although considered controversial by some chemists of the old school. One of them was Arthur Smithells, a professor of chemistry at Leeds, who expressed his worries over 'the invasion of chemistry by mathematics and, in particular, from the sudden appearance of the subject of radio-activity ... with its accompaniment of speculative philosophy'.[58] He was tempted to see so-called radiochemistry as 'a chemistry of phantoms', a view shared by Mendeleev. According to traditionalists like Smithells and Mendeleev, the word radiochemistry was a misnomer as radioactivity was incompatible with the true meaning of chemistry. While radiochemistry flourished, after about 1935 some workers in the field began using the new term 'nuclear chemistry', which first appeared in print in 1932. The two terms are roughly synonymous and since World War II they have been used with approximately the same frequency (*OED*). Nuclear chemistry, an interdisciplinary branch of chemistry and physics, was officially sanctioned when Frédéric Joliot-Curie was appointed professor of 'Chimie Nucléaire' at the Collège de France in 1937.

As Röntgen's name appeared in a unit, so did the names of Becquerel and Curie. The curie unit (Ci) was introduced as early as 1910, at a meeting of the International Congress of Radiology and Electricity where it was agreed that 'the name Curie, in honour of the late Prof. Curie, should ... be employed for a quantity of radium'.[59] Thus, the unit referred only to Pierre Curie, who had died four years earlier in a traffic accident, and not to Marie Curie. Marie was excluded, not because of her sex but because she was still alive: it was an unwritten rule that units should be named only after deceased scientists. The unit named after Becquerel dates from 1975 and is defined as the activity of a radioactive material in which one nucleus of radium decays per second. While the becquerel (Bq) is part of the present SI system, the curie is not and therefore no longer used. The same is the case with the rutherford unit (Rd) from 1946 which in 1975 was replaced by the becquerel. The not widely used rutherford was proposed by Edward Condon and Leon F. Curtiss as an alternative or supplement to curie.[60]

No unit was ever named after the French physicist René Blondlot, and fortunately not. A highly reputed experimental physicist at the University of Nancy, in 1903 Blondlot concluded that he had found a new kind of radiation emitted from discharge tubes.[61] Further experiments indicated that the rays were also emitted from a variety of other sources such as gas burners, the Sun, and metals in states of strain. Blondlot christened the new and exciting phenomenon in honour of the city in which it was discovered—*N-rays* for Nancy rays, hence a toponym. The new rays soon attracted massive attention among French physicists and in particular after they had been detected to emerge also from the human nervous system. Physicists outside France were more sceptical as they were unable to confirm the existence of N-rays.

[58] Smithells (1907).
[59] Rutherford (1910).
[60] Condon and Curtiss (1946). Alexander (1946).
[61] For details, see Nye (1980).

The eponym 'Blondlot rays' was used in at least one paper denying the reality of the rays.[62]

For a couple of years, it was uncertain whether the rays existed. In the years 1903–6 some 300 papers were published on the subject and a substantial part of them accepted the claims of Blondlot and his allies. However, as late as 1908 it was evident that the rays did not exist and that affirming observations were due to psycho-physiological and psycho-sociological effects. Like Mesmer's 'animal magnetism' a century earlier (Section 1.5), Blondlot's N-rays only survived as an instructive example of flawed science. Both terms can be found in the *Oxford English Dictionary*.

4.3 Quantum languages

Max Planck's famous hypothesis of energy quantization dating from 1900, and the equally famous eponymous constant of nature h associated with it, marks the beginning of the quantum revolution. The word *quantum* is old and had for long been used to denote, for example, a definite amount of some quantity. For physicists and chemists in the pre-quantum era it could be used to express a unit of matter or electricity, but with the advent of quantum theory it acquired a new meaning and was used in extended forms. The noun 'quantization' and the verb 'quantize' made their entry in the physicists' vocabulary. These neologisms had a quite different, more specialized meaning than the old words 'quantification' and 'quantify'. Planck did not originally speak of a quantum of energy but used the term 'energy element' (*Energieelement*) in the restricted sense of a body absorbing and emitting discrete portions of radiation energy. The first to use 'quantum' in its current sense may have been Einstein in his famous 1905 paper on the nature of light.

During the first decade of the twentieth century key terms such as 'quantum theory' and 'Planck's constant' were rarely used. Planck's new theory was largely restricted to the study of blackbody radiation and a few other subjects such as the specific heats of solids. Only with Niels Bohr's epoch-making atomic theory of 1913 did it turn out that quantum concepts were also immensely fruitful, indeed indispensable, when it came to understanding the structure of atoms and molecules. Like some earlier atomic models, Bohr's pictured the atom in analogy with the solar system, the nucleus playing the role of the Sun and the surrounding electrons the role of the planets. He originally assumed that atoms were planar structures just like the solar system. However, he was keenly aware that metaphors of this kind could not encompass the novelties of his theory. Thomas Kuhn, not only an innovative philosopher of science but also an expert in the history of the Bohr atom, wrote as follows:

> Bohr and his contemporaries supplied a model in which electrons and nucleus were represented by tiny bits of charged matter interacting under the laws of mechanics and electromagnetic theory. That model replaced the solar system metaphor but not, by doing so, a metaphorlike process. Bohr's atom model was intended to be taken

---

[62] McKendrick and Colquhoun (1904).

only more-or-less literally; electrons and nuclei were not thought to be exactly like small billiard or Ping-Pong balls.[63]

In the same year that Bohr proposed his atomic model, the German physicist Johannes Stark announced that spectral lines are split in strong electric fields, a discovery that greatly stimulated quantum theory and in 1919 resulted in a Nobel Prize for its discoverer. Bohr immediately offered an interpretation of the phenomenon that supported his theory of atomic structure. The 'Stark effect', as it was and is still called, is an eponym but perhaps not the right one. At least, some Italian physicists quickly pointed out that their compatriot Antonino Lo Surdo had independently and virtually simultaneously made the same discovery and that 'Stark–Lo Surdo effect' was therefore a more appropriate name.[64] Of course, Stark disagreed. Although Lo Surdo is essentially forgotten, in Italian papers from the early period 'Stark–Lo Surdo' was commonly used. Given that Stark published his discovery a few months earlier than Lo Surdo, the case cannot be considered an example of Stigler's law of eponymy.

Another eponymous effect that played a most important role in early quantum theory was the magnetic splitting of spectral lines called the 'Zeeman effect' after the Dutch physicist Pieter Zeeman. Physicists were particularly worried that although they could account for the 'normal' Zeeman effect, where a line is split into three, the old quantum theory was unable to explain the more complicated pattern in the anomalous effect. Physics papers in the period up to the mid-1920s are filled with references to the 'anomalous Zeeman effect', which in a sense is a misnomer insofar that it is more common than the normal type of magnetic splitting. This was known at the time, but for historical reasons the physicists kept to the confusing terminology. Or perhaps they were aware that the word anomalous is not only synonymous with 'unusual' or 'uncommon' but can also refer to what is unexplained and unexpected.

Bohr based his 1913 theory of the hydrogen atom on two basic assumptions of which the first postulated that the electron can only move around the atom in certain *stationary states* or orbits characterized by different *quantum numbers* (he did not employ the latter term, though). He called the lowest energy state of an atom its 'permanent state', a term which soon became known as the 'ground state'. The higher energy states are called 'excited' states, but this term was also absent from Bohr's papers of 1913. The words 'stationary state' and 'quantum number' were essentially neologisms although as usual they can be found in different contexts in the earlier literature. Thus, in an important paper of 1900 Max Planck spoke of a system's stationary state in the sense corresponding to its maximum total entropy. Two years later we find quantum number in Lord Kelvin's ill-fated and entirely classical theory of so-called electrions from 1902 (Section 2.2). Kelvin wrote that the different nature of atoms belonging to various chemical elements 'may be partially due to the quantum-numbers of their electrions being different'.[65] He obviously used quantum in the sense of 'total quantity'.

---

[63] Kuhn (1993). See also Petruccioli (1993).
[64] Leone, Paoletti, and Robotti (2004).
[65] Planck (1900), p. 110. Thomson (1902), p. 259.

There are a few more terms related to Bohr's theory that deserve attention. First, for a hydrogen atom in its ground state he calculated the distance from the nucleus to the electron in terms of known constants of nature. Numerically, his result was 0.55×10^{-8} cm, which is in excellent agreement with the modern value for what became known as the 'Bohr radius'. This quantity is still counted as an important unit in the SI system. Second, Bohr deduced from his theory of 1913 that atoms in outer space might exist in states given by high quantum numbers and therefore be very large compared to ordinary atoms. Monster atoms of this kind have been observed and since the early 1970s been named 'Rydberg atoms'. In this case the eponym is questionable as the Swedish physicist Janne Rydberg never referred to atoms of this kind but only hypothesized the corresponding frequencies. 'Rydberg–Bohr atoms' or 'Bohr–Rydberg atoms' would seem to be more appropriate. There are thousands of papers including the term 'Rydberg atoms' but none with Bohr's name appended. Clearly, in this case Stigler's law is relevant.

There is yet another word which in the public mind is an essential feature of Bohr's atomic theory, and of quantum mechanics generally, namely the popular term *quantum jump*. As Bohr stated in his second postulate, when an atom is in an excited energy state it will spontaneously pass into a lower one and end up in the ground state. As a result, light will be emitted with a frequency given by the energy difference divided by Planck's constant, that is, $\nu = \Delta E/h$. Bohr spoke of transitions between the states but never used the metaphor of electrons jumping from one state to another. Nor was the metaphor widely used by other physicists in their writings. The first time it turned up was probably in an article of 1915 in which Max Planck expressed his reservations with regard to Bohr's idea of 'a jump of the oscillating electron from one stationary orbit to another stationary orbit'.[66] Nine years later the American physical chemist Richard Tolman explicitly referred to 'quantum jumps' in a paper discussing the lifetime of excited molecules.[67]

The abrupt and uncaused quantum transitions, jumps or not, were controversial also after they were carried over into quantum mechanics. Erwin Schrödinger, who much disliked them, published in 1952 a paper with the rhetorical question 'Are There Quantum Jumps?' as its title reads. As it has happened with other words with an origin in the world of science, quantum jump later became adopted in the public sphere with inevitable changes of meaning. Today, quantum jump and quantum leap are routinely used to denote a drastic change in society, to brand a new commercial product, or to characterize a sudden advance in technology. In this sense, the terms relate to something big and revolutionary—a kind of Big Bang (Section 6.5)—in stark contrast to the atomic quantum jumps which are minute, scarcely noticeable, and happen all the time. In 1961 the journal *Science* carried an article about the communication satellites of the near future, informing its readers that they 'promise quantum jumps in overseas capacity at low cost'.[68] To mention but one later example, the Duracell company marketed in the 1990s a new battery as 'a revolutionary advancement in battery technology ... a quantum leap in battery power'.

---

[66] Cited in Kragh (2012), p. 149. In Bohr's theory an electron was not 'oscillating'.
[67] Tolman (1924).
[68] Meckling (1961).

Quantum mechanics as we know it today built on but also radically changed the framework of the earlier quantum theory due to Bohr, Sommerfeld, and others. The new theory was originally created by the young German theorist Werner Heisenberg, who in the early autumn of 1925 wrote a pioneering paper with 'quantum mechanics' hidden in the text and without the author paying attention to the word. Later the same year Heisenberg and his collaborators Max Born and Pascual Jordan published a comprehensive work generally regarded as the first full exposition of the new quantum theory. Its title was 'Zur Quantenmechanik II' (On Quantum Mechanics II). While the term was thus firmly introduced in 1925, it was not completely new. It appears in the physics literature at least three times before it became more than just a word.

The first, to my knowledge, who used the word quantum mechanics was the British physicist Charles G. Darwin in a discussion session on quantum theory at the 1921 meeting of the British Association for the Advancement of Science. 'The spectrum theory is far the most interesting branch of the quantum theory, as it has led and is still leading to the extensions of quantum mechanics', he said. In a paper on superconductivity from 1922, Einstein stated that 'our ignorance of quantum mechanics of composite systems' made it unlikely that superconductivity would soon be understood (see also Section 2.5). Finally, in 1924 Born wrote a paper in *Zeitschrift für Physik* with the title 'On Quantum Mechanics' in which he critically reviewed the problems of the existing quantum theory. Born envisaged 'a formal passage from classical mechanics to a "quantum mechanics".[69]

What was generally referred to as quantum mechanics in late 1925 was also known as 'matrix mechanics' or 'Göttingen mechanics', the first name reflecting the mathematical formalism of the theory and the second a toponym, the German university city where it was principally created. In Heisenberg's original paper there appeared arrays of physical symbols with the strange property that they did not satisfy the commutative law of multiplication (that is, xy differed from yx, see below). His professor, the mathematically trained Max Born, recognized that Heisenberg's symbolic quantities were *matrices* and that the new theory could be formulated in terms of the *matrix* calculus with which he (in contrast to Heisenberg) was acquainted. Hence, matrix mechanics became just another name for the Göttingen quantum mechanics.

The term *matrix* (plural: matrices) is itself neither a quantum word nor a science word, but was used much earlier in the meaning of 'source' or 'origin' as located in the womb of a mother. It is based on the Latin *mater* (mother). Johnson's definition of 'matrix' in his *Dictionary* from 1755 was 'Womb; a place where any thing is generated or formed'. Since the eighteenth century the term was also used in geology and mineralogy to denote, for example, a rock in which a fossil or gemstone was embedded and matured. To the extent that the word matrix is commonly known today, it is not as an anatomical, geological, or mathematical term but more likely from the innovative science fiction film from 1999 called *The Matrix*. Another area from which the word is commonly known is the 'matrix printer' marketed since the late 1950s.

---

[69] Darwin (1922), p. 473. Einstein, 'Theoretical Remarks on the Superconductivity of Metals', available at https://arxiv.org/abs/physics/0510251. Born (1924), translation in Van der Waerden, (1968), pp. 181–98.

The term entered mathematics in the 1850s, when it turned up in papers by the British mathematicians James J. Sylvester and Arthur Cayley. In a remarkably long (148 pages) paper in *Philosophical Transactions*, Sylvester defined the word matrix as 'a square or rectangular arrangement of terms in lines and columns'.[70] Sylvester was a scientific wordsmith almost as productive if not as successful as the older William Whewell. At the end of his 1853 memoir, he appended a glossary of 'New or Unusual Terms', which not only included matrix but also more bizarre words such as 'bezoutoid', 'covariant', 'dialytic', 'endoscopic', 'rhizoristic', 'syrrhizoristic', and 'syzygetic'. Some of his many mathematical words remained protologisms or survived only with different meanings such as did 'endoscopic', which is primarily a medical term referring to the use of the instrument called an endoscope. Matrix and a few other terms such as 'covariant' came into general use among mathematicians and physicists. Although Sylvester coined matrix in its mathematical sense, he credited Cayley with the discovery of matrix algebra.[71]

Apart from 'quantum mechanics' and 'matrix mechanics', by the late spring of 1926 yet another name made its entry, the *wave mechanics* of Erwin Schrödinger. The Austrian physicist also spoke of his theory as 'undulatory mechanics'.[72] From a later point of view, the terms quantum mechanics and wave mechanics are synonyms, just two words for the same theory, but this was not originally the case. Schrödinger disliked the Göttingen quantum mechanics and thought of his own wave mechanics as a superior alternative to it, not least before it was more *anschaulich*, that is, visualizable. The German noun *Anschaulichkeit* was a key term in the early phase of quantum mechanics. It is commonly translated as 'visualizability' but also has the more metaphorical connotation 'intelligibility'. When a person understands something, he or she will often say 'I see'.

Although 'wave mechanics' appeared in several papers and books, by the late 1920s 'quantum mechanics' had won general acceptance such as illustrated by the first generation of textbooks from George Birtwistle's *The New Quantum Mechanics* (1928) to Dirac's influential *The Principles of Quantum Mechanics* (1930). To learn what quantum mechanics was all about, Dirac advocated a symbolic and nonverbal approach. The new theory, he said in his textbook, was based on 'physical concepts which cannot be explained in terms of things previously known to the student, which *cannot even be explained adequately in words at all*'.[73] Nonetheless, the theory had of course to be formulated in words. While some of the words were taken from the old vocabulary, others were new and coined for the purpose. Some of them were neologisms, others were known words with new meanings, and others again were hybrid German–English words.

To the first category belonged Arthur Eddington's whimsical nonce word *wavicle* invented to describe the wave–particle duality as exhibited by a particle built up of Schrödinger's quantum waves: 'We can scarcely describe such an entity as a wave or as

---

[70] Sylvester (1853). Silver (2017).
[71] See Jammer (1966), pp. 205–6, for the priority controversy concerning the discovery of the matrix theory.
[72] Schrödinger (1926).
[73] Dirac (1930), p. v. Emphasis added.

a particle; perhaps as a compromise we had better call it a "wavicle". On a more serious note he also discussed in 1928 what is usually known as Heisenberg's *uncertainty principle*, a much-discussed cornerstone in the conceptual foundation of quantum mechanics. Recognizing its importance, Eddington suggested calling it the *principle of indeterminacy*, a term he possibly coined.[74] The American physicist Arthur Ruark introduced the almost identical 'indetermination principle' in a brief note from the same year—in which he suggested that Heisenberg's principle was wrong![75]

Heisenberg did not state his relations as a 'principle' in his paper of 1927. Bohr, in an important address published in 1928 where he introduced the *complementarity principle*, wrote throughout of 'uncertainty' and not of 'indeterminacy', but without using the phrase 'uncertainty principle'.[76] This phrase, or rather 'principle of uncertainty', may have made its entry in a 1928 paper by another American physicist, Gregory Breit.[77] It may seem of no consequence whether one speaks of the uncertainty or the indeterminacy principle, but philosophers of science often discriminate between them and still discuss whether one of the terms is more appropriate than the other. Uncertainty is generally regarded as relating to knowledge, or rather to lack of knowledge, whereas 'indeterminacy' denotes a characteristic of nature.

The complementarity principle became a key element in the so-called *Copenhagen interpretation* of quantum mechanics which refers to a coherent cluster of views regarding this theory and its philosophical basis. The much-used label is of rather late date as it was coined by Heisenberg in a 1955 festschrift in honour of Bohr. He introduced the term in order to defend his own and Bohr's view against the criticism raised by other physicists. Although the name 'Copenhagen interpretation' soon became entrenched in the literature, it is widely agreed that the expression is unclear as it covers several different views. According to Bohr's close collaborator Léon Rosenfeld, writing in 1960, 'it would be better to discard such an ambiguous expression as "Copenhagen interpretation", were it only because it falsely suggests that there could be other possible interpretations of quantum theory'.[78]

Bohr developed his complementarity principle in close connection with Heisenberg's uncertainty principle. Although Bohr's more general principle had its roots in quantum physics, he thought that it might also be applicable to fields outside physics including biology, psychology, anthropology, and linguistics. He was particularly fascinated by the concept of language and its relation to physics and science generally. According to Bohr, the essence of scientific knowledge, as well as any other kind of genuine knowledge, is that it allows unambiguous communication in terms of words. On one occasion he explained to his assistant Aage Petersen that quantum physics is at bottom an abstract description of phenomena that must be expressed in ordinary words: 'It is wrong to think that the task of physics is to find out how nature is. Physics

---

[74] Eddington (1928), pp. 201 and 220. The book was based on the series of Gifford Lectures that Eddington gave in January–March 1927. 'Wavicle' appears in *OED* and has been used a few times by later physicists and in popular presentations of quantum physics.

[75] Ruark (1928).

[76] Bohr (1928). For details, see Cassidy (1988) with comments on the terminology in France (by Lévy-Leblond and F. Balibar) and Italy (G. Battimelli), pp. 279–80.

[77] Breit (1928).

[78] Rosenfeld (1960). For details on the origin of the Copenhagen interpretation and its further use among physicists and philosophers, see Howard (2004).

concerns with what we can say about nature.... We are suspended in language.'[79] This often-cited phrase presumably encapsulates what Bohr thought, but neither I nor others have been able to find it in his published works. In any case, his deep interest in language did not focus on particular words and the terminology of physical concepts and objects.

David Bohm and some other physicists have argued that our language utterly fails to meet the demands of the new physics and that, consequently, a new form of language is required to describe the quantum reality in particular. The influential British linguist Michael Halliday disagreed. In a paper first published in 1987, he contended that 'natural language—not as it is dressed up in the form of a scientific metalanguage, but in its common-sense, everyday, spontaneous spoken form—does in fact "represent reality" in terms of complementarities'. He affirmed that, in his view, 'our natural languages do possess the qualities needed for interpreting the world very much as our modern physicists see it'.[80]

Halliday's view seems closer to Bohr's than to Bohm's. In his *Physics and Beyond*, Heisenberg recalled conversations on language that he had with Bohr during a skiing holiday in 1933. According to Heisenberg, Bohr said:

> It is one of the basic presuppositions of science that we speak of measurements in a language that has basically the same structure as the one in which we speak of everyday experience. We have learned that this language is an inadequate means of communication and orientation, but it is nevertheless the presupposition of all science.

Referring to the philosophy of William James, Bohr further observed (again in Heisenberg's rendering):

> We never know what a word means exactly, and the meaning of our words depends on the way we join them together into a sentence, on the circumstances under which we formulate them, and on countless subsidiary factors. ... Though our minds may seem to seize on only the most important meaning of a word we hear spoken, other meanings arise in its darker recesses, link up with different concepts and spread into the unconscious. That ... applies to the language of science as well. Particularly in atomic physics, nature has taught us that some of our most trusted concepts have a strictly limited application.[81]

Quantum mechanics was primarily due to German-speaking physicists, which is reflected in many of the words used by their colleagues in England, the United States, and elsewhere. German words were often combined with English words or they were simply adopted in their German forms at least temporarily.[82] In the early quantum

---

[79] Petersen (1963). See also Favrholdt (1993) and, for a critical view, Lévy-Leblond (1999).

[80] Halliday (2003), pp. 125 and 129. On Bohm's unsuccessful attempt to create a new language of physics, see Bohm (1980). See also Ford and Peat (1988) for a development of Bohm's proposal of a close connection between language and thought.

[81] Heisenberg (1971), pp. 130 and 134–5.

[82] Patterson and Knorr (1933).

theory of atoms and molecules some scientific papers in English contained words such as *Aufbauprinzip* (construction principle), *Tauchbahn* (penetrating orbits), *Leuchtelektron* (light electron), and *Rumpfelektron* (inner electron).

According to quantum theory, the energy of a particle can never be zero. There inevitably remains a minimum energy known as the zero-point energy, a concept introduced by Planck as *Nullpunktsenergie*. In the case of a harmonic oscillator, $E_n = (n + \frac{1}{2}) h\nu$, hence $E_0 = \frac{1}{2}h\nu$ and not $E_0 = 0$. Although the German term was initially used by a few British and American physicists, it soon became known in English as either 'zero-point energy' or 'zero-point-energy'. The translation may seem straightforward and unobjectionable, but according to Darwin it was 'clumsy and ugly' because it was borrowed literally from German. 'The translation of *Nullpunkt-senergie* is a poor one, because we do not translate *Nullpunkt* as *zero-point*, but simply use the English term *zero* ... It does seem a pity not to create an English technical term and speak of *residual energy*, etc., which could be done without ambiguity.'[83] Darwin further objected that in English it is not allowed to use a noun to qualify an adjective and for this reason he characterized 'quantum-theoretical' and 'quantum-mechanical' as unwanted Germanisms. As an alternative he coined the adjective *quantal* as an antonym of classical. Although the term was used on some occasions, it never became a common word in quantum physics or elsewhere.

In 1930 Schrödinger deduced from Dirac's relativistic quantum theory that a free electron would exhibit a peculiar kind of oscillatory motion which he called *Zitterbe-wegung* (trembling motion).[84] American workers adopted the German word and so did other physicists. The decidedly un-English word is still part of the language spoken by the physics community. The same is the case with *bremsstrahlung* (braking radiation), which refers to the electromagnetic radiation produced when electrons or other charged particles are slowed down in matter as, for example, in an X-ray tube. This term even appears in the *OED*, whereas zitterbewegung (with small letter z) does not. There are several other and much better known German words in English language, witness *wunderkind*, *zeitgeist*, *wanderlust*, and *kindergarten*. Biochemists are familiar with the term *zwitterion* where the first syllable refers to the German noun *Zwitter* meaning hermaphrodite or hybrid. An amino acid with a positively charged amino group (NH_3^+) and a negative carboxyl group (COO^-) is a zwitterion.

The new quantum language included a series of hybrid words that made sense only to physicists and mathematicians, such as the *eigen*- words eigenvalue, eigenfunction, eigenvector, and eigenstate.[85] These were largely adopted from Schrödinger's German terminology, but the words *Eigenwert* and *Eigenfunktion* were originally coined by the German mathematician David Hilbert in a paper of 1904. British and American physicists kept *eigen* and combined it with an English end syllable. *Eigenfunction* seems to have been coined by an Englishman, namely Dirac, in an important paper of 1926 dealing with radiation theory. Writing the general solution to the Schrödinger equation as $\psi = \sum c_n \psi_n$ he stated that 'the ψ_n's are a set of independent solutions,

---

[83] Darwin (1936).
[84] Kragh (1992).
[85] https://jeff560.tripod.com/e.html. The German 'eigen' literally means 'own' as in 'its own.'

which may be called eigenfunctions'.[86] A paper from the same early period written by the American physicist Karl Darrow illustrates the impact of German terminology. Darrow wondered about what words to use:

> The permitted values of m are known in German as the *Eigenwerte* ... The English term would be 'characteristic values'; but it is long and has many meanings, and I think it preferable to borrow the German word ... To each *Eigenwert* of m there corresponds a value of the vibration frequency $mu/2\pi$, which in German is called an *Eigenfrequenz*; but here we may as well keep to the English term *natural frequency*. To each *Eigenwert* there corresponds a solution to the differential equation, an *Eigenfunktion*.[87]

Note that Darrow used the full German words, such as *Eigenfunktionen* in plural, rather than the hybrid terms preferred by Dirac and most other English-speaking physicists. In a voluminous textbook on quantum physics from 1930, two American authors commented: 'The German word for characteristic value is Eigenwert; the characteristic function ψ_n belonging to the Eigenwert E_n is called an "Eigenfunktion". We shall often use these words in English sentences without apologetic quotation marks.'[88] At least one reviewer of *Principles of Quantum Mechanics*, the American mathematician Bernard Koopman, expressed reservation with regard to Dirac's use of eigen- words: 'We object to the coining of the words "eigenvalue" and "eigenfunction", particularly since usage provides the terms "characteristic number" and "characteristic function", as the precise rendering of the German "Eigenwert" and "Eigenfunktion".'[89] In his textbook of 1930, Dirac used both 'eigenfunction' and 'eigen-ψ'.

While eigenfunction and eigenvalue were in part imported from German, *eigenstate* has a different story as it was one of Dirac's new word constructions, this time from the second edition of his textbook on quantum mechanics, where he added in a footnote: 'The word "proper" is sometimes used instead of "eigen", but this is not satisfactory as the words "proper" and "improper" are so often used with other meanings.'[90] In German, eigenstate is *Eigenzustand* (plural—*zustände*). Many English-speaking authors began writing *proper* instead of *eigen*, with the result that words such as 'proper function' and 'proper value' appeared frequently in the physics literature. And yet the pure English words never dominated. The hybrid words, objectionable as they may have been in the eyes of purists, eventually became commonly accepted. While the ratio of scientific papers with eigenvalue to those with proper value was about 1:1 in the 1930s, in the 1970s it was approximately 20:1. In a 1967 book, the American mathematician Paul Halmos wrote: 'For many years I have battled for proper values, and against the one and a half times translated German-English

[86] Dirac (1926a).
[87] Darrow (1926).
[88] Ruark and Urey (1930), p. 526.
[89] Koopman (1931).
[90] Dirac (1935), p. 32 and in later editions of the book.

hybrid that is often used to refer to them. I have now become convinced that the war is over, and eigenvalues have won it.'[91]

One of Dirac's lasting contributions to the quantum dictionary was to replace the frequently occurring quantity $h/2\pi$ (h = Planck's constant), with a new symbol $\hbar$ sometimes called 'Dirac's h' or the 'Dirac constant'. He introduced the new 'universal constant having the dimensions of action' in his 1930 textbook, first in an equation with no explanation, which only came subsequently: 'In order that the theory may agree with experiment, we must take $\hbar$ equal to $h/2\pi$, where h is the universal constant that was introduced by Planck.'[92] It is unknown why Dirac chose the unusual symbol $\hbar$ (pronounced h-bar) as he never explained the choice. It may just have been due to its similarity to h, but it is also possible if not likely, in my opinion, that he may have been inspired by the astrological symbol for the planet Saturn (♄) which looks somewhat similar.

Another but later notational innovation was the 'bracket' formalism, which Dirac introduced in a brief paper of 1939, coining the words 'bra-vector' and 'ket-vector' and their corresponding symbols < and >. A vector or wave function corresponding to the quantum state A was symbolized as |A> and that of the conjugate vector A∗ as <A|. The scalar product of a bra vector <A| and a ket vector |B> was written as <A|B>. 'As names for the new symbols < and > to be used in speech, I suggest the words *bra* and *ket* respectively', he ended the paper.[93] In a memorial article two years after Dirac's death, two prominent British physicists, Richard Dalitz and Rudolf Peierls, commented in jest: 'He was probably not aware of the colloquial meaning of "bra".'[94]

At first Dirac's new notation was not much noticed—he used it himself for the first time in 1943 and five years later in the third edition of his textbook. Eventually it gained acceptance, and today the bra(c)ket formalism is a common, powerful, and much admired notation in quantum mechanics. Combining the bra-ket terminology with the *eigen-* prefix, in the third edition of his textbook (on p. 31) Dirac introduced the neologisms *eigenket* and *eigenbra*, as in the sentence 'The eigenvalues associated with eigenkets are the same as the eigenvalues associated with eigenbras.' These words have not been widely adopted by quantum physicists.

As mentioned, in Heisenberg's first paper on quantum mechanics, variables representing two different physical quantities x and y do not in general satisfy the law of *commutation*. While classically $xy - yx = 0$, in quantum mechanics there are cases where $xy - yx \neq 0$. For example, if x denotes the position of a particle and y its momentum, $xy - yx = ih/2\pi$. To express in words that two quantum variables satisfy the commutative law, physicists sometimes said they were *commutable*, but this term was never commonly used.[95] Mathematicians and logicians had long been interested in commutation relations and also introduced the term *commutator* for the quantity $xy - yx$. The words are derived from the Latin *commutare* meaning 'to exchange' or 'to change altogether'. With regard to the noun commutator it was better known from an

[91] Halmos (1967), p. x. The accepted French word for eigenvalue is 'valeur propre'.

[92] Dirac (1930), p. 96.

[93] Dirac (1939).

[94] Dalitz and Peierls (1986), p. 169. A woman's 'bra' as short for 'brassiere' was not commonly used in the 1930s (*OED*).

[95] Birtwistle (1928), p. 63: 'In cases where $ab = ba$, a and b are said to be "commutable".'

entirely different area, namely as a device transforming a direct electrical current into an alternating current in a generator. Commutators of this kind were familiar to electrical engineers and therefore also to Dirac, who was originally trained in electrical engineering.

When Dirac formulated his own version of quantum mechanics in papers of 1925 and 1926 he found it convenient to coin a verb associated with the commutability or non-commutability of two variables: 'When $xy = yx$ we shall say that x commutes with y', he stated.[96] In his textbook published four years later, he repeated: 'In the special case when $\alpha_1\alpha_2$ is equal to $\alpha_2\alpha_1$, we say that α_1 *commutes* with α_2 or that α_1 and α_2 *commute*.'[97] Late in life Dirac returned to what he thought was a neologism, a verb coined in analogy with the noun commutator and the adjective commutative:

> The mathematicians who had been handling noncommutative algebra said that u permutes with v. It seemed to me that the word 'permute' was not really very appropriate. One thinks of permutations as rearranging the order of several quantities, and here we are concerned only with two quantities. So I invented the word 'commute'. I do not think it had been previously used in mathematics. I said that, when $uv = vu$, u and v commute with each other. That again is a notation which everyone has accepted since then.[98]

However, although Dirac may have been the first to use the word commute in the contexts of quantum mechanics, it was not quite as new as he thought. It appeared in a few mathematical memoirs on matrices written in the nineteenth century. Moreover, electrical engineers used phrases such as 'the current may be commuted'. In present parlance commute typically refers to a person who travels daily or regularly from one's home to a job in a city. In this well-known sense, the words commute and commuter (not commutator!) were not common at the time when Dirac and others pioneered quantum mechanics.

Other words with an origin in or associated with the early quantum era include tunnelling, spin, spinor, observable, and degeneracy. *Tunnelling* is borrowed from everyday life, as making a tunnel through a mountain, but in quantum mechanics it may refer to a positive particle escaping from the nucleus despite the fact that the particle's energy is less than the potential barrier. It has to 'tunnel' its way through a mountain of potential energy. The metaphoric term was not used in early papers on the 'tunnel effect' from 1927[8] but seems to have entered the physics vocabulary in the mid-1930s. The first English use of the term was possibly in Yakov Frenkel's monograph *Wave Mechanics* from 1932, which included a section on 'Transition through a Potential Energy Mountain (Tunnel Effect)'.[99]

The hugely important concept of *spin* was in a quantum context introduced by the Dutch physicists Samuel Goudsmit and George Uhlenbeck when they discovered from spectroscopic studies in the fall of 1925 that the electron has an extra

[96] Dirac (1926b), p. 562.
[97] Dirac (1930), p. 27.
[98] Dirac (1977), p. 129.
[99] Frenkel (1932), p. 111.

degree of freedom. They originally interpreted the new spin quantum number in the pictorial sense of an electron endowed with an intrinsic rotation but without referring to either spin or spinning.[100] These terms became common only in the early months of 1926 and at first with spinning more popular than spin. The two physicists possibly adopted the metaphor from Arthur Compton, who in 1921 speculated that 'the electron itself, spinning like a tiny gyroscope, is probably the ultimate magnetic particle'.[101] Compton did not refer to this quantity as the electron's spin, such as did later quantum physicists. Although it was quickly realized that spin is a purely quantum-mechanical property with no analogue in classical physics, the figurative terms remained. No one seems to have proposed an alternative name for what is widely acknowledged to be a misnomer.

With regard to the term *observable*, in itself it is neither a quantum word nor one of recent origin. While in a phrase like 'the observable part of the universe' it is an adjective, the quantum observable is a noun. Heisenberg introduced his seminal paper of 1925 by declaring that the new quantum mechanics was founded exclusively upon relationships between quantities which in principle are observables. Physical quantities that can be measured, such as momentum, spin, and energy, are such observables. More technically, they are eigenvalues of certain quantum mechanical operators called 'Hermitian', an eponym named in honour of the French nineteenth-century mathematician Charles Hermite. If the quantities do not satisfy this condition, they are unobservables—cannot be observed. This usage is of course very different from how the same words are used in common language. While we have no problem with observing the position of some macroscopic body, in quantum mechanics the position of an electron is not observable.

If two or more different quantum states give the same value when measured, they are said to be *degenerate*, a term used by Heisenberg, Schrödinger, and other physicists in the early stage of quantum mechanics. As stated in the Born–Heisenberg–Jordan paper, 'Degenerate [*entartete*] systems will be characterized by the fact that multiple eigenvalues occur'.[102] The same terminology was used in the old Bohr–Sommerfeld quantum theory but there referring to the eccentricity of electron orbits and not to eigenvalues. An atomic system with several different orbits corresponding to the same total energy was said to be degenerate.[103] Bohr used the same terminology in a lengthy paper of 1918 in which he defined the term in a more general and abstract sense. In a different context, the term can be found even earlier, probably first in a paper of 1914 in which Walther Nernst described the state of gases at very low temperatures as degenerate. The more gas deviated from the law of ideal gases, the higher its degree of degeneracy.[104]

Ten years after Nernst introduced the term, Schrödinger referred to 'gas degeneracy' (*Gasentartung*) in the title of a paper. Also in 1924, Fermi and Einstein used it in their works on the statistical theory of gases which led to so-called Fermi–Dirac and

---

[100] Uhlenbeck and Goudsmit (1925).

[101] Compton (1921).

[102] Born, Heisenberg, and Jordan (1925). Translation in van der Waerden (1968), pp. 321–86, on p. 352.

[103] Sommerfeld (1922), p. 246, who attributed the term *Entartung*, used in a more general sense, to a 1916 paper by Karl Schwarzschild (p. 685).

[104] Nernst (1914).

Bose–Einstein statistics. Einstein began his paper as follows: 'In a treatment which has appeared recently in these proceedings ... a theory of the "degeneracy" of ideal gases has been given.'[105] However, it was only when the gas theory appeared in a new quantum-mechanical dressing that *degeneracy* became an essential concept in quantum physics. A paper by two American physical chemists, Gilbert N. Lewis and Joseph Mayer, reveals the German heritage. Discussing 'what has been called degeneracy or Entartung', they commented: 'These words now have their etymological meanings, for when two or more particles are in exactly the same state they lose their ... identity, being no longer distinguishable from one another.'[106]

A related meaning of degenerate gas turned up in the late 1920s when physicists realized that in matter at very high density the electrons will exert what was called a 'degeneracy pressure'. This outward pressure of an exotic state of matter might cause a certain type of high-density stars to neutralize the inward push of gravity and leave them as white dwarf stars. The term 'degenerate matter' may first have been used by Eddington in 1931. This new kind of degeneracy played a crucial role in astrophysics and was the central theme of the controversy between Eddington and Subrahmanyan Chandrasekhar on white dwarfs in the mid-1930s. White dwarfs and neutron stars are examples of 'degenerate stars'. As regards the name *white dwarf* it was introduced by the Dutch-American astronomer Willem Luyten in 1922 and soon popularized by Eddington in a series of works.[107]

Of course, although words such as *degeneration, degeneracy,* and *degenerate* are today well-known terms in physics, they are metaphors and much better known from other and more sombre contexts. The adjective degenerate is from the Latin *degenerere*, a falling off from the generic or natural state into a lower one, whereas degeneration is the condition of being degenerate. If the referent is applied to a human being—and not to an atom or a molecule—the words are inherently pejorative. The double meaning of the term was light-heartedly alluded to in a letter from Léon Rosenfeld to Chandrasekhar referring to the controversy with Eddington: 'The story of Eddington's degeneracy (if I may use such an ambiguous expression) takes the shape of the *Iliad*, with the various gods and heroes coming in.'[108]

The old idea that non-white races are degenerate received support from social Darwinism and eugenics in the late nineteenth century, and a few decades later—at the time when quantum mechanics was developed—the Nazis began rallying against *entartete Kunst* (degenerate art). The quantum physicists used the term in a purely technical sense and apparently they did not connect it with the ideological and political connotations which must have been so visible at the time.

In a paper published shortly before the advent of quantum mechanics, the American theoretical physicist John Slater introduced the term 'physically degenerate'. As he explained, 'a system is physically degenerate if its variable has had time to go through only a part of a cycle'.[109] The same year, 1925, Adolf Hitler used the very same phrase in the first volume of his influential *Mein Kampf*, a work where degenerate

---

[105] Einstein (1924). Bose did not refer to the term degenerate.
[106] Lewis and Mayer (1929).
[107] Holberg (2007), p. 119.
[108] Miller (2005), p. 117.
[109] Slater (1925).

and degeneration appear abundantly. For example, Hitler said about the intellectual classes that they were 'physically degenerate' because their education made them 'unfit for life's struggle.'[110] It goes without saying that Slater and Hitler used the phrase in entirely different meanings. As far as I know, there is only one indication in the quantum literature of a leading physicist worrying over the technical use of degeneracy. In the second edition of *Principles of Quantum Mechanics* published in 1935, Dirac stated in a footnote:

> A system with one stationary state belonging to each energy-level is often called *non-degenerate* and one with two or more stationary states belonging to an energy-level is called *degenerate*, although these words are not very appropriate from the modern point of view.[111]

Exactly what Dirac meant with 'the modern point of view', is unknown, but he presumably had in mind the political and cultural meaning of degenerate.

The scientific use of the term can be further illuminated if we consider the terminology of medicine, biology, and genetics rather than that of physics. In fact, the term appeared in pathological and other medical texts decades before physicists adopted it.[112] In the 1950s George Gamow was much interested in the genetic code and of course familiar with the physicists' technical use of degenerate. According to one source, 'In quantum physics, degeneracy is a specialized term for different stationary states corresponding to the same energy level. Borrowing the word from quantum physics, Gamow applied the term "degeneracy" in biology to refer to different structures that could variably achieve the same outcome.'[113] Although Gamow's ideas about genetic coding turned out to be untenable, they inspired Francis Crick of double-helix fame who later referred to an unpublished paper of 1955 in which he discussed Gamow's theory. This paper, Crick said, 'pointed out that in Gamow's scheme several different triplets [of nucleotides] could code one amino acid, and it introduced the word degeneracy to describe this.'[114]

Regardless of whether the word in this context was due to Gamow or Crick, eventually, if with some hesitation, it was adopted by geneticists and molecular biologists. It generally denotes the ability of biological elements that are structurally different to perform the same function or yield the same output. Thus, there is a superficial similarity between the ways that physicists and biologists use the term. In modern systems biology degeneracy plays an important role but the terminology is often considered controversial because of the term's historical association to eugenics and racial prejudices. It has been suggested to hyphenate the word in order to distinguish it from its former ideological connotations: '"De-generacy" with a hyphen may satisfy the need for a more precise vocabulary and distance the word from its historical baggage.'[115]

---

[110] English translation, http://gutenberg.net.au/ebooks02/0200601.txt
[111] Dirac (1935), p. 172. The footnote also appeared in the third and fourth editions of 1947 and 1958, respectively, but not in the original 1930 edition.
[112] Lawrence (2010).
[113] Mason (2015). For Gamow's important contributions to cosmology, see Section 6.5.
[114] Crick (1966).
[115] Mason (2015).

Physicists are familiar with the adjective *virtual* which appears frequently in quantum words such as 'virtual state' and 'virtual particle'. However, most people will undoubtedly think of the popular term *virtual reality* which dates from the late 1970s and is often abbreviated VR. Virtual reality may seem to be a curious word construction, perhaps an oxymoron, insofar that the two words are traditionally taken to be antonyms. Apart from 'real', other antonyms to virtual include 'actual', 'genuine', 'concrete', and 'observable', whereas 'potential' is sometimes considered a synonym. According to *OED*, since about 1600 virtual has been used in the sense 'relating to essential, as opposed to physical or actual, existence'. Johnson's *Dictionary* from 1755 similarly gives the meaning as 'having the efficacy without the sensible or material part'.

Although the quantum meaning of the polysemic word virtual belongs to the twentieth century, the term was used in the terminology of optics and mechanics long before. John Harris' *Lexicon Technicum* of 1704 refers to 'virtual focus' and 'virtual image', but not to similar terms outside optics. The first time that virtual appeared significantly in a quantum context was in a radiation theory proposed in 1924 by Niels Bohr and his two collaborators Hendrik Kramers and John Slater (also known as the BKS theory).[116] According to this short-lived theory, any atom communicated all the time with other atoms by means of 'virtual fields' originating in 'virtual oscillators'. After the emergence of quantum mechanics, physicists began to speak of the 'virtual state' of an atomic nucleus and also, in the late 1930s, of 'virtual particles'. When the Japanese physicist Hideki Yukawa introduced his important meson theory (Section 3.2), he conceived the meson as a virtual particle constantly emitted and absorbed by the nucleons, the real constituents of the nucleus. Generally, virtual particles exist only temporarily as they are extremely short-lived (about 10^{-24} second for a pion) and exchanged between the ordinary particles which have a lifetime long enough that they can be observed experimentally.

In an interesting paper of 1988, one of the very few on the subject, the French theoretical physicist Jean-Marc Lévy-Leblond reflected on the language of quantum physics. 'A novel theory, when it appears, cannot but use old words to label new concepts', he said. 'Words are rarely, if ever, invented *ex nihilo*. They are borrowed from other fields, other languages: their meaning is displaced and stretched to fit their new context.'[117] Although this is largely correct, as we have seen at several occasions there are many and important cases where new words have been coined as genuine neologisms (although not ex nihilo in the strict sense).

With respect to the term quantum mechanics Lévy-Leblond argued that it is an unfortunate name because it unduly stresses the 'mechanical birth scars' and ignores that it is as much or even more a non-mechanical field theory. For the whole area he suggested the term *quantics* and for all quantum entities whether particles or not the term *quantons*. Both words have been adopted by some commercial companies and a few philosophers, but none of them by the community of quantum

---

[116] A special issue of the journal *Perspectives on Science* to be published in 2024 will be devoted to various aspects of virtual in science, technology, and culture.
[117] Lévy-Leblond (1988). Lévy-Leblond (2003).

physicists. As Lévy-Leblond admitted, he borrowed the word *quanton* from the Argentine-Canadian physicist and philosopher Mario Bunge, who in a paper of 1966 explained that '"quanton" is a name for any entity satisfying, say, Schrödinger's equation.'[118]

Historians of twentieth-century theoretical chemistry may have come across another rare word that apparently belongs to the quantum family, namely the *quanticule*, meaning something like little quantum. The term was introduced in 1943 by Kasimir Fajans, who over the next thirty years developed his 'quanticule theory' of the chemical bond. However, Fajans' theory was not based on quantum mechanics but meant to be an alternative to it as far as molecules and ionic compounds were concerned.[119] Only a handful of scientists accepted the concept of quanticules and today the term is no longer alive.

While arguing that some quantum words are misleading, among them such standard terms as 'observable', 'indeterminism', and 'wave–particle duality', Lévy-Leblond found it suitable to propose a few more neologisms of his own such as *pantopy* (from Greek 'all places') instead of the commonly used non-locality. In contrast to the trend in modern scientific nomenclature he advocated the traditional method of basing new words on Greek and Latin roots. 'While this strategy may run counter to the temptation of public advertising and media temptation', he stated, 'it has the merit of *not* capturing too easily the lay minds by the use of concrete or pseudo-intuitive wordings and to stress the real difficulty of new scientific concepts.'[120] In other words, he recommended a quantum vocabulary designed in such a way that it does not easily invite metaphoric use.

A similar elitist view was advocated by the famous American polymath Charles Sanders Peirce many years before the birth of quantum mechanics. In a letter to his compatriot, the philosopher William James, he made some comments on scientific nomenclature. As Peirce argued, it was no less than a moral obligation of scientists to agree upon and use terms that did *not* appeal to the common man:

> It is an indispensable requisite of science that it should have a recognized technical vocabulary, composed of words so unattractive that loose thinkers are not tempted to use them ... It is vital for science that he who introduces a new conception should be held to have a *duty* impressed upon him to invent a sufficiently disagreeable series of words to express it. I wish you would reflect seriously upon the moral aspect of terminology.[121]

Peirce's radical-aristocratic view with regard to scientific terminology may not have been very controversial in his own time, but it definitely is today.

---

[118] Bunge (1966). 'Quanton' is accepted as an English word by *OED*.
[119] Hurwic (1987).
[120] Lévy-Leblond (1999).
[121] Letter of 3 October 1904, quoted in Ketner (1981).

4.4 From solid-state to condensed matter physics

Many physicists in the fin-de-siècle period were engaged in research on magnetic properties, the crystalline state of matter, and the thermal and electrical conductivity of metals. They were solid-state physicists *avant le mot*. Together with his brother Paul-Jacques, 21-year-old Pierre Curie demonstrated in 1880 the so-called piezo-electric effect, namely an electric voltage produced by the compression of certain crystals. The noun *piezoelectricity* and the adjective *piezoelectric* refer to the Greek name *piézō* (πιέζο) meaning to squeeze or press. Fifteen years later, shortly before the discovery of radioactivity, Pierre Curie established that certain forms of magnetism undergo a drastic change when the material is heated. The temperature at which this happens became known as the *Curie temperature* and the quantitative dependence of magnetism on temperature as *Curie's law*. The more famous Marie Curie (or Sklodowska—she married Pierre in July 1895) has no share in these eponyms.

Another French specialist in magnetism, Pierre Weiss, introduced at about 1910 the *magneton* as a unit for the magnetic moment of elementary magnets. According to Weiss, 'as the electron has symbolized the new ideas on the discontinuous nature of electricity, so the magneton marks an analogous revolution in our representation of magnetic phenomena.'[122] Was the magneton merely an experimentally based unit or might it be conceived as a kind of magnetic particle, such as suggested by the suffix *-on*? A few researchers, including the British physicist Samuel McLaren, speculated that Weiss' magneton was indeed a kind of magnetic analogue to the electron, but the idea—which can perhaps be seen as an anticipation of Dirac's later hypothesis of magnetic monopoles—turned out to be unfruitful.[123] In unpublished notes from 1913 Niels Bohr calculated a value for the magneton much larger than Weiss's, namely $eh/4\pi mc$ with m the mass of the electron. The quantity soon became known at the Bohr magneton, an eponymous label introduced by Wolfgang Pauli in 1920.

Michael Faraday was the first to notice that a small class of materials, such as silver sulphide Ag_2S, have unusual electrical properties, but his observation was isolated and did not result in the recognition of a category of materials between metallic conductors and insulators. The existence of *semiconductors* only attracted attention at about 1880 and it took until 1911 before the word was coined in its German version *Halbleiter*. With the advent of quantum mechanics semiconductors became an important part of solid-state physics, eventually resulting in the discovery of the transistor effect in 1948 and all the marvels that followed from it (Section 2.6). In this period a number of new words appeared, many of them eponyms relating to the new study of matter in its solid state. Examples of such words which have long been part of the physicists' vocabulary are Brillouin zones (Léon Brillouin, France), Ising model (Ernst Ising, Germany), Bloch states (Felix Bloch, Switzerland), and Meissner effect (Walther Meissner, Germany).

Whereas the magneton does not exist as a physical entity, but only as a unit, the superficially similar word *magnetron* refers to something real, namely to a vacuum

---

[122] Quoted in Hoddeson et al. (1992), p. 387.
[123] Hendry (1983).

tube device widely used in early radar systems. Generally, the suffix -*on* signifies a particle and -*tron* an instrument of some kind, but there are several exceptions to the rule. As mentioned in Section 3.1, the end syllable of the words electron, positron, and neutron is really -*on* and not -*tron*, whereas the now obsolete 'mesotron' (meson) had -*tron* as its suffix.[124] As regards the word *radar*, it was originally a code name or acronym for '*ra*dio *d*etection *a*nd *r*anging' adopted by the U.S. Navy in November 1940. It appeared in public as early as 2 October 1941 in the *New York Times*. Some thirty years later the word began being used figuratively as an expression for a person's awareness or intuition of something. To say that a thing or subject is 'off the radar' indicates that it is unnoticed or considered unimportant. Phrases such as 'it went under my radar' and 'I kept my radar tuned to it' are commonly heard today.

To return to the magneton and the magnetron, another superficially similar physics word is *magnon*. However, this entity has a different story and relates to the early days of solid-state physics and the concept of what is called *quasiparticles*. The concept of a name often precedes the name by many years, as was the case with the magnon, which was known but not named so in the early 1930s. The name, which only appeared in English in the late 1950s, has its origin in Russian physicists' discussions of the quantum theory of magnetism almost twenty years earlier. It first appeared in print in a 1941 paper written by the Russian theorist Isaak Pomeranchuk, who credited the name to his colleague and compatriot Lev Landau.[125] Pomeranchuk also gave name to the 'pomeron', an important quantity in the early theory of strong interactions called quantum chromodynamics or QCD (Section 3.5). However, the pomeron is not a real elementary particle as it can be assigned neither charge, spin, nor mass.

The term *quasi*, a borrowing from Latin, means 'in appearance only'. It is mostly used as a prefix to word X, meaning that the combined term quasi-X is a kind of X or that it resembles X but is not quite the same as X. A quasi-religion is a belief system which has elements in common with traditional religions but is nonetheless different. New Age beliefs are sometimes labelled quasi-religious.

In the early 1960s astronomers identified a rare new kind of enigmatic celestial objects which emitted an enormous amount of electromagnetic energy in the radio region. Since the objects superficially looked like stars, they were called 'starlike objects', 'quasi-stellar radio sources', or 'quasi-stellar objects' (QSOs). According to the Dutch astronomer Maarten Schmidt, the principal discoverer of QSOs, at the 1963 Texas Symposium on Relativistic Astrophysics there was 'half a session spent on finding a name for the confounded thing. And all the names had been unfortunate. Nobody liked any of the names proposed there.'[126] In May 1964 the quasi-stellar objects were renamed *quasars*, a name that caught on and is today very popular not only in astronomy and physics but also beyond. It appeared in a book title with Fred Hoyle's *Galaxies, Nuclei, and Quasars* published in 1965. The inventor of

---

[124] Munns (2017).
[125] Walker and Slack (1970).
[126] Interview with Schmidt of 29 July 1977, see https://www.aip.org/history-programs/niels-bohr-library/oral-histories.

the neologism, the Taiwanese-American astrophysicist Hong-Yee Chiu, justified the name as follows:

So far, the clumsily long name 'quasi-stellar radio sources' is used to describe these objects. Because the nature of these objects is entirely unknown, it is hard to propose a short, appropriate nomenclature for them so that their essential properties are obvious from their name. For convenience, the abbreviated form 'quasar' will be used throughout this paper.[127]

Although quasar soon came into general use, it took several years before the term was accepted by the leading research periodical *Astrophysical Journal* whose editor, Chandrasekhar, did not like it. When Schmidt in 1970 wrote a paper in the journal with 'quasar' in the title, Chandrasekhar added a note saying 'The *Astrophysical Journal* has until now not recognized the term "quasar"; and it regrets that it must now concede ... [and that] the term can no longer be ignored.'[128]

One of the few objections to quasar came from Nicholas Kurti, a distinguished Hungarian-born British professor of physics and a specialist in low-temperature physics. At a Royal Society conference in 1965, he expressed reservation with regard to 'the disturbing rate of increase in new scientific and technical terms'. As Kurti pointed out, 'The very act of coining a new word implies a theoretical assessment, almost an acknowledgment of the fact that a concept or phenomenon, or particle has a well defined identity.' He suggested that, in general, when a neologism was justified, it should be 'expressive, well-sounding and not too offensive to the purist'. Kurti disliked acronyms and abbreviations or, as he expressed it, 'telescoping of two words to save a few letters, or a syllable'. This is where quasar came in:

A good example is 'quasi-stellar', which at least means something to a layman, changed to 'quasar'. Moreover, it seems that 'quasar' is used in the sense of 'quasi-stellar radio source' and this means that one uses the abbreviation of the adjective in the expression for the whole thing. To be logical one should either talk about 'quasar radio source' or about Quasars (singular, being the abbreviation of **QUA**si-**S**tell**A**r **R**adio Source). The latest cosmological abbreviation is Q.S.G. (quasi-stellar galaxy). Before long we may be talking about Quasarxies.[129]

The quasi prefix can also be used in a somewhat different and more critical sense. Thus, the term 'quasi-science' has been suggested for something masquerading as science but not living up to accepted scientific standards. In this pejorative meaning, quasi-science may function as another word for pseudoscience or at least flawed science.[130]

A more recent and most important addition to scientific *quasi-* terminology are the so-called *quasicrystals* discovered by the Israeli researcher Dan Shechtman in

---

[127] Chiu (1964).

[128] *Astrophysical Journal* **162** (1970): 371–9.

[129] Kurti (1965). The quasi-stellar galaxies discovered in 1964 resembled quasars except that they emitted light and not radio waves. For Kurti and terminology, see also Section 4.1.

[130] Bourdillon (2015).

1982 but first reported in public two years later. While ordinary crystals have a periodic structure, quasicrystals resemble in some respects amorphous solids by having an aperiodic structure. Shechtman's discovery claim was controversial and it took more than a decade before it was accepted and eventually earned him the 2011 chemistry Nobel Prize. The name quasicrystal was introduced in 1984 by the American physicist Paul Steinhardt who supported Shechtman's claim with theoretical arguments and suggested that quasicrystals might exist in nature, as indeed they do.[131] With about 10,000 papers on the subject, today the study of quasicrystals is a vibrant field of interdisciplinary science. Whereas quasicrystal was essentially a neologism of the 1980s, the adjective 'quasicrystalline' had been used earlier.

While the concept of a name often precedes the name itself, in other cases it is the opposite: the name comes first, sometimes in the form of a protologism. In a paper of 1926 the British physicist Ralph Fowler referred to light quanta or photons as 'quasi-particles', with which he meant that although photons are particles of a kind, still they are different from the ordinary material particles.[132] The later concept with the same name, quasiparticles, arose when physicists in the 1930s applied quantum mechanics to condensed forms of matter. These entities—essentially quantum excitations of a collective kind that occur inside a solid body—behave in many respects like ordinary particles and are generally considered to be real things although of an ontological order different from that of the particles in the standard model. In contrast to electrons and other elementary particles, they cannot exist in vacuum or in isolation from a collection of matter particles. To stress the similarity to ordinary particles they are given names with the termination -on. Apart from magnon, examples are phonon, exciton, polaron, roton, and plasmon (for the latter, see Section 2.6).

The prominent Russian quantum theorist Yakov Frenkel coined the name 'phonon' (sound particle) for a quantum of elastic waves in 1932, making it known to Western physicists in his book *Wave Mechanics*: 'It is possible to associate the acoustical waves with certain particles which we shall call "phonons" and to replace the study of the heat oscillations forming these waves by the study of the motion of the corresponding "phonons".' In a footnote he added that he did not want to convey 'the impression that such phonons have real existence.'[133] Four years later Frenkel suggested 'exciton' for a wave quantum propagating through a crystalline medium, a concept he had initially called an 'excitation packet'. He wrote: 'Just as the positive hole can be pictured as a collectivized positron, the excited state can be pictured as a kind of particle which we shall call a (collectivized) exciton.'[134]

In an earlier paper, his contribution to the 1927 physics congress held in Como, Italy, Frenkel more explicitly used metaphorical terms borrowed from the recent history of Soviet Russia. His subject was conventional, the quantum theory of metallic conduction, but his language was not:

---

[131] Steinhardt (2018).
[132] Fowler (1926). In contrast to ordinary particles, photons are massless.
[133] Frenkel (1932), p. 267.
[134] Frenkel (1936).

The only type of freedom electrons could obtain … is, so to speak, the freedom to change their master, or the atom to which they belong. While in the gaseous state each electron belongs to its proper atom, in the liquid or solid state it becomes 'the slave of the collective' formed by all atoms, and enjoys a rather relative freedom of constant transition from 'hand to hand', i.e. from one atom to another.[135]

Many of the quasiparticles were introduced and named by physicists in the Soviet Union under Stalin's regime. As argued by Alexei Kojevnikov, a historian of science, the way those physicists spoke and thought of the quasiparticles reflected aspects of the communist ideology in the 1930s and in particular its emphasis of collectivism. 'The development of a new fundamental language in physics and of some of its highly sophisticated mathematical models was enabled by the collectivist conception of freedom', Kojevnikov writes. 'The transfer of metaphors and concepts between scientific and political discourses can thus play an important productive role not only in biological sciences, where it has been studied extensively, particularly in the case of Darwinism, but also in a mathematized hard science like physics.'[136]

To the family of quasiparticles belongs also the *skyrmion*, an object originally proposed in 1961 by the British physicist Tony Skyrme but not, of course, with this name. The eponymous 'skyrmion' was first used in 1979, at a time when Skyrme was still alive. Since then, more than 4,000 research papers have included the word. Yet another new quasiparticle, or at least the name of one, appeared in 1982 when Frank Wilczek, a future Nobel laureate, invented the unusual word *anyon* for a hypothetical particle that can only exist in two dimensions. While the Faraday–Whewell 'anion' was coined from impeccable Greek (Section 2.2), Wilczek's whimsical 'anyon' was coined for fun. In an article of 2017, he recalled:

There are many more … categories of particles beyond bosons and fermions. I coined the word 'anyon' to describe the quarticles whose motion is restricted to two space dimensions, and which are neither bosons nor fermions. I meant this humorously, to suggest 'anything goes', but of course that implication should not be taken literally.[137]

The word 'quarticle' is not a misprint but a name that Wilczek proposed for a quantum particle.[138] Although the two-dimensional anyons may appear to be unphysical and cannot be observed in the usual sense, in a formal sense they turn up in quantum computing and areas of experimental physics. The exotic quasiparticles are taken seriously by both theorists and experimentalists, and the word anyon is accepted by *OED* and other dictionaries.

Although much work was done on solid-state physics in the 1930s, the field only evolved into a distinct discipline and community of physicists after World War II. During the war years, physicists in the United States proposed to form a new

---

[135] Quoted in Kojevnikov (2004), p. 60.
[136] Kojevnikov (1999), p. 300. See also Kojevnikov (2004), pp. 52–71.
[137] Wilczek (2017). See also Section 3.5 for Wilczek's coining of axion and anything-goes names.
[138] The same nonce had earlier been proposed by Nick Huggett, a philosopher of physics. See Huggett (2003).

division within the American Physical Society to take care of materials science and related fields oriented more towards applied physics than those already existing. The question of what to call the division was much discussed and eventually, after 'metal physics' had been considered, it was agreed that 'solid-state physics' was appropriate. Although the name soon became popular, initially it was met with resistance in some quarters of physics. As one physicist expressed his reservation, 'solid state physics sounds kind of funny'.[139] Funny or not, from about 1950 solid-state physics as a discipline developed at a phenomenal rate with its own textbooks, graduate courses, conferences, specialized journals, etc.

The term solid-state physics is still popular but no longer the accepted name for the research area dealing with complex matter in its many varieties. In the 1960s a new term appeared, first in Europe and subsequently in the United States. 'Condensed Matter Physics' was the title of a journal founded in West Germany in 1963 as *Physik der kondensierten Materie*. Fifteen years later, the American Physical Society decided to rename its large Division of Solid State Physics as Division of Condensed Matter Physics. This may seem to be just a new name for an old one, but it was more than that. For one thing and as reflected in the name change, the new discipline was broader in scope as it encompassed not only metals and other solids but also liquids, plasmas, and other forms of complex matter. It even had significant connections to elementary particle physics.

Moreover, and no less importantly, condensed matter physics was oriented more towards fundamental research and less towards industrial applications. The new field has been described as a 'deep, subtle, and intellectually coherent discipline', words which one would not use for traditional solid-state physics.[140] According to Joseph Martin, who has examined the transition from solid-state physics to condensed matter physics in an American context, 'the terminological shift reflected a change of professional ideology brought about as solid state physicists confronted challenges to their intellectual prestige and funding for explorative research'.[141] Despite the shift in terminology, the term 'solid state' appears much more frequently in the scientific literature than 'condensed matter'.

In 1929 the linguist Otto Jespersen wrote about 'a highly linguistic trick that has lately come into fashion in many countries, namely that of coining terms from the initials of a composite expression, which are read either separately ... or pronounced together'.[142] He had *acronyms* in mind but did not use the term, which had not yet been invented and only appeared in English in 1940 (*OED*). Whether in science or elsewhere, abbreviations in the form of acronyms were practically nonexistent before the twentieth century. They only entered significantly after World War II and have since then steadily increased in number and variation, leading to a veritable *acronymania*.[143]

According to a recent study based on millions of articles, the proportion of acronyms in titles in scientific papers has increased from 0.7 per 100 words in 1950 to

---

[139] Quoted in Martin (2015), p. 9.
[140] Zangwill (2021), p. 3.
[141] Martin (2015), p. 21.
[142] Jespersen (1929), p. 95.
[143] Subramanyam (1979). The term 'acronymania' was introduced in the journal *New Scientist* in 1968.

2.4 per 100 words in 2019. In cardiological trials alone, acronyms increased from 250 in 1992 to nearby 4,200 ten years later. Only few of the thousands of acronyms were used regularly and three-letters words of this kind were more popular than words with two or four letters. 'New acronyms are too common, and common acronyms are too rare', the authors comment.[144] Today it is widely recognized that obscure acronyms are overused in scientific papers and that they tend to hinder understanding by non-specialists in particular.

Acronyms in medicine were originally built from the first letters of words but later any letter or letters appearing in a word would do. They may even be made of other acronyms as in TAPS = *TPA APSAC Patency Study*, where TPA = *Tissue Plasminogen Activator* and APSAC = *Anisoylated Plasminogen Streptokinase Activator Complex*. According to two American professors of medicine writing in 2003, the development of acrymania had gone too far: 'The goal seems to be finding an acronym that is cuter and wittier than the previous one. . . . We have reached the point where investigators are selecting a colorful acronym, and then dreaming up a suitable study to match it.'[145]

Whereas acronyms are often useful, sometimes they are employed for no good reason, as when particle physicists acronymize the straightforward term 'long-lived particle' to LLP. It made more sense to abbreviate dichlorodiphenyltrichloroethane as just DDT, such as was done from about 1942. The organic compound was synthesized in 1874 but became socially and militarily important—and later controversial—only in the early 1940s when it was found to be a potent insecticide. With the rising importance of the substance, the systematic name was no longer used outside chemical nomenclature. DDT or initially D.D.T. became the standard term.

By far the most popular of the numerous acronyms is, however, the initialism DNA (*deoxyribonucleic acid*), which dates from about the same time as DDT and since about 1980 has appeared in the book literature with a frequency of 45 per million words.[146] Many of the other very popular acronyms are from the biomedical sciences such as HIV (*human immunodeficiency virus*) and the four-letter mRNA (*messenger ribonucleic acid*). As illustrated by laser and DNA, some of the widely used acronyms are effectively independent words or, as they have been called, pseudo-neologisms. They have entered common language and are used by speakers or writers without knowing of or caring about their origin.

There are also a few noteworthy examples of acronyms that relate to the history of solid-state and condensed matter physics. The American physicist Charles Townes developed in the early 1950s a method for stimulated emission of microwaves which in 1955 resulted in the invention of 'a very high resolution microwave spectrometer'. Having announced his new device, he contemplated with his students what to call it.

> They started with Greek and Latin words, but nothing worked, so they tried creating a descriptive acronym. They finally settled on Microwave Amplification by

[144] Barnett and Doubleday (2020). https://elifesciences.org/articles/60080. The authors define an acronym as 'a word in which half or more of the characters are upper case letters', which excludes, for example, commonly used terms such as 'laser' and 'radar'.

[145] Fred and Cheng (2003).

[146] *OED* refers to a paper of 1944, but there is at least one earlier paper with both DNA and RNA (ribonucleic acid), namely Cohen (1942).

the Stimulated Emission of Radiation, which spelled out the name that would stick, *MASER*.[147]

Whereas the maser was a scientific instrument with only limited use outside science, its further development into the *laser* became a great success also on the commercial and military markets. This well-known name was an acronym directly borrowed from the maser, only with 'light' substituting 'microwave': laser = *l*ight *a*mplification by *s*timulated *e*mission of *r*adiation. For a short while the instrument was also known as an 'optical maser'. The new acronym was first used in public by Gordon Gould, one of several inventors of the laser concept, at a conference in 1959. Since then laser has become a household word.

To the extent that *squid* is known at all, it is probably from the fried culinary dish known as calamari prepared from squids, an abundant species of molluscs with eight arms and two long tentacles. However, to condensed matter physicists a SQUID (in capital letters) is an instrument based on superconductivity used for measuring small magnetic fields with extreme sensitivity: SQUID = *S*uperconducting *Q*uantum *I*nterference *D*evice. The first SQUID was constructed in 1964, but without any special name for it. The acronym appeared in print two years later: 'A magnetometer utilizing a superconducting quantum interference device (SQUID) as a magnetic flux sensor is described.'[148] It was just introduced as a convenient abbreviation and neither the authors nor other physicists noted that nature had invented squids millions of years ago. The group of physicists who had invented the device at first found the acronym unacceptable and instead referred to 'Josephson Weak Link Devices', a reference to the so-called Josephson effect named after British physicist Brian Josephson.

---

[147] Hecht (2005), p. 25.
[148] Forgacs and Warnick (1966). Silver (2006).

5

Worlds and words of chemistry

The historical lineage of chemistry differs substantially from those of mathematics, astronomy, medicine, and physics. Still in the seventeenth century 'chymistry' (as it was often called) was not recognized as a proper science. Nor had chemistry cut all ties to the earlier alchemy. The later development of chemistry and its many interdisciplinary connections in the form of, for example, physical chemistry and biochemistry can be followed through the words associated with the various chemical sciences. The so-called chemical revolution in the last quarter of the eighteenth century is particularly interesting from a linguistic perspective as considerations concerning nomenclature and language played a central role in the revolution. Lavoisier was not only a great reformer of chemistry but also a reformer of scientific language.

Section 5.2 is of a less historical and more general nature. It focuses on eponyms in chemistry and other sciences and also on how words in a particular area of science undergo semantic shifts when reused in—or migrate to—either common language or the technical language of another science. Surprisingly many of our commonly used words and phrases have roots in chemical concepts and experiments and as such they are examples of polysemy.

Since the days of Lavoisier and Dalton chemistry has been based on the notion of chemical elements and their combinations in the form of numerous molecular or ionic compounds. The naming of the elements was considered important at the time of their discovery and later chemists and historians have investigated the subject in considerable detail. There is a story behind the names of all the 118 elements known so far. Until fairly recently the names were informally agreed upon by the involved chemists, but since about 1950 official names and discovery criteria have increasingly become the business of international nomenclature committees under the wings of the IUPAC, the International Union of Pure and Applied Chemistry. This is illustrated in particular by the names designated to the artificially produced elements heavier than uranium, the heaviest of which are called superheavy elements. As discussed in Section 5.5, the names have often been controversial and associated with priority disputes.

5.1 Aspects of chemistry

Chemistry is an old word, the ultimate origin of which is not known with any certainty, but it is generally accepted that the word alchemy, and later chemistry, were derived from the Arabic 'al kimiya'.[1] The Arabic term probably came from a Greek

---

[1] Brock (1993), pp. 28–9.

word, which has been claimed to have its origin in an even older Egyptian word for 'blackness'. However, this is a speculation.

The word 'chymia' turned up in the literature in the seventeenth century, when it was eventually separated from 'alchymia' and mostly restricted to pharmaceutical and technological manipulations of matter. During the following century 'chemistry' often appeared in the spelling *chymistry* and its practitioners were known as *chymists* rather than chemists. There were entries on both terms in the 1707 edition of Blount's *Glossographia*, whereas 'chemistry' and 'chemist' were absent. Blount defined chymistry as 'the Anatomy of natural Bodies by Fire', and said about *alchymy* that it was 'the sublimer Part of Chymistry, which teaches the Transmutation of Metals, and the Philosopher's Stone according to their Cant'. The term chemist only became common in the second half of the eighteenth century. Still at that time the corresponding English word 'physicist' had not yet been coined (Section 1.3).

As emphasized by leading historians of chemistry, the 'chymistry' of the early seventeenth century was in no way just an old name for what we call chemistry and nor did it signify a clean break with the alchemical tradition. To translate 'chymist' as 'chemist' is in part anachronistic because the translation ignores the persisting influence of alchemy. The great Robert Boyle is often portrayed as the father of scientific chemistry, but in his celebrated *The Sceptical Chymist* from 1661 he did not discriminate between the words 'chemistry' (or chymistry) and 'alchemy'. The problem of language has been aptly summarized as follows:

> This linguistic development is not merely an impermanence or incommensurability of terms by historians and philosophers, namely, that words change their meanings and referents over time and particularly over changes in theoretical systems—e.g., the physics of the late Middle Ages is not the physics of the late 1990s. Here instead, there existed a set of referents which was for a long time equally well denominated by two synonymous terms—'alchemy' and 'chemistry'; at a given point in time that set of referents was divided, segregating out a certain class of members to which the term 'alchemy' was the assigned, and its former synonym 'chemistry' applied to the remainder. Synonyms became non-synonymous.[2]

The learned German writer Andreas Libavius supposedly introduced the distinction between chemistry and alchemy in what is sometimes called the first textbook on chemistry as an independent science. And yet, ironically, his Latin book of 1597 bore the title *Alchemia* (and not Chemia). Libavius was much concerned with terminology and in an earlier work, *Rerum Chymicarum Epistolica*, he attacked Paracelsus and his disciples on this issue: 'The true chemist does not like neologisms. The Paracelsian likes nothing more. To hide in a perplexity of words like the Delphic Demon is the best that these artful dodgers are capable of.' On the other hand, he admitted that there was more to chemical terminology than authentic or Latinized Greek words: 'This should not be taken to absurd extremes and [those words] which have been

---

[2] Newman and Principe (1998), p. 41.

taken over from the Arabs and the Egyptians have been made more pliable now by frequent use. Let the latin-speaking practitioner speak more latinly when he can.'[3]

Chemistry only developed slowly, and in the early eighteenth century it was not yet a respected scientific discipline taught at the universities. The chemists studied nature, but so did the natural philosophers later to be named physicists, and the latter group enjoyed much higher academic recognition. In 1699 Bernard Fontenelle, perpetual secretary of the Royal Academy of Sciences in Paris, gave voice to the low status of chemistry as compared to what he conceived to be physics:

> Through its visible operations, chemistry resolves bodies into a certain number of crude tangible principles; salts, sulfurs, etc. while through its delicate speculations, physics acts on the principles as chemistry acts on bodies, resolving them into other even simpler principles, small bodies fashioned and moved in an infinite number of ways. ... The spirit of physics is clearer, simpler, less obstructed, and, finally, goes right to the origins of things, while the spirit of chemistry does not go to the end.[4]

Nearly a century later, in his *Metaphysische Anfangsgründe der Naturwissenschaft* (Metaphysical Foundations of Natural Science) from 1786, Immanuel Kant argued that chemistry could never be a genuine science on par with physics because its subject matter was intractable to mathematization and systematic deduction from laws of nature. According to the great philosopher, chemistry was nothing more than a systematic art or experimental doctrine.

The meanings of the words chemistry and physics, their domains and the relationship between the two sister sciences, have changed significantly over time. The same holds for the territories of nature thought to belong to either chemistry or physics. To mention but one example, in the early part of the nineteenth century it was customary to classify topics such as electricity, magnetism, heat, and light as chemical rather than physical sciences. Half a century later, these topics had become topics of physics or what was often called 'chemical physics'.

According to an American textbook with the title *Introduction to Chemical Physics* published in 1874, 'Chemistry is ... usually divided into two portions. The first treats of the Chemical Agents, Heat, Light, and Electricity, and is commonly called Chemical Physics; the second, of the chemical properties and relations of the various kinds of matter.'[5] The book does not mention either 'ion' or 'ionic' and also the term 'thermodynamics' is absent. Moreover, except in its title there is no mention of the term physics, which throughout is replaced by the older 'natural philosophy' as the preferred synonym.

The interdisciplinary term 'chemical physics'—not to be confounded with 'physical chemistry'—reappeared in the early 1930s, but then with an entirely different meaning, namely with a focus on quantum chemistry and molecular spectroscopy. These areas made up the main content of the *Journal of Chemical Physics* founded in

---

[3] Cited in Hannaway (1975), p. 119.
[4] Bensaude-Vincente (2009), p. 373. Fontenelle thought of Cartesian and not Newtonian physics.
[5] Pynchon (1874), p. 21. Online as Google Book.

1933 with Harold Urey as its first managing editor. In a similar vein, in the fin-de-siècle period there were attempts to create a 'mathematical chemistry',[6] a name and concept which was only nominally the same as used in the modern branch of so-called 'computational chemistry'. Identical terms may imply identical concepts, but generally they do not.

The changed physics–chemistry relations and the gradual emancipation of chemistry as an autonomous science can be followed by looking at the content and titles of scientific textbooks and journals. Nineteenth-century journals were often media for both physicists and chemists, such as illustrated by *Journal für Chemie und Physik* and *Annalen der Physik und Chemie* in Germany, and *Annales de Chimie et de Physique* in France. Another example, but this time of the close relations between chemistry and pharmacy, is *Annalen der Chemie und Pharmacie* edited by Justus von Liebig. Journals wholly devoted to chemistry did exist, but they were few and came to dominate the market only in the late part of the century. The British *Journal of the Chemical Society* began publication in 1862 and the *Journal of the American Chemical Society* in 1879.

Like in most other sciences, the progress in chemistry gave rise to a confusingly large number of specialities and subdisciplines, some of which carry names which are noteworthy from a historical and linguistic point of view. One of the interdisciplinary fields was geochemistry, which has been dealt with in Section 1.5. According to the consensus view at about 1840, the term *astrochemistry*—the chemical study of the stars and other celestial bodies—was close to being an oxymoron. For how could chemists possibly gain access to whatever chemical processes occurred in the distant stellar realm? Or acquire knowledge about the elements of which the stars were supposedly composed? The French philosopher Auguste Comte published his ambitious and influential work *Cours de Philosophie Positive* (Course on Positive Philosophy) in six volumes from 1830 to 1842. In the second of the volumes from 1835 he proclaimed that the stars could never be the subjects of 'physical, chemical, physiological, and social [*sic*] research'.[7] And yet, as a result of the invention of the marvellous spectroscope, terms such as 'astrochemistry' and 'astrochemist' entered the scientific language in the late 1870s. The English astronomer Norman Lockyer published in 1886 a book with a title that would have shocked Comte: *Chemistry of the Sun*.

Comte is today mostly known as the founder of positivism and also, perhaps, for having coined the word *sociology* for the emerging science of society or what he previously had called 'social physics'. As to the term 'sociology', he apologetically called it a convenient barbarism because it mixed Latin (*socius*) with Greek (*logia*). He is also credited for the word 'altruism' (Latin *alter* meaning 'other'), which he introduced as an antonym to 'egoism' known from about 1800.

In the twentieth century astronomers, physicists, and chemists went on to study the molecules and ions floating around in the atmospheres of planets and stars or elsewhere in the universe. The result was a neologism coined by the German-American astronomer Rupert Wildt in a programmatic paper of 1940: 'The question can be raised whether it would not be a timely expansion of the notion of geochemistry

---

[6] Van Laar (1901).
[7] Comte (1835), p. 9.

to adopt the term cosmochemistry to designate the science which shall deal with matter under all cosmic conditions.'[8] With the founding of the journal *Geochimica et Cosmochimica* in 1950 the new science of *cosmochemistry*, also called chemical cosmology, received broad professional recognition and is today a flourishing interdisciplinary field of research.

If astrochemistry would be considered an oxymoron by 1840, so would the term *bioinorganic chemistry* (or 'inorganic biochemistry'), which is the name for a large modern research field principally studying the role of metals in biological processes. The reason is that according to the traditional view, organic chemistry was about matter derived from living organisms, plants, and animals, whereas inorganic chemistry was strictly limited to the dead nature. While the term bioinorganic chemistry would have seemed oxymoronic, another modern chemical discipline, namely *bioorganic chemistry*, would have seemed a redundant term.

In 1806 Berzelius defined organic chemistry as 'the part of physiology which describes the composition of living bodies together with the chemical processes taking place in them.'[9] As an alternative to 'physiology', he suggested to use the term *organology*, which he may have borrowed from the phrenologists among which he counted himself (Section 1.5). The same word was on a few occasions used for 'the doctrine of the life of the whole plant, and of its particular organs', as defined by the German nineteenth-century botanist Matthias Schleiden.[10] Later, in musicology, it came to signify the science of old musical instruments. In contrast to this word, others of Berzelius' many neologisms were adopted by the international chemical community, including household terms like 'isomer', 'polymer', 'catalysis', 'halogen', and 'protein'. He also named the elements cerium and selenium, both of which he discovered, the first in 1803 and the second in 1817.

The word *protein* first appeared in print in papers from the autumn of 1838 written by Gerardus Mulder, a Dutch professor of chemistry. According to Mulder, there existed in 'the foodstuff of the whole animal kingdom' a substance 'probably formed only by plants', and this substance 'we designate *protein* from πρωτειος, *primarius*.'[11] However, there is convincing evidence in a letter from Berzelius to Mulder of 10 July 1828 that protein was actually coined by Berzelius, who wrote him: 'The name protein which I propose to you for the organic oxide of fibrin and of albumin, I wanted to derive it from πρωτειος, because it seems to be the original or principal substance of animal nutrition.' In his publications, Mulder did not acknowledge Berzelius for having proposed the name and the Swedish chemist never stressed his priority. The word protein became one of the most commonly used science words. It typically occurs about eighty times per million words in modern English, which makes it more popular than e.g. DNA, laser, and nuclear (*OED*).

While the existence of a vital force in organic matter was taken for granted by Berzelius and most of his contemporaries, with the syntheses of organic compounds directly from inorganic substances this 'force' disappeared. In 1845 Hermann Kolbe

---

[8] Wildt (1940). See Kragh (2001) for early cosmochemistry.
[9] Wentrup (2022).
[10] Schleiden (1849), p. 454, Google Books.
[11] Quoted in Vickery (1950), which is also the source of the following quotation. See Tanford and Reynolds (2001) on the history of proteins.

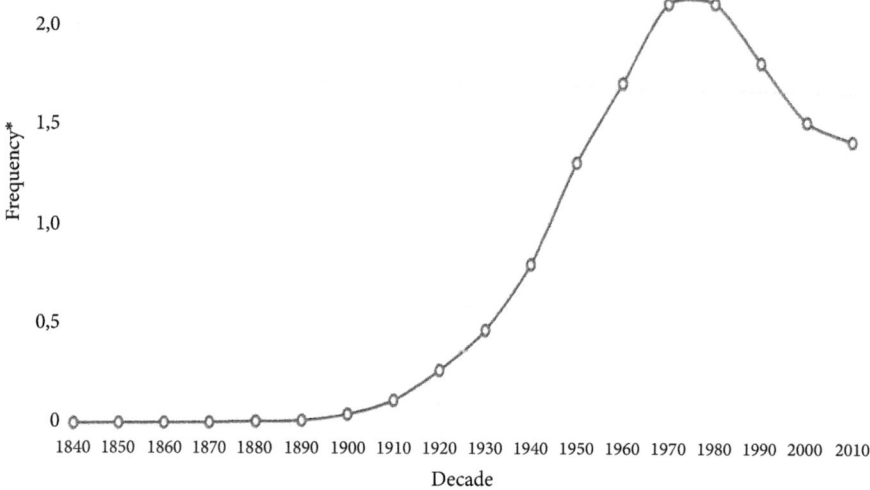

Figure 5.1 The frequency in ppm (parts per million) of the word 'biochemistry' in written English 1840–2010.

Reproduced from *Oxford English Dictionary*.

in Germany was able to produce the organic compound acetic acid (CH_3COOH) purely by nonbiological means. A decade or so later, organic chemistry simply became the area of chemistry investigating carbon compounds (with a few exceptions such as carbon dioxide CO_2 and soda ash Na_2CO_3). 'Inorganic biochemistry' first appeared in a scientific publication in 1969 and the synonymous 'bioinorganic chemistry' in 1972. 'Bioorganic chemistry' emerged at about the same time as a merger of organic chemistry and biochemistry.[12]

Looking back on the stormy development of twentieth-century biochemistry, in 2005 the German-American biochemist Helmut Beinert recalled that when he was a student in the 1930s, the term bioinorganic chemistry would have been considered 'outright ridiculous'. He further noted that the classical definition of the terms organic and inorganic had always been problematical: 'Because metals of many kinds are found in living matter, e.g. sodium, potassium, and calcium in considerable quantities, and because all metals are subjects of inorganic chemistry, there must then, by definition, always have been an inorganic component of biochemistry. Thus, according to this reasoning, "inorganic biochemistry" and "bioinorganic chemistry" certainly are no new subjects; rather they may only be new words.'[13]

Biochemistry in its traditional all-organic version emerged as a scientific discipline in the early twentieth century, when it replaced and extended what was often called 'physiological chemistry' (Figure 5.1). One of biochemistry's early research areas was the study of the 'accessory food factors' now known as vitamins. The very

---

[12] For the history and prehistory of bioorganic chemistry, see Morris, Travis, and Reinhardt (2001).
[13] Beinert (2002).

term *vitamin*, but in the spelling vitamine, was introduced in 1912 by the Polish-born biochemist Casimir Funk, who believed that all vitamins were amines, that is, included nitrogen groups such as $-NH_2$ and $> NH$. 'It is now known that all these diseases ... can be prevented and cured by the addition of certain preventive substances, which are of the nature of organic bases, we will call "vitamines".'[14] Funk spoke of the 'scurvy vitamine' and the 'beri-beri vitamine', later to become vitamins C and B_1, respectively.

When it was realized in the 1920s that not all vitamins are amines, the final *e* was dropped and vitamine became vitamin. In a brief paper of 1920, the British biochemist Jack Drummond noted the confusing variety of names that had been proposed for the nutrient factors—not only 'vitamine', but also 'nutramine', 'auxinone', and a few more. Rather than inventing a new word, he suggested to slightly revise Funk's broadly adopted term:

> The criticism usually raised against Funk's word Vitamine is that the termination '-ine' is one strictly employed in chemical nomenclature to denote substances of a basic character, whereas there is no evidence which supports his original idea that these indispensable dietary constituents are amines. ... The suggestion is now advanced that the final '-e' be dropped, so that the resulting word Vitamin is acceptable under the standard scheme of nomenclature adopted by the Chemical Society, which permits a neutral substance of undefined composition to bear a name ending in '-in'.[15]

Drummond also recommended a revised nomenclature that included the currently very well-known terms vitamin A, vitamin B, etc. 'This simplified scheme should be quite sufficient until such time as the factors are isolated, and their true nature identified', he wrote. When the Danish biochemist Henrik Dam in 1935 discovered a new fat-soluble vitamin that prevented bleeding, he called it vitamin K, for as he said in his Nobel lecture from 1946, 'The letter K was the first one in the alphabet which had not, with more or less justification, been used to designate other vitamins, and it also happened to be the first letter in the word "koagulation" according to Scandinavian and German spelling.'[16]

5.2 Eponyms and meaning transfers

The history of chemistry and its neighbour sciences offers numerous examples of eponyms, some of which have previously been mentioned (Section 1.2). Mineralogy, a science with strong historical links to inorganic chemistry, comprises thousands of minerals of which a substantial part is named after scientists or other people. We are not surprised to find in the list 'berzeliite' (after Berzelius) and the hard-to-pronounce

---

[14] Funk (1912).
[15] Drummond (1920). Rosenfeld (1997).
[16] https://www.nobelprize.org/prizes/medicine/1943/dam/lecture/

'lotharmeyerite' (after Lothar Meyer, co-inventor of the periodic table), but *armalcolite* probably does not ring a bell. It is a well-hidden acronymic eponym named after Neil *Arm*strong, Buzz *Al*drin, and Michael *Col*lins, the three astronauts who visited the Moon in 1969 and there found the mineral. As there are hidden eponyms in mineralogy, so there are hidden toponyms. The clay mineral *kaolinite* is the main component of kaolin and both names are derived from an older European spelling of the Chinese village Gaoling in the Jiangxi Province.

Apart from chemical elements, of which more later in this chapter, there is a large number of instruments, reactions, and compounds named after more or less famous chemists.[17] All chemistry students have prepared 'Tollens' reagent', an alkaline solution of silver nitrate and ammonium hydroxide, and used it to distinguish between the groups of organic compounds called aldehydes and ketones. A positive test resulting in a beautiful 'silver mirror' indicates that the solution is an aldehyde. The students probably have no idea that the name of the test refers to the German chemist Bernhard Tollens, who described it in a paper of 1882. The ubiquitous conical Erlenmeyer flask routinely used in modern chemical and biological laboratories has its name from the German nineteenth-century organic chemist Emil Erlenmeyer, and the no less ubiquitous Petri dish used by microbiologists and medical scientists owes its name to yet another German, Richard Julius Petri, who invented the simple glass container in 1887. It could just as well have been the 'Koch dish' after the great bacteriologist Robert Koch.[18]

Yet another of the highly successful instruments from the nineteenth century was the Bunsen burner invented by the famous Heidelberg chemist Robert Bunsen in 1857. This type of burner was a crucial ingredient of the spectroscope which Bunsen a few years later constructed in collaboration with his colleague in Heidelberg, the physicist Robert Kirchhoff. However, it has been questioned whether Bunsen really invented the iconic burner named after him. The final design was made by one of his laboratory technicians, Peter Desaga, who, aware of the selling-power of the name Bunsen, produced and sold the apparatus as a 'Bunsen burner'.[19] One more example of Stigler's law of eponymy?

As regards the many eponymous chemical compounds, one of the earliest and best known is not generally recognized to be an eponym. Nicotine with the empirical formula $C_{10}H_{14}N_2$ was first isolated in 1828 and its formula dates from 1843. It is a naturally produced alkaloid and an ingredient in the tobacco plant *Nicotiana tabacum* so named by the influential Swedish botanist Carl Linnaeus. Behind this name and therefore also *nicotine* hides the Frenchman Jean Nicot, a diplomat and ambassador to Lisbon, who introduced the tobacco plant to France in 1560 and promoted its use for medical purposes.[20] It was originally known as the ambassador's herb or *nicotiana*, which later became nicotine.

More than 400 years later, a team of five chemists discovered a strange molecule of molecular weight 720 and yet made up of carbon atoms only. Since the atomic

---

[17] Wagner (1951) refers to several hundreds of such names, most of which are no longer used.
[18] Shama (2019). Petri served as an assistant to Koch, who can equally well be regarded as the inventor of the bacteriological culture dish.
[19] Russell (1999).
[20] Charlton (2004).

weight of carbon is 12, the molecule consisted of no less than 60 carbon atoms. What was first called C60 proved by closer inspection to have a remarkably symmetric three-dimensional structure resembling the 'geodetic domes' constructed by the American architect and author Richard Buckminster Fuller. When the five chemists published their discovery in *Nature* in 1985, the title of their paper read 'C60: Buckminsterfullerene', admittedly an unusual name for a molecule:

> We are disturbed at the number of letters and syllables in the rather fanciful but highly appropriate name we have chosen in the title to refer to this C60 species. For such a unique and centrally important molecular structure, a more concise name would be useful. A number of alternatives come to mind (for example, ballene, spherene, soccerene, carbosoccer), but we prefer to let this issue of nomenclature be settled by consensus.[21]

Yet another fanciful name for C60 was suggested by another researcher, who the following year published a paper with 'footballene' in its title. 'For convenience this molecule is called here "footballene" ("soccerballene" in the US)', he wrote.[22] As more or less expected, 'buckminsterfullerene' did not stick (and nor did 'footballene'), but the shorter and still eponymous 'fullerene' did and so did the popular term 'buckyball'. The systematic name, should one care to know, is

$$\{(C_{60} - I_h)\,[5,6]\,\text{fullerene}\}.$$

In 1996, three of the discoverers, Harold Kroto, Robert Curl, and Richard Smalley, were awarded the Nobel Prize in chemistry for their work. The unusual name proposed in the 1985 *Nature* paper may have been due to Kroto in particular. 'He thought it tripped off the tongue. The -ene ending was euphonious as well as chemically correct ... [and] a remarkably economical descriptive term.'[23]

As pointed out in other sections, several of the names originating in scientific contexts are found in everyday language with a meaning different from the original one. They are used figuratively and metaphorically. Examples are the verbs electrify, galvanize, and mesmerize referred to in Section 2.1. Yet another word in this category is *dinosaur* as used in an expression like 'he is a dinosaur' about a person who is old-fashioned and outdated. Moreover, the term is widely used in Britain and some other countries as a metaphor for trade union. At about 1980 'dinosaur' was used as a slang term for 'an ageing rock musician, usually of the 1960s generation and resolutely impervious to suggestions that they should hang up their guitars'.[24] It is frequently stated that the scientific term dinosaur was coined by the leading British naturalist and palaeontologist Richard Owen in a report read to the British Association in July 1841. However, Owen did not mention the term in his oral report but only in the

---

[21] Kroto et al. (1985).
[22] Haymet (1986).
[23] Aldersey-Williams (1995), p. 77.
[24] Green (1991), p. 75. For other metaphoric uses of dinosaur, see Haste (1993) and the 'dinosaur element' in Section 5.5.

published version of the following year. He found 'sufficient ground for establishing a distinct tribe or sub-group of Saurian Reptiles, for which I would suppose the name of *Dinosauria*'. And in a footnote: 'Gr. δεινὸς, fearfully great; σαύρος, a lizard.'[25]

As mentioned in Section 4.3, the French physicist Lévy-Leblond reflected thoughtfully on how the language of science relates to the insights of quantum mechanics. In this context he wrote:

> Many words initially created for and to be found in professional scientific discourse slowly leak out to find their way in common parlance, where they take on original meanings which cannot but come back within the scientific discourse to give it new colours ('energy' and 'entropy', 'electricity' and 'magnetism' are cases in point).

He was worried over 'the simplistic borrowing of picturesque but misleading common words, as in expressions like "big bang", "coloured quarks" or "butterfly effect"'.[26] The latter term goes back to a 1972 paper by the American meteorologist and mathematician Edward Lorenz, a founder of modern chaos theory, but it only entered common usage with James Gleick's popular book *Chaos* published in 1987.

Lévy-Leblond's worries were earlier given voice by Lancelot Hogben: 'Assimilation of technical terms in everyday speech—especially by mass media—exposes them to the process of semantic erosion responsible for the multiplicity of meaning conveyed by other words in daily use.'[27] Maxwell may have had something similar in mind when he, nearly a century earlier, warned against the transfer of scientific words and phrases to other areas where they were used illegitimately and merely served to make a view appear justified by the authority of science. In a lecture of 1871 to physics students, he said: 'Such indeed is the respect paid to science, that the most absurd opinions become current, provided they are expressed in language, the sound of which recals [*sic*] some well-known scientific phrase.'[28]

In addition to science words migrating *to* ordinary language there is, about equally frequently, a migration or transfer in the opposite direction, that is, common words which at some stage have been adopted for scientific and more specialized purposes. The names of basic mechanical concepts such as *work*, *force*, and *power* are borrowed directly from everyday language but in science they are strictly defined and provided with meanings that may contradict those used in ordinary speech. A person may work very hard—with all his power and getting very tired—and yet do no work at all in the sense of physics. This is what Theodore Savory, a biology teacher and writer on scientific language, referred to in a critical comment on the words used by physicists:

> The mathematical physicist is guilty of linguistic rape of a family of related words— force, work, power and weight. In mechanics, force does not mean strength, as it does when the ordinary man says that he is perhaps impressed by the force of an argument. ... Work gives even more trouble, because a physicist has decided that a

---

[25] Owen (1842), p. 103.
[26] Lévy-Leblond (1999), pp. 78–9.
[27] Hogben (1970), p. 145.
[28] Maxwell (1965), part 2, p. 242. Maxwell did not indicate which absurd opinions he had in mind.

force works, or does work, only when it moves something. I may push and pull in vain at an immovable object, make myself hot and tired by my efforts, and find that mathematically I have done no work.[29]

Such double or multiple semantic shifts of words, in whatever direction, is a general phenomenon and not one specifically related to the sciences. The linguist Carolynn van Dyke speaks of *recycled terms* in the sense of 'words whose roots are all in a contemporary dictionary but were defined without reference to the word's new subject-areas'.[30] As she points out, this is not a one-way transfer since 'Recycled scientific and technological terms can easily return to the nontechnical semantic realms from which many of them arose.' Generally, semantic shift or change refers to how the meaning of words changes over the course of time. According to some linguists, a word is new not only if it is a neologism but also if it is a *neosemanticism* introduced with a new meaning. 'Neosemanticisms ... are words or groups of words already in the language that aquire fresh meanings by use in new situations.'[31]

The term *computer* is derived directly from the Latin *computus*, which in the early Middle Ages was a system used for the calculation of the date of Easter and generally for calendrical calculations. Combining mathematics with astronomy, computus was for many centuries after about 400 the principal form of scholarship or, anachronistically, science. While computus was a method, later on those who made calculations for either ecclesiastical or astronomical purposes became known as 'computers'. Still in the early part of the twentieth century computers were human and often women employed at astronomical observatories.[32] In its modern sense of a programmable electronic device, 'computer' dates from about 1945. It is well known that recently the jargon of computer science has thoroughly influenced the common language. We may complain that a household equipment is not 'user-friendly', that we are missing 'input' to make a decision, or that we need to 'google' a certain word. Again, people may say that 'the mind is the software of the brain' and refer to the material constitution of our body as 'human hardware'. Although the terms 'software' and 'hardware' were known and used long before the computer age, today they are firmly associated with computing in which sense they may first have turned up in the late 1950s. They appear in a 1958 paper, characteristically in inverted commas: 'Today the "software" comprising ... aspects of automative programming are at least as important to the modern electronic calculator as its "hardware" of tubes, transistors, wires, tapes and the like.'[33]

Somewhat similarly, the term *virus* was well known in the medical world long before it entered computer language in the 1980s. Of Latin origin, the word refers to poison and other noxious liquids and it was commonly used in the late nineteenth century for an agent that causes infectious diseases. The reuse of the word in computing language was largely due to the twenty-eight-year-old American computer scientist Fred Cohen, who in 1984 introduced a paper as follows: 'We define a

[29] Savory (1967), p. 64.
[30] Van Dyke (1992).
[31] Maurer and High (1980).
[32] Grier (2005).
[33] Tukey (1958).

computer "virus" as a program that can "infect" other programs by modifying them to include a possibly evolved copy of itself.[34] Today, many people probably think of a computer virus and not a biological virus if confronted with the term. Referring to an early incident of 1988, a communication expert writes:

> With the 'virus' as vehicle, many aspects of the tenor could be elucidated using related terms. News stories explained that about six thousand computers were *infected* as the 'virus' proved to be *virulent* and *highly contagious*. NASA *isolated* its computers from the *infected* network and *quarantined* them. Attempts were made to *sterilize* the network. Programmers struggled to develop a *vaccine*, and to *inoculate* against new attacks. The 'virus' proved to be a master metaphor.[35]

In computer language, the popular term *mouse* does not refer to a small brownish or black rodent, but to a handheld device by means of which the pointer on a monitor screen can be moved. In contrast to the rodent, the computer mouse in plural can be both 'mice' and 'mouses'. The invention of the device is usually credited the American engineer and Internet pioneer Douglas Engelbart who, together with his collaborator William English, constructed a prototype in 1964. Three years later, Engelbart filed a patent, not for a mouse but for what he called a 'position indicator control'. The name mouse first appeared in print in a report of 1965 with English as its lead author, but it took most of a decade before it caught on. In an interview of 2006, Engelbart stated: 'I didn't give it a name when I was doing all these experiments. I didn't call it a "mouse". It was so successful we were sure it would go to the rest of the world, and they'd give it a dignified name. We referred to it as the XY positioning indicator or something.'[36] The nickname mouse—not a particularly dignified one—was used internally early on, but it is unknown who first coined the term. In another interview, Engelbart said:

> Somebody, I can't remember who, attached the name mouse to it. You can picture why, because it was an object about this big, and had one button to use for selection, and had a wire running out the back. 'It looks like a one-eared mouse!' someone said. Soon all of us just started calling it a mouse.[37]

In the remainder of this section I look at a few word migrations relevant to the history of chemistry and biochemistry. But first an instructive example from the history of medicine. The concept and name of *allergy* (Allergie in German) was introduced in 1906 by the Austrian medical doctor Clemens von Pirquet, who based the neologism on the Greek *allos* (other, different) and *ergon* (work, action). It took about a decade before the term and its associated concept gradually won acceptance as a name for hypersensitive reactions of the body. The *Journal of Allergy* was established in 1929. With the popularity in medical circles followed popularity in the general public:

---

[34] The paper was only formally published in 1987. See Cohen (1987), who acknowledged his teacher Leonard Adleman for having suggested the name 'computer virus'.

[35] Gozzi (2017).

[36] Atkinson (2007).

[37] Interview of 2003, in Moggbridge (2006), p. 28.

Soon, allergy escaped from scientific and clinical arenas and went on the streets, where people began to use it to express any apparent adverse reaction to anything. Furthermore, people used the word to express antipathy, rejection or aversion. 'Allergy' appeared in newspapers, novels and songs with both medical and nonmedical significances ... 'Allergy' became a fashionable word.[38]

Of course, the term coined by Pirquet, and especially its adjectival form 'allergic', still has this double meaning. People with no symptoms of allergy can be allergic to hamburgers, baseball, romantic novels, or philosophical debates. The Greek *allos* enters several other scientific names of which *allotropy* may be the best known (*tropos* = form). This is the phenomenon where a chemical element in the same state can exist in two or more different physical forms. Diamond, graphite, and the fullerenes are allotropic modifications of the same element, carbon.

As we have seen, identical words with multiple meanings, either homonyms or polysemes, are in many cases found in both scientific language and ordinary nonscientific language. It often happens that a familiar word turns up in a scientific context with a specialized and very different meaning, which may cause confusion to science students. As an American chemistry teacher mused: 'Thus, "gas" is not what you put in your car, nor is an "Ideal Gas"' one which gives good mileage; "precipitation" refers to the formation of solids and not to rain; "acids" are not psychedelic drugs and "basic" does not mean fundamental; not all pleasant-smelling chemicals are "aromatic".'[39]

There are several other words in this category, random examples being 'culture', 'mole', 'solution', 'spontaneous', and 'radical'. Apart from its general meaning, since the 1880s *culture* also refers to the production of bacteria or other microorganisms in the biological laboratory. A *mole* is a small animal primarily living off earthworms but is also a chemical unit given by Avogadro's number of molecules or other particles. The *molar* concentration is the number of moles per unit volume, an adjective, whereas the noun molar refers to a tooth. When we think to have found the answer to a problem, we have found its *solution*, but in chemistry the term typically refers to a substance dissolved in water or some other liquid. The common meaning of *spontaneous* is an immediate and unconstrained action or thought ('her answer was spontaneous'). On the other hand, chemical reactions are said to occur spontaneously even though they may be extremely slow. For the different meanings of *radical*, see later in this section.

It is not unusual that the same word appears with different technical meanings in two or more areas of science, such as 'molar' in dentistry and chemistry. In botany the term *pyrene* (or pyrena) denotes a fruitstone within a drupe, whereas in chemistry the same word refers to an organic compound consisting of four benzene rings with the stochiometric formula $C_{16}H_{10}$. The botanical pyrene is derived from Greek 'pyren' or 'pyreno' (πυρήν) meaning fruitstone, whereas the chemical pyrene refers to 'pyr' or 'pyro' (πύρ) meaning fire. There are even examples of different plants, animals, and chemical compounds which for a limited period of time have been

[38] Igea (2013).
[39] Ryan (1985).

designated by the very same name. 'Virus' belongs to the category of homonyms and so do 'plasma' and 'fission' (Section 2.6).

In the case of *hybrid* and *hybridization*, the terms appear in half a dozen sciences, most notably in biology (genetics), linguistics, and chemistry, but also in computing, geology, and particle physics (Section 3.5). The words are usually associated with the offspring resulting from cross breeding, say a mule from a horse and a donkey, but as mentioned several times, they are also used in linguistics when a hybrid word is formed by hybridization. It is less well known that they have a significant place in the chemists' vocabulary as terms describing the bonds that keep atoms together in a molecule.

Chemists speak of atomic and molecular orbitals that can be mixed to produce hybrids or hybridized orbitals. This concept from early quantum chemistry was introduced by the eminent chemist Linus Pauling and it may first have appeared in print in a paper of 1931 written by another eminent chemist, Robert Mulliken (both were awarded the Nobel Prize in chemistry). Pauling was most likely inspired by the biologists' use of the terms. In 1931 he was invited to give a seminar to a group of biologists, and, according to a historian of science: 'Wide-ranging readings in biology soon spilled over in his thinking about the chemical bond. The "changed quantisation" of 1928 and 1931 became widely known as "hybridisation" by the late 1930s.'[40]

As yet another example, consider the word *vector*. A mathematically educated person will understand 'vector' in one sense and one trained in biology or medicine will understand the same word in another sense. A vector in the first meaning is a quantity that has a magnitude as well as a direction in space and can be resolved into spatial components. The physical quantity velocity **v** is a vector given by (v_x, v_y, v_z), where the components are so-called scalars defined by the magnitude only. According to the second meaning used in epidemiology in particular, a vector is 'any insect or other arthropod, rodent or other animal of public health significance capable of harboring or transmitting the causative agents of human disease, or capable of causing human discomfort and injury'.[41] Malaria is a vector-borne disease transmitted by mosquitoes. Obviously the two meanings have nothing in common except the word vector itself.

The term vector first turned up in astronomical language, namely in connection with Kepler's second planetary law as formulated in terms of what is called the *radius vector*. Harris' *Lexicon Technicum* from 1704 defined the term vector as follows: 'A Line supposed to be drawn from any Planet moving round a Center, or the Focus of an Ellipsis, that that Center or Focus, is by some Writers of the New Astronomy, called the Vector; because 'tis that Line by which the Planet seems to be carried round its Center, and with which it describes proportional areas in proportional Times.' Something like this is still the standard textbook formulation of Kepler's second law even though the radius vector is not a vector in the mathematical sense. It was in part for this reason that the Irish mathematician and astronomer William Rowan Hamilton— after whom the 'Hamiltonian' is named (Section 1.2)—in the 1840s introduced a new

---

[40] Nye (2000).
[41] https://www.ocvector.org/what-is-a-vector

terminology that included vector and scalar. In a paper published in the *Proceedings of the Royal Irish Academy*, Hamilton wrote as follows:

> In the new mode of speaking which it is here proposed to introduce, and which is guarded from confusion with the older mode by the omission of the word '*radius*', the *vector* of the sun *has* (itself) *direction*, as well as length. It is, therefore *not* sufficiently characterized by *any single number* ... but *requires*, for its *complete numerical expression*, a *system of three numbers*; such as the usual and well-known rectangular or polar co-ordinates of the Sun or other body or point whose place is to be examined. ... A *vector* is thus (as you will afterwards more clearly see) a sort of *natural triplet*.[42]

It took several decades before the vector notation was generally adopted by mathematicians and physicists. As regards the biological and medical use of vector, it came considerably later and apparently without being related in any direct way to the mathematical term. The 1922 edition of *Encyclopaedia Britannica* writes about the 'intermediary hosts of the parasite actually causing the disease' that they are 'known as "carriers" or "vectors"' (*OED*). The two meanings of vector only have in common that the word is derived from the Latin *vehere* meaning 'carrier' or 'to carry'. In the original astronomical sense, it was the Sun or its gravitational force that carried the planets, and in the later epidemiological sense it was small animals that carried the disease.

One of the simplest and most common experiments in elementary chemistry is the *litmus test* in which the acidic or basic character of a solution is tested by means of a litmus paper. If the paper turns blue, the solution is basic (pH > 8.3) and if it turns red the solution is an acid (pH < 4.5). Litmus, a compound extracted from various lichens, was first used as an indicator for acidity in the early fourteenth century and eventually, in the handy form of litmus paper, it became a standard remedy in school chemistry. As a figurative term litmus test only entered language in the second half of the twentieth century, typically in the meaning of a crucial indication that something is actually the case or has succeeded. 'The passing of the bill was a litmus test for the new government', a newspaper may report. Or we may be told that with the solar eclipse observations of 1919 Einstein's theory of gravitation passed a litmus test. Such phrases are very common in today's plain language.

In the eighteenth century, strong sulphuric acid (H_2SO_4) was known as 'oil of vitriol' or 'vitriolic acid' because it was manufactured from sulphates or 'vitriols', a name that comes from the Latin word *vitrum* for glass. The sulphate crystals looked like pieces of coloured glass. Iron sulphate was called 'green vitriol' and earlier 'vitriol of Mars', and copper sulphate was 'blue vitriol' or 'vitriol of Venus'. Stannous sulphate alias tin(II) sulphate $SnSO_4$ was known as 'vitriol of Jove', a reference to the planet Jupiter or the Roman god of the same name. As mentioned in Section 2.1, the electric fluid generated by the friction of glass was known as vitreous electricity.

Another common substance was the 'caustic soda' or what later became sodium hydroxide (NaOH). While sulphuric acid is a strong acid and sodium hydroxide a

---

[42] Quoted in https://mathshistory.st-andrews.ac.uk/Miller/mathword/

strong base, the two substances have in common that they are highly corrosive and therefore dangerous. The words are used figuratively in approximately the same sense, namely to denote speech or behaviour which is bitterly critical, harshly condemnatory, or sarcastic in an unkind way. A person may launch 'a vitriolic attack' against some other person by making use of a 'caustic rhetoric'. More recently the metaphoric phrase 'oxygen of publicity' has crept into the language as an expression for how the mass media may indirectly boost questionable or harmful causes. In a speech of 1985, Britain's prime minister Margaret Thatcher said that 'we must try to find ways to starve the terrorist and the hijacker of the oxygen of publicity on which they depend'.[43]

There is more. An individual cannot have a 'good chemistry', but two persons can, meaning that they easily come along, enjoy being together, or have an intuitive feeling of what the other is thinking. If Johnny tells Linda that 'Well, there just wasn't any chemistry between us', Linda knows what he means. An American newspaper reported about a basketball team: 'Injuries have hurt our chemistry, and team chemistry is such a delicate thing.'[44] It is not obvious what chemistry has to do in phrases like these, but perhaps there is help to find in George Bernard Shaw's play *You Never Can Tell* from 1897. Gloria says to Valentine, 'I hope you are not going to be so foolish—so vulgar—as to say love', to which Valentine replies:

No, no, no. Not love: we know better than that. Let's call it chemistry. You can't deny that there is such a thing as chemical action, chemical affinity, chemical combination—the most irresistible of all natural forces. Well, you're attracting me irresistibly—chemically.[45]

Affinity is the key word. Apart from its many other connotations, this concept has played a crucial role in the history of chemistry, where since the mid-eighteenth century it was conceived as a kind of force causing some substances to combine and others not. Later attempts to turn the elusive affinity into a measurable quantity resulted in elaborate electrical and thermal theories of affinity until it was largely replaced by the 'free energy' of current chemical thermodynamics.[46] There is another common expression with scientific and semantic roots similar to that of 'good chemistry'. If two or more people understand each other well or can easily cooperate—are 'of the same mind'—they are said to be 'on the same wavelength'. The idiom, which has been in use since the 1920s, alludes to the reception of radio programmes at a particular wavelength sent through 'the ether'.[47] It would make no sense before the invention of radio.

The adverb *hermetically* is still used to denote a tube, can, or vessel so closely sealed that it prevents air entering it. Hermetic cans and plastic bags are all around in food markets. The name refers to the Greek god Hermes but more specifically to the 'hermetic' tradition in the Renaissance, when alchemists, in their vain search

---

[43] https://www.margaretthatcher.org/document/106096. *OED*.
[44] Watkins (1989).
[45] https://www.gutenberg.org/files/2175/2175-h/2175-h.htm
[46] Levere (1971).
[47] *OED*. Morse (1927).

for the philosopher's stone, experimented with heated substances isolated from the surrounding air. They claimed to be the successors of the mythical Hermes Trismegistus (Hermes the Thrice-Greatest) who supposedly had collected all ancient wisdom in the writings called *Corpus Hermeticus*.

Dynamite is a neologism coined by Alfred Nobel in 1867 when he obtained a patent for mixing the highly explosive liquid nitroglycerine synthesized twenty years earlier with an inert absorbent. As an absorbent Nobel used kieselguhr, a porous siliceous earth that easily absorb liquids and the name of which is of German origin. The Greek *dynamis* means power or strength (dynamo, dynamics) and the suffix *-ite* 'belonging to' or 'connected with'. At the time the suffix was used to designate numerous minerals eponymously, for example gadolinite after Johan Gadolin. With the industrial success of dynamite, the name came to be used figuratively, first as an expression of something potentially dangerous or explosive as in 'the released document is political dynamite'. Later the term was also associated with positive connotations such as 'excellent' and 'impressive'.

To chemists, a bromide is a salt containing the negative bromide ion Br^- just as a chloride contains the ion Cl^-. Sodium bromide is NaBr, potassium bromide KBr. But in literary usage the same word signifies a cliché, a banality, or a feel-good phrase, for example 'time heals all wounds', 'boys are boys', and 'you don't look a day over fifty'. Any connection between the two very different meanings? Yes, there is one.

In the mid-nineteenth century it was discovered that potassium bromide in particular had calming effects and could be used as a sedative and even, so it was claimed, to treat epilepsy and forms of hysteria. 'Of all the sleep-producing agents at our disposal, the bromide of potassium is most deserving the name of hypnotic', reported *Scientific American* in 1869.[48] Although the detrimental effects of the substance had become clear by the 1880s, its popularity and overuse continued for a decade or two. Soon bromide was being used figuratively for anything or anyone that might put one to sleep because of commonness or plain dullness. As early as 1906, an American author, Gelett Burgess, published a humorous piece with the title 'Are You a Bromide?' in which he presented a large number of 'bromidioms'. One of them was 'This world is such a small place, after all, isn't it?' and another was 'Now, this thing really happened!'[49]

The word *catalysis* and associated terms like 'catalyse' and 'catalyst' go back to Berzelius, who in a paper of 1835 noted that small amounts of a substance might drastically increase the reaction rate without being consumed in the reaction. He ascribed it to a 'new force, up till now unknown, [which] is common to organic and inorganic nature', and said,

> it will be more convenient to designate the force by a new name. I will therefore call it the 'Catalytic Force' and I will call 'Catalysis' the decomposition of bodies by

[48] Quoted in Lamb (2018).
[49] Gelett Burgess, 'Are You a Bromide?', available at https://www.gutenberg.org/ebooks/10870. On Burgess as a prolific but unserious coiner of words, see Keyes (2021), pp. 171–3.

this force, in the same way that we call by 'Analysis' the decomposition of bodies by chemical affinity.[50]

The word is derived from Greek *katalysis* (κατάλυις) meaning 'dissolution' or 'to dissolve'. As it turned out only many years later, the mere presence of a catalyst is not enough for its action. It does take part in the reaction it catalyses, but is reformed before the reaction is over.

Berzelius' terms soon became very important in chemistry and biochemistry, and from about 1940 they also gained a footing in common language, typically as something or someone causing an event to happen. One can read, for example, that the assassination of Archduke Franz Ferdinand on 28 June 1914 was the catalyst that caused World War I. The figurative meaning is close to but not quite the same as the chemical notion of catalysis, where a catalyst speeds up the reaction but does not cause it. The antonym to catalyst is *inhibitor*, a term used in biochemistry since about 1914. Whereas the verb 'inhibit' is a common English word, the noun inhibitor is rarely used in a nonscientific context. A biological catalyst is known as an *enzyme*, a word coined by the German physiologist Wilhelm Kühne in 1877: 'In order to avoid misunderstandings and cumbersome circumlocutions, the presenter proposes to designate as *enzymes* the unformed or not organized ferments, whose action can occur without the presence of organisms and outside of the same.'[51]

The second part of enzyme is the Greek root *zyme* (ζύμη) meaning leaven or ferment. It was used by the German chemist Georg Stahl, better known for his influential phlogiston theory (Section 5.3), to construct the word 'zymotechnia' as it appeared in his *Zymotechnia Fundamentalis* from 1697. Much later the term was popularized and promoted by the Danish plant physiologist and entrepreneur Alfred Jørgensen for the industrial use of fermentation processes. The 'zymotechnology' of the fin-de-siècle era was a predecessor of what came to be known as *biotechnology*, a term which, in its German equivalent 'Biotechnologie', was coined by the Hungarian agricultural scientist and engineer Karl Ereky in 1919.[52]

There are a few words that occur prominently in both chemistry and the social and political spheres without there being any obvious connection between the two areas. One of them is *radical*, a well-known term for a person with extreme political views or for a corresponding ideology. Sympathizers of the views of Islamic State (IS) are often said to have been radicalized. An economic reform may be radical in the sense of being thorough or even revolutionary. Einstein's theory of relativity was considered too radical by many contemporary physicists. The term has its origin in the Latin word *radix* for 'root', which is reflected in one of the meanings of radical, namely the root or base of something. We have the word in 'radish' and also in the verb 'eradicate' meaning to root something out.

When modern chemists speak of a radical, they are referring to an atom, molecule, or ion with one or more unpaired valence electrons (e.g. Cl and CH_3). Such particles are highly reactive and can exist under normal conditions only for a very short time. However, the chemical concept of radicals has changed drastically over time since it

---

[50] Cited in Linström and Pettersson (2003).
[51] Kühne (1877), p. 190.
[52] Bud (1993), pp. 6–26.

was introduced in 1782 by the French chemist Guyton de Morveau, who based the name on the Latin *radix*.[53] Morveau's notion was entirely different from ours, as he essentially used it to denote the oxygen-free part of an acid. Radicals later became a central concept in organic chemistry, but in a meaning which has little in common with modern usage. In chemistry there is no concept of 'radicalization'—the process of making or becoming increasingly more radical—such as there is in the political sphere.

The meaning of the Greek root *organon* (οργανον) is an instrument or tool. From this root we have words such as organ, organism, and organize, and also the adjective *organic* as in organic chemistry, a term that Aristotle would neither have accepted nor understood. But organic also has another and quite different meaning which in large parts of the world is presumably more popular than the one of the chemists'. 'Organic food' has since about 1970 come to mean food produced with no use of synthetic chemicals whether or not these are organic in the chemical sense of the term. Such food (or other products, say cosmetics) is often marketed as being 'free of chemicals', which is strictly speaking absurd. From a chemical point of view, 'organic food' is close to being a tautology and 'non-organic food' an oxymoron.

What in English is organic food was and still is 'ecological food' in some European countries. The term *ecology* was coined by the leading German Darwinian zoologist Ernst Haeckel, who in a book of 1866 derived it from the Greek *oikos* (οἶκος) for house or habitation: 'By ecology, we mean the whole science of the relations of the organism to the environment including, in the broad sense, all the "conditions of existence".'[54] To Haeckel, ecology was thus a new holistic science of biology and natural geography. Only in the 1960s did ecology become a fashionable word associated with 'green' lifestyle and political movements concerned with the harmful effects of human activity on 'mother Earth'. It has been argued that Henry Thoreau, the famous American naturalist and poet, used the word ecology as early as 1858, but the suggestion rests on an instructive mistake, namely a misreading of the word 'geology' in one of Thoreau's letters.[55]

While the confusing use of organic as practically synonymous with 'natural' is only indirectly related to chemistry, *deoxyribonucleic acid* better known as DNA provides a recent example of a (bio)chemical term which has successfully entered common language as a powerful metaphor. There is little doubt that had it not been for the abbreviated form, an acronym, this would not have happened. Nor would word formations such as 'DNA fingerprint' and 'DNA profiling' have been possible. Today DNA is often used figuratively as signifying the essence of a person or a thing, as in 'it is part of his DNA' or in an advertisement saying 'we build good cars because it's in our DNA'. It is hard to imagine an advertisement with the alternative 'we build good cars because it's in our deoxyribonucleic acid'. The term is no less successful among scientists. It appears in the title of nearly half a million scientific papers.

In many cases words and phrases do not migrate between the social and scientific spheres but within the latter, that is, between different fields of science. It is not uncommon that a technical term originating in one scientific discipline is

---

[53] Constable and Housecroft (2020).
[54] Haeckel (1866), p. 286.
[55] McIntosh (1975).

subsequently adopted with a different meaning in another discipline, such as did the words *plasma* and *metabolon* (Sections 2.6 and 4.1). An interesting case of what may be called internal word migration is how characteristic phrases in Darwinian evolution theory found their way to chemistry and physics—and sometimes even to astronomy—in the second half of the nineteenth century. In 1872, thirteen years after Darwin's bombshell *On the Origin of Species*, the British-Australian physicist Morris Pell wrote about an original 'warfare among the molecules' governed by the principle of natural selection. He argued that the fittest of the molecules would conquer those who were too weak to survive in the struggle for existence.[56]

Other scientists who explicitly invoked Darwin's biological theory in more than just a rhetorical sense included the Austrian physical chemist Leopold Pfaundler and the Danish thermochemist Julius Thomsen. According to Pfaundler, 'Darwin's principles are valid also in the molecular world.'[57] Thomsen too made use of Darwinian metaphors when trying to explain that only certain atomic weights were represented in the elements while others, such as 8, 10, and 13, were missing. His answer:

> Just as the biologist supposes that the right of the fittest has manifested itself in the evolution of the species ... the chemist has shown that the atomic weights of the elements do not form a successive series of numbers. ... The biologist believes that evolution from one species to another occurs through a series of generations ... and the chemist must adopt a similar hypothesis if he is to suggest the mechanism by means of which the transformation or development of an element leads to another element.[58]

The reputed British chemist William Crookes was yet another advocate of 'inorganic Darwinism', a phrase he coined in a spirited address to the British Association for the Advancement of Science in 1886. According to Crookes, the known elements were 'the gradual outcome of a process of development, possibly even a "struggle for existence"'.[59] Like a few other chemists in the period, he considered the periodic system to be a close analogy to the evolutionary trees of organic beings. Crookes stated that 'The array of the elements cannot fail to remind us of the general aspect of the organic world', and as an example he mentioned 'the Monotremata [platypus, echidnas] of Australia and New Guinea, and among the elements the metals of the so-called rare earths'.

5.3 A revolution in chemical nomenclature

A reformed technical language was an essential part of what is often known as the 'chemical revolution' in the last decades of the eighteenth century. The chief architect was Lavoisier, who called himself a *physicien* (physicist) and not a *chimiste* (chemist).

---

[56] Pell (1872).
[57] Snelders (1977).
[58] Kragh (2016), p. 279.
[59] Brock (1985), pp. 195–7.

According to him, it was as much a revolution in physics as in chemistry. At the time the borders between the two allied natural sciences were fluid and, as mentioned, they would remain so for several decades.

The term *chemical revolution* (often written with first letters capitalized) or something close to it goes all the way back to Lavoisier, who used 'revolution' in the new meaning of a drastic change and not in its traditional meaning of a phenomenon that occurs repeatedly (see Section 6.1). Antoine Fourcroy characterized in 1782 the new chemistry as a 'great revolution', and eight years later—at the time of the French political revolution—Lavoisier wrote in a letter to another of his chemical allies, Jean-Antoine Chaptal, that 'All young people adopt the new theory and I conclude that the revolution in chemistry is complete.'[60] The name chemical revolution was permanently fixed on the historical record a century later with the publication of the eminent chemist Marcelin Berthelot's book entitled *La Révolution Chimique: Lavoisier.*

Dissatisfaction with existing chemical language was a key element in Lavoisier's revolution, and it was far from new. More than a century earlier, Boyle indignantly criticized chemical authors for their obscure language and ambiguous terminology. 'Even Eminent Writers', he wrote in his *Sceptical Chymist,*

> will now and then give divers things, one name; so they will oftentimes give one thing, many Names; and some of them (perhaps) such, as do much more properly signifie some Distinct Body of another kind; nay even in Technical Words or Termes of Art, they refrain not from this Confounding Liberty; but will, as I have Observ'd, call the same Substance, sometimes the Sulphur, and sometimes the Mercury of a Body.[61]

Whether belonging to chemistry or physics, the revolution that occurred in the period from about 1770 to 1790 relied on progress in so-called *pneumatic chemistry*. As far as theory was concerned, the most important element was the rejection of the central concept of *phlogiston* that had dominated chemical thinking for most of the century. The word *pneuma* (πνεῦμα) is an ancient Greek name for a space-filling substance or spirit which played an important role in the philosophy of nature expounded by Chrysippos and other members of the Stoic school. Much later, the word came to refer to the new gases discovered in the eighteenth century such as carbon dioxide, hydrogen, oxygen, ammonia, and chlorine. Pneumatic chemistry was thus what we would call the chemistry of gases. Today *pneumatics* is an important branch of engineering which makes use of pneumatic devices operated by compressed air or other gases.

The term *gas* was coined by the Flemish natural philosopher Jan Baptista van Helmont, who in 1648 emphasized that it was a neologism: 'For want of name, I have called that vapour, Gas, being not far severed from the Chaos of the Auntients

---

[60] Siegfried (1988). See also Cohen (1985), pp. 229–36.
[61] *Sceptical Chymist*, Part IV. https://www.gutenberg.org/files/22914/22914-h/22914-h.htm. Golinski (1987).

[ancients]'.[62] According to van Helmont, there were several gases, none of which were similar to air or ordinary vapours. A gas, he said, was a space-filling principle that could neither be retained in a vessel nor be reduced to a visible form. Chemists in the eighteenth century rarely used the term, preferring to speak of different 'airs' or 'elastic aeriform vapours' rather than gases. Although Samuel Johnson included 'gas' in his dictionary of 1755, he had his reservations about this 'word invented by the chymists'. He complained: 'It is used by Van Helmont and seems to signify, in general, a spirit not capable of being coagulated; but he uses it loosely, in many senses, and very unintelligibly and inconsistently'.[63] When the Scotsman Joseph Black in 1756 announced that he had discovered a new type of gas, what came to be known as carbon dioxide, he called it 'fixed air'. By heating *magnesia alba* (magnesium carbonate) Black had identified a gas that differed from atmospheric air. The new gas or air was fixed in the sense that it was part of a solid substance and could be released by heating it ($MgCO_3 \rightarrow MgO + CO_2$).

The chemists tried to understand the airs in terms of the popular concept of *phlogiston*, a term introduced but not coined by the German chemist Georg Ernst Stahl in 1718 for the principle or substance supposedly contained in all combustible bodies. It is based on Greek *phlogistos* (φλογιστός) meaning inflammable. The term phlogiston had been used before Stahl as an adjective by van Helmont and others, but it was only with Stahl that it became an important chemical term.[64] Johnson briefly referred to two meanings of the term, which could either denote 'a chemical liquor extremely inflammable' or 'the inflammable part of any body'.

In a nutshell, the followers of Stahl argued that in a combustion process the flammable body X released its phlogiston (φ) to the surrounding atmospheric air, thus $(X + \varphi) \rightarrow X + \varphi$. In a closed vessel the combustion will cease because the air has been 'phlogisticated', that is, saturated with phlogiston. When an 'earth' or 'calx' (an oxide or carbonate) is heated with charcoal rich in phlogiston, it will be reduced to a metal, a process interpreted as $X + \varphi \rightarrow$ metal. It should be pointed out that during most of the eighteenth century chemists used the term earth as a designation of substances later identified as metallic oxides, e.g. the earth of iron being Fe_3O_4. In contrast to other meanings of the polysemic word earth—as used by geologists and astronomers, for example—the chemists used it in both the singular and plural forms. As a reminiscence of the past we still speak of the 'rare earths' but now as a group of metallic chemical elements largely identical to the lanthanide series in the periodic table starting with atomic number 57 (lanthanum) and ending with 71 (lutetium).

To proceed, when it was discovered that 'inflammable air' (hydrogen) produced water droplets when subjected to an electric spark, some chemists interpreted it as evidence that water is a compound body consisting of 'vital air' (oxygen) and phlogiston. For a while Henry Cavendish in England tended to identify phlogiston with the inflammable air and later he suggested that this air was water saturated with

[62] Wothers (2019), p. 91. Partington (1961), pp. 227–32.
[63] Johnson (1755), who repeated Harris' description in *Lexicon Technicum* from 1704.
[64] For historical and philological details, see Partington (1961), pp. 667–8.

phlogiston. By conceiving the vital or 'highly respirable' air as 'dephlogisticated'—water without phlogiston—he explained what we call the synthesis of water as a redistribution of phlogiston:

$$(\text{water} + \varphi) + (\text{water} - \varphi) \rightarrow \text{water}.$$

What matters here is that the terminology for substances and processes to a large extent reflected the generally accepted phlogiston theory. As the vital air or our oxygen was called dephlogisticated air and the foul air or our nitrogen was often known as phlogisticated air, so Carl Wilhelm Scheele in Sweden wrote of 'dephlogisticated nitrous acid' (HNO_3). He likewise described the chlorine gas he had discovered as 'dephlogisticated acid of salt'. In the process we write as

$$MnO_2 + 4HCl \rightarrow Cl_2 + 2H_2O + MnCl_2,$$

Scheele saw a transfer of phlogiston from the acid to the manganese.

The aim of the radical reform of chemical nomenclature proposed in France in the 1780s was in part to get rid of terms based on the phlogiston theory, but it was much broader in scope and started before Lavoisier's crusade against phlogiston. In 1782 Guyton de Morveau, at the time still accepting the general framework of the phlogiston theory, published his *Mémoir sur les Dénominations Chimique* in which he formulated a number of general principles for a new chemical language. Among them were that 'denominations should be as much as possible in conformity with the nature of things' and the anti-eponym recommendation that 'the names of discoverers of substances should be excluded from a general nomenclature'. He further recommended that new designations should be formed only from Greek and Latin roots, 'so that the words can easily be found again from the sense and the sense from the word'.[65]

Morveau was aware of and inspired by the Swedish chemist and mineralogist Torbern Bergman, who was a pupil of Carl Linnaeus, the great systematizer of botanical nomenclature. Strongly influenced by Linnaeus, Bergman suggested that the names of chemical compounds should be in Latin and follow a binomial nomenclature similar to that used in botany. For example, for the salts that came to be known as magnesium sulphate ($MgSO_4$) and silver nitrate ($AgNO_3$) Bergman introduced *magnesia vitriolata* and *argentum nitratum*. The latter salt was traditionally known as 'lunar caustic', a name that reflected the affinity of silver to the Moon but gave no clue to the composition of the substance. Bergman favoured Latin as the common language of chemical and mineralogical nomenclature for two reasons. First, Latin was known to the whole learned world and, second, it had the advantage that it was a dead language and therefore not liable to innovation and change. According to Bergman in a work of 1784:

[65] Cited in Crosland (1978), p. 158. For the influence of ancient Greek on modern chemical terminology, see Loyson (2009).

> In establishing entirely new names, I desire that their origins be Latin. This language is, or at least was, the vernacular of the learned: now it is dead and is not subject to constant changes. Therefore ... chemical language can attain general agreement in all places, which promises no small benefit not only in reading foreign works, but also in translating.[66]

In 1787 Morveau, who by then had converted to the anti-phlogistic theory, joined forces with Lavoisier, Claude Berthollet, and Antoine Fourcroy in the writing of *Méthode de Nomenclature Chimique*, a book which has been hailed as nothing less than 'a momentous contribution to the world-wide vocabulary of Western science' (Figure 5.2).[67] The collaborative work was translated into English in 1788 as *Method of Chymical Nomenclature* and in 1793 a German translation followed with the more explicit title *Methode der chemischen Nomenclatur für das antiphlogistische System*. It also appeared in Spanish and Italian translations.

Expectedly, the recommendations in *Méthode* initially met with resistance both in France and abroad. Some critics objected that chemical nomenclature should be neutral with regard to theory, which the new system was not as it rested to some extent on Lavoisier's anti-phlogistic chemical system. Others found it wrong to name chemical substances after their composition, since the true composition might not yet be known with certainty. To Jean-Claude de la Métherie, a French mineralogist, the nomenclature of Lavoisier and his allies was unacceptable for aesthetic and euphonious reasons alone. As he objected, it made use of 'harsh and barbaric words that shock the ear and are not at all in the spirit of the French language'.[68] Thomas Jefferson, the third president of the United States, was not only a statesman and diplomat but also interested in science and technology. He was among the first Americans to comment on the new nomenclature, which he did not like at all. In a letter from Paris dated 19 July 1788, he said about chemistry that although it was a useful art, as a science it was still in an embryonic state:

> It is probably an age too soon to propose the establishment of system. The attempt therefore of Lavoisier to reform the chemical nomenclatugere is premature. One single experiment may destroy the whole filiation of his terms, and his string of sulfates, sulfites, and sulfures may have served no other end than to have retarded the progress of the science by a jargon from the confusion of which time will be requisite to extricate us. Accordingly it is not likely to be admitted generally.[69]

Despite the initial opposition, within one or two decades the new nomenclature of Lavoisier and his allies was by and large adopted in almost all European countries.

On behalf of all four authors, in the spring of 1787 Lavoisier read a paper to the Paris Academy of Science titled 'Memoir on the Necessity for Reforming and Improving Chemical Nomenclature' in which he presented an outline of the new system as

---

[66] Quoted in Gordin (2017), p. 47.
[67] Hogben (1970), p. 28. Lefèvre (2018).
[68] Quoted in Golinski (1992).
[69] Duveen and Klickstein (1954). Jefferson stayed in France from August 1784 to October 1789.

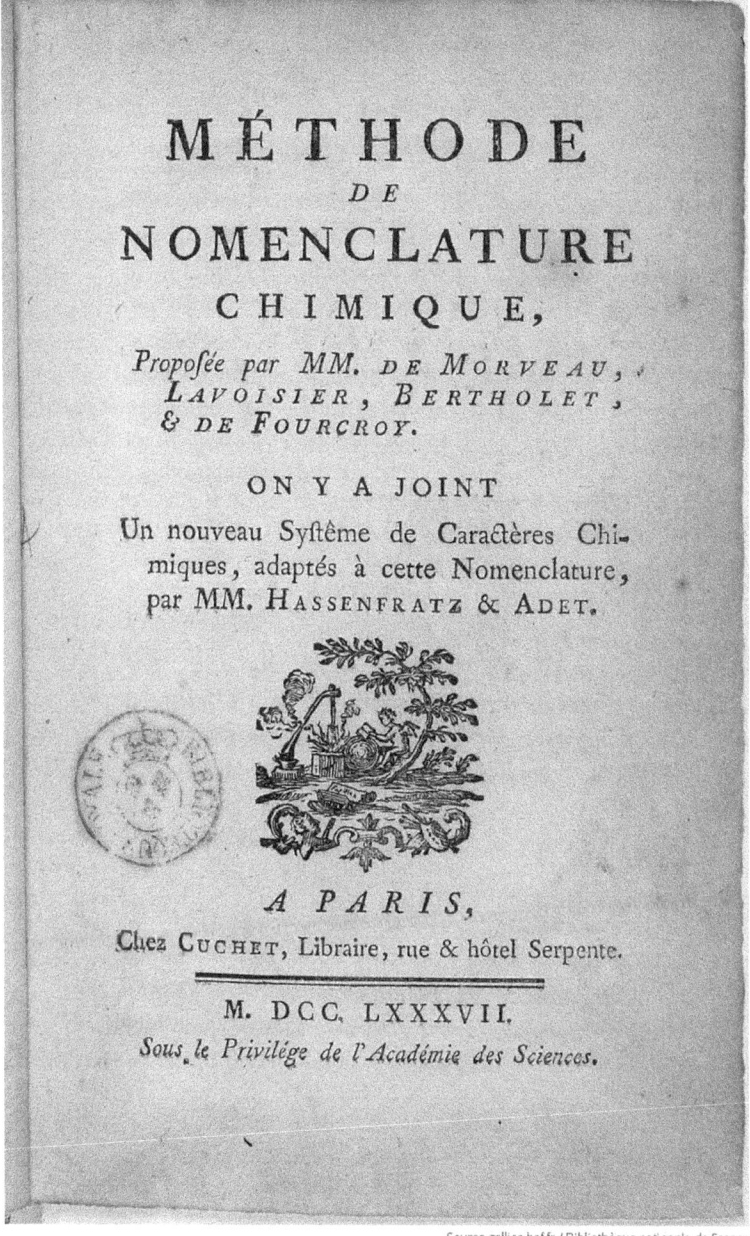

Figure 5.2 *Method of Chymical Nomenclature* published by Lavoisier and his three co-authors.

Courtesy of Gallica, Bibliothèque National de France, https://gallica.bnf.fr/ark:/12148/bpt6k1050402r.image (accessed on 8/11/2023).

part of a more comprehensive programme of linguistic reform. About languages in general he said that they are 'not only intended, as is commonly supposed, to express ideas and images by signs, but also are real analytical systems by means of which we advance from the known to the unknown, and to a certain extent in the manner of mathematicians'. If languages, he continued, 'really are instruments fashioned by men to make thinking easier, they should be of the best possible kind; and to strive to perfect them is indeed to work for the advancement of science'. Lavoisier proceeded:

> A well-composed language adapted to the natural and successive order of ideas will bring in its train a necessary and immediate revolution in the method of teaching and will not allow teachers of chemistry to deviate from the course of Nature; either they must reject the nomenclature or they must irresistibly follow the course marked out by it. *The logic of the sciences is thus essentially dependent on their language ...* We shall have three things to distinguish in every physical science: the series of facts that constitute the science; the ideas that call the facts to mind; and the words that express them. The words should give birth to the idea; the idea should depict the fact ... it follows that the science can never be brought to perfection, if the language be not first perfected.[70]

Morveau's recommendation to avoid eponymous names as far as possible was repeated in *Méthode*, where Lavoisier and his co-authors objected to the widespread use of traditional names such as 'Glauber's salt', 'Scheele's fire air', and 'Boyle's fuming liquor'.[71] As they pointed out, these and many similar names, whether eponyms or not, conveyed nothing at all about the composition of the substances or the underlying principles of chemistry.

In the preface to his influential and innovative textbook *Traité Élémentaire de Chimie* from 1789, Lavoisier stressed that he did not use the term 'element' to express 'those simple and indivisible atoms of which matter is composed, [for] it is extremely probable [that] we know nothing at all about them'. Elsewhere in the preface, and earlier in *Méthode*, he quoted extensively from the French Enlightenment philosopher Étienne Bonnot de Condillac whose *La Logique* of 1780 and other works evidently served as a major inspiration.[72] According to Condillac, the progress of the human mind depended closely on progress in the use of language. Words and other linguistic signs were producers of new ideas and not merely a neutral medium in which the ideas could be conveniently expressed. Language, he said, was an epistemological instrument through which the philosopher could proceed from the known to the unknown and as such it was no less important than the laboratory's material instruments.

In full accordance with Condillac, Lavoisier asserted that nomenclature was not extrinsic to science, but part and parcel of it. As he wrote:

---

[70] Quotes from McKie (1952), pp. 264–7. Emphasis added. See also Surman (2016).

[71] For these and numerous other obsolete chemical terms from the eighteenth century, see https://www.chemteam.info/Chem-History/Obsolete-Chem-TermsTOC.html. See also Senning (2019), pp. 451–96.

[72] On Lavoisier's debt to Condillac's philosophy, see Roberts (1992) and Crosland (1978), pp. 170–1. See also Anderson (1985) for a critical analysis of Lavoisier's theory of language.

The impossibility of separating the nomenclature of a science from the science itself, is owing to this, that every branch of physical science must consist of three things; the series of facts which are the objects of the science, the ideas which represent these facts, and the words by which these ideas are expressed. Like three expressions of the same seal, the word ought to produce the idea, and the idea to be a picture of the fact.[73]

Thus, inspired by Condillac, Lavoisier thought that the proper choice of words would generate new ideas and that words were therefore important to scientific creativity. As noted in Section 1.2, this was a view that had earlier been criticized by John Locke according to whom there was no natural connection between words and ideas. On the other hand, Maxwell stated in 1870 that well-chosen metaphorical words could work as generators of science.

To jump further ahead in time, Einstein's view about words and creative thinking was closer to Locke than to Lavoisier. 'The words or language, as they are written or spoken, do not seem to play any role in my mechanism of thought', he told the French mathematician Jacques Hadamard, who likewise described his own mathematical thinking as largely wordless.[74] In conversations with his friend, the psychologist Max Wertheimer, Einstein elaborated: 'These thoughts [leading to the theory of relativity] did not come in any verbal formulation. I very rarely think in words at all. A thought comes, and I may try to express it in words afterwards.'[75]

The different views of Lavoisier and Einstein with respect to the importance of words can perhaps be explained, at least in part, by the difference between late eighteenth-century chemistry and early twentieth-century theoretical physics. While Lavoisier was forced to think in more descriptive terms, in Einstein's abstract thinking about relativity and motion words were of limited significance. Although Einstein seems to have been unconcerned with specific words, on a few occasions he reflected on the language of science which he suggested was international and different from ordinary language. 'Everything depends on the degree to which words and word-combinations correspond to the world of impression', he said in an address of 1941. Moreover:

Is there no thinking without the use of language, namely in concepts and concept-combinations for which words need not necessarily come to mind? ... We may conclude that the mental development of the individual and his way of forming concepts depend to a high degree upon language. This makes us realize to what extent the same language means the same mentality. In this sense thinking and language are linked together.[76]

---

[73] Lavoisier (1965), p. xiv.
[74] Hadamard (1954), p. 142.
[75] Quoted in Holton (1988), p. 387. See also Jakobson (1997), written by the famous linguist Roman Jakobson, who writes about 'Einstein's personal and primary inclination to attribute to the act of thinking complete independence from language' (p. 142).
[76] Einstein (1982), p. 336.

Einstein or rather his theory of relativity had some impact on the later development of linguistic theories. The physician Max Talmey, who was Einstein's friend and a contributor to constructed languages, thought that the problem with comprehending relativity theory was essentially linguistic and that it could be avoided by using his own favourite model language called *Arulo*, a word formed from Auxiliary Rational Universal Language.[77] More importantly, the linguist Benjamin Lee Whorf referred to Einstein's theory when formulating his ideas of so-called linguistic relativity.[78] There is no indication that Einstein knew about Whorf's theory and its controversial claim that language determines what we know about nature (Section 1.2). Had he known about the theory, he most likely would have rejected it.

And now from Einstein back to Lavoisier. The often-reproduced table of chemical elements in Lavoisier's famous textbook started with a select group of 'simple substances belonging to all the kingdoms of nature, which may be considered as the elements of bodies'. Apart from light and caloric, the group comprised 'oxygen', 'hydrogen', and 'azote', the latter an older name for what Jean-Antoine Chaptal proposed to call nitrogen (nitrogène) in 1790. *Azote* is based on a Greek word for life (*zōos*) combined with the prefix *a-* (not, without, lacking) and thus means 'without life'. As a German name for azote or nitrogen the prominent chemist and pharmacist Sigismund Hermbstädt first suggested the dramatic term *Tödlicherstoff* (deadly substance), soon to be replaced by *Stickstoff* (suffocating or extinguishing substance), which is still the name used in German. The Danish name is the equivalent *kvælstof*.

For a couple of decades, azote and nitrogen were both used in France and England—Dalton preferred azote in his atomic theory—but by 1830 the gas was mostly known as nitrogen. As late as 1878 the French scientist Charles Leturneau wrote of azote with chemical symbol Az, stating the formula of ammonia to be AzH_3 and that of ammonium chloride AzH_4Cl.[79] Although azote disappeared from the chemists' lexicon, the word element *azo* has remained in use for certain organic compounds containing nitrogen. Azo compounds such as azobenzene $C_{12}H_{10}N_2$ contain the group $-N=N-$. The inorganic liquid hydrazine H_2NNH_2 is not an azo compound but the second part of its name still refers to azote.

As Lavoisier commented, oxygen was 'the base of vital air' and hydrogen 'the base of inflammable air'. The wording reveals that he did not think of oxygen gas as an element in our sense but as a compound of elemental oxygen and caloric. He likewise conceived the hydrogen gas to be a combination of elemental hydrogen with caloric. It is well known that Lavoisier coined the word *oxygen* from the Greek *oxys* (ὀξύς) meaning sharp or acidic because he mistakenly believed that it was the characteristic principle in all acids. 'I declare', he said in an earlier treatise of 1783, 'that this pure and highly respirable air, is the constitutive principle of acidity; that this principle is common to all acids'.[80] Of course, his association of acidity with oxygen did not imply that all oxides were acidic. Although Lavoisier's oxygen was a misnomer,

[77] Talmey (1932), who believed that relativity theory, when expressed in Arulo and not in a natural language, 'becomes free of all mysteriousness and can be made comprehensible to the educated layman'. He did not claim that Arulo facilitated scientific creativeness.

[78] Gordin (2022). Heynick (1983).

[79] Letourneau (1878), p. 6.

[80] Wothers (2019), p. 108.

it was hugely important and kept after chemists recognized that it is hydrogen (or the hydrogen ion H^+) and not oxygen which is the 'constitutive principle of acidity'. The German name for oxygen is *Sauerstoff*, in Dutch it is *Zuurstof*, and in Swedish it is *syre*, all of which names refer to the same term (acid) as in the word invented by Lavoisier.

Lavoisier wrote about the muriatic acid also called marine acid (our hydrochloric acid, HCl) that although he had not been able to decompose the substance, 'we cannot have the smallest doubt that it, like all other acids, is composed by the union of oxygen with an acidifiable base'.[81] Dalton was no less sure. In his *New System of Chemical Philosophy* from 1808 he confidently wrote about the 'oxymuriatic' gas that, 'All experience shews that it is a compound of muriatic acid and oxygen.'[82] However, a few other chemists doubted what Lavoisier thought was undoubtable. In the early years of the new century his former collaborator Berthollet came to the conclusion that oxygen was not a necessary component of all acids and that Scheele's gas (chlorine) was not after all oxymuriatic acid formed by the oxidation of muriatic acid such as Lavoisier thought. And yet neither Berthollet nor other chemists in the early post-phlogiston era regarded chlorine as an element.

That only came later, namely with Humphry Davy's recognition that neither water nor oxygen was present in oxymuriatic acid gas. In a paper read to the Royal Society in November 1810, he asserted that the greenish gas could not possibly be decomposed and would thus, in the light of Lavoisier's pragmatic definition of an element, be of elementary nature. He did not claim to have discovered a new element, just to have ascertained its proper status with which followed a need for a new name:

> To call a body which is not known to contain oxygen, and which cannot contain muriatic acid, *oxymuriatic acid*, is contrary to the principles of that nomenclature in which it is adopted; and an alteration of it seems necessary to assist the progress of the discussion and to diffuse just ideas on that subject. ... After consulting some of the most eminent chemical philosophers of this country, it has been judged most proper to suggest a name founded upon one of its obvious and characteristic properties—its colour, and to call it *Chlorine* or *chloric gas*.[83]

The name was derived from Greek *khloros* (χλωρός), meaning green or greenish, to which was added the suffix -*ine* which came to be characteristic for the group of halogens. *Chlorofyll*, isolated and named in 1817, also refers to the green colour, but in this case combined with the Greek *phyllon* for 'leaf'.

Davy tried, albeit unsuccessfully, to isolate the element hidden in fluoric acid (HF) based on the mineral fluorspar (or fluorite, CaF_2) first described by Agricola in 1529. In letters to Davy of 1812, Ampère suggested that as muriatic acid contained no oxygen, so was it the case with fluoric acid. As to the name of the hypothetical element, Ampère proposed either 'fluore' or 'phtore'. Davy settled on a version of the first

[81] Lavoisier (1965), p. 71.
[82] Dalton (1808), p. 302.
[83] Cited in Crosland (1978), p. 221.

proposal, namely *fluorine* in analogy with chlorine. Several years later, the influential Scottish chemist Thomas Thomson adopted Davy's name, adding in a footnote:

> Ampere has given it the name Phtorine, (Phtore) from the Greek word φθόρος, *destructive*. But it is quite evident that this new name cannot be adopted. There would be no end to names if every person at pleasure could coin new ones. ... Davy informs us that Ampere himself originally suggested the name *fluorine*.[84]

The dangerously reactive fluorine gas F_2 was only isolated in 1886, when the French chemist Henri Moissan was able to collect and identify it. It was in part for this feat that he was awarded the Nobel Prize in 1906.

In his *Elements of Chemical Philosophy* of 1812, Davy gave much consideration to the nomenclature of new elements and their compounds. He proposed to use particular suffixes to denote the amount of chlorine in combinations with other substances, namely *-ane* for the smallest proportion of chlorine and *-anea* for the larger. The copper chlorides were thus called *cuprane* and *cupranea*, corresponding to our copper(I) chloride CuCl and copper(II) chloride $CuCl_2$. Davy's proposal was unsuccessful and criticized by Whewell, amongst others: 'In this nomenclature, common salt would be *Sodane*, and Chloride of Nitrogen would be *Azotane*. This suggestion never found favour. ... But Davy's terms were bad; for it does not appear that Chlorine enters, as Oxygen does, into so large a portion of chemical compounds, that its relations afford a key to their nature, and may properly be made an element in their names.'[85]

According to Lavoisier and his followers in early nineteenth-century chemistry, an 'oxidation' of a substance meant that it literally combined with oxygen. For example, the element sulphur could be 'oxidized' to the acid sulphur dioxide (SO_2), and calcium sulphite ($CaO \cdot SO_2$) likewise to calcium sulphate ($CaO \cdot SO_3$). After it had been accepted that some acids and salts contain no oxygen, the term oxidation (and the converse term reduction) continued to be applied as if oxygen were present. It took a long time before rhetorical references to the element oxygen disappeared from what was—and still is—called oxidation and reduction processes.[86] The important and very practical concept of 'oxidation number' used in balancing chemical reaction schemes was introduced in the 1930s. Although it is a purely formal concept, nominally it still refers to oxygen. Beginners in chemistry may be puzzled when told that in the synthesis of hydrochloric acid, $H_2 + Cl_2 \rightarrow 2HCl$, hydrogen is 'oxidized' and chlorine 'reduced'. Since an oxidation of one element implies a reduction of another element and vice versa, chemists formed the term *redox* by blending of the two words. The term introduced in a 1928 paper by the German biochemist Leonor Michaelis and his American co-author Louis Flexner has for long been accepted as a staple part of the chemists' vocabulary.[87]

---

[84] Thomson (1831), p. 89. Online as Google Book.
[85] Whewell (1840), p. cxvi.
[86] Jensen (2007b).
[87] Michaelis and Flexner (1928).

5.4 Naming the elements

Phosphorus is the first chemical element known to have been discovered in the usual sense of the term. Credit for the chance discovery in 1669 is attributed to the German merchant and alchemist Hennig Brand but it only became publicly known several years later and then under a variety of names all including 'phosphorus', e.g. *phosphorus hermeticus* and *phosphorus smaragdinus*. The word *phos* is ancient Greek for 'light' and *phoros* (or phorus, phore) means 'bringing' or 'bearing'.[88] The new substance was thus a bearer of light. The same term was previously used for the 'morning star', that is, the planet Venus when first appearing on the sky (Section 6.2). By the early eighteenth century phosphorus was well known but thought to cover several substances. Harris' *Lexicon Technicum* of 1704 contained a detailed entry on phosphorus, describing it as 'A Chymical Preparation, which being exposed to the Light or Air, will shine in the Dark [and of which] there are several kinds'. That the several kinds were all versions of the same element was first recognized by Lavoisier in a paper of 1777 although on this occasion he did not explicitly refer to phosphorus as a chemical element.

Lavoisier's table of simple substances included manganese among its metallic bodies together with, for example, copper, zinc, and iron. It also referred to 'magnesia', but this was classified as an 'earthly substance', meaning that the elemental metal had not yet been isolated from the earth. Manganese, later placed as number 25 in the periodic system, was first isolated in an impure form by the Swedish chemist Johan Gottlieb Gahn in 1774. Only in 1808 did Humphry Davy, in a series of famous experiments based on the new art of electrolytic decomposition, succeed in isolating and thereby discovering the light metal (magnesium) hidden in the magnesia earth. In slightly earlier experiments he had been able to decompose potash and soda electrically and in this way to isolate two other metals, potassium and sodium also known as kalium and natrium.

Davy had for long been greatly interested in the relationship between words and things. In a notebook of the late 1790s he wrote, 'Science and knowledge is the association of a number of ideas, with some idea or term capable of recalling them to the mind in a certain order'.[89] Now, reporting on his important electrolytic discovery, he suggested names that were theory-neutral and based only on empirical facts:

It will be proper to adopt the terminology which, by common consent, has been applied to other newly discovered metals, and which, though originally Latin, is now naturalized in our language. Potassium and Sodium are the names by which I have ventured to call the two new substances: and whatever changes of theory, with regard to the composition of bodies, may hereafter take place, these terms can

---

[88] Weeks and Leicester (1968), pp. 110–30. Flood (1960), p. 143. The *semaphore* invented in the 1790s was an optical telegraph system, a bearer (phore) of signals or signs (sema). Earlier mentioned words with *-phore* as suffix are electrophore (electrophorus) in Section 2.1 and odoriphore and osmophore in Section 3.1.

[89] Cited in Batchen (1993).

scarcely express an error; for they may be considered as implying simply the metals produced from potash and soda.[90]

With regard to the name of the new metal obtained from the magnesia earth, Davy wrote:

These new substances will demand names; and on the same principles as I have named the bases of the fixed alkalies, potassium and sodium, I shall venture to denominate the metals from the alkaline earths barium, strontium, calcium, and magnium; the last of these words is undoubtedly objectionable, but magnesium has been already applied to metallic manganese, and would consequently have been an equivocal term.[91]

As Davy was aware, in a classification from 1775 the prominent Swedish chemist Torbern Bergman, when reporting on Gahn's discovery, had used different word-endings to distinguish between what he confusingly called *magnesium* (our manganese, Mn) and *magnesia* (our magnesium oxide, MgO). Despite Davy's proposal and reasonable argument, the name magnesium persisted for the light metal, although 'magnium' is still used in some languages in Scandinavia and Eastern Europe.

Neither manganese nor magnesium is a magnetic metal, so why are they named as if they were? The three terms—manganese, magnesium, and magnetism—are etymologically related and the reason is found in the historical development of the terms, starting with the mineral pyrolusite (manganese dioxide, MnO_2), which originally was confused with lodestone. Flood's glossary of scientific terms from 1960 offers a summary account of the complex history:

In the Middle Ages lodestone was called *magnes* (masculine) and pyrolusite *magnesia* (feminine). Further confusion resulted in the eighteenth century when another mineral (now called magnesium carbonate) was discovered and, for some strange reason, believed to be related to magnesia. It now became necessary to distinguish the two kinds of magnesia by calling pyrolusite *magnesia nigra* (black) and magnesium carbonate *magnesia alba* (white). From this tangle we have obtained not only the word *magnet* but also the names of two chemical elements: *magnesium* (directly from *magnesia*) and *manganese* (a corruption of *magnesia*).[92]

In the English translation of Lavoisier's table of 1789 there appeared another earth called 'argill' which the author suggested might be a metallic element in alum saturated with oxygen. It was not a term coined by Lavoisier, for *argille* was well known in France, and in 1782 Bergman included *argilla* among the simple earth substances. The words were derived from *argillos* (ἀργιλλος), meaning clay in ancient Greek. In

[90] Cited in Wothers (2019), p. 147.
[91] Davy (1808), p. 347. Weeks and Leicester (1968), pp. 163–9 (manganese) and pp. 495–502 (magnesium).
[92] Flood (1960), p. 104. Wothers (2019), pp. 159–70.

another classification scheme from the same year, Morveau suggested *alumine* as a better term. In Lavoisier's original table from 1789 he preferred this term and only referred to *argille* as an older name.

While Davy failed to isolate the metal by means of electrolysis, at least his efforts resulted in a name. He first called it 'alumium' and later, in his *Chemical Philosophy* of 1812, he modified it to 'aluminum', which is still the standard American practice. In Britain and most other countries the alternative 'aluminium' was quickly adopted. An anonymous reviewer of Davy's book referred in 1812 to 'aluminium, for so we shall take the liberty of writing the word, in preference to aluminum which has a less classical sound'.[93] The *-ium* suffix conformed to the spelling of elements like sodium and potassium, but other metals known at the time ended in *-um*, such as was the case with platinum and molybdenum. Aluminium may first have appeared in a report on one of Davy's unsuccessful experiments in which it was said that 'this experiment is not wholly decisive as to the existence of what might be called *aluminium* and *glucinum*'.[94] Chemists, industrialists, and the general public have for long lived with different spellings of the element, and it changed nothing when IUPAC in 1990 decided that the spelling aluminium should be officially adopted.

As there is and have been two different spellings of element 13, so it is the case with its neighbour element 14 in the periodic table. Lavoisier assumed the element to be present in the earth or ground stone called *silex* (*silice* in the French version), and in 1808, in the same paper in which he suggested 'alumium', Davy coined the name 'silicium' for it: 'Had I been so fortunate to have ... procured the metallic substances I was in search of, I should have proposed for them the names of silicium, alumium, zirconium, and glucium.' Davy's name was revised nine years later by the Scottish chemist Thomas Thomson, who argued: 'The base of silica has been usually considered as a metal, and called *silicium*. But there is not the smallest evidence for its metallic nature, and as it bears a close resemblance to boron and carbon, it is better to class it along with these bodies, and to give it the name *silicon*'.[95] The element was finally isolated by Berzelius in 1824 who kept to 'silicium' despite recognizing Thomson's point. But the Swedish chemist suggested that Davy's name was nonetheless justified since 'there is no sharp borderline between metalloids and metals'.[96] Thomson's 'silicon' is the accepted name in English, whereas Davy's 'silicium' is used in several other languages such as German, French, Dutch, and Danish.

Although Berzelius did not coin the term *metalloid*, he was the first to use it widely for the varied class of non-metallic elements.[97] He divided the non-metals into three groups of which one (which he called the 'real metalloids') comprised elements such as phosphorus, carbon, boron, and silicon. In part because of Berzelius' authority, the usage of 'metalloids' as more or less synonymous with 'non-metals' remained popular well into the twentieth century. However, as pointed out by several chemists, Mendeleev among them, metalloids might better be categorized as

---

[93] *The Quarterly Review*, 1812, quoted in Williams (1993).
[94] *Critical Review: Or, Annals of Literature* **22** (January 1811), p. 9.
[95] Thomson (1817), p. 252. See also Whewell (1847), vol. 2, p. 562, who proposed to rename phosphorus *phosphur* in analogy with sulphur.
[96] Trofast (2016), p. 86.
[97] Goldsmith (1982).

intermediate elements having both metallic and non-metallic properties. In 1970, IUPAC recommended: 'Because of the inconsistent uses in different languages of the word "metalloid", its use should be abandoned. Elements should be classified as metals, semi-metals and non-metals.'[98] Nonetheless, metalloid continued to be used and in fact increasingly so, whereas semi-metal was not nearly as popular. In the period 1970–2010 the first term appeared in English literature with a frequency approximately ten times higher than that of the second term (*OED*; Google Ngram).

The term 'glucinum' or 'glucinium' relates to an earth called 'glucine' identified in the gemstone beryl in 1798 by the French chemist Nicholas Louis Vauquelin, who called it by that name because of the sweet taste of its salts. Three years later the German mineralogist Martin Heinrich Klaproth criticized the name on the grounds that there were other mineral earths with a sweet taste. He proposed *beryllina* as an alternative. The words 'glucos' or 'glycos', based on the Greek γλυκύς for sweet wine, can be found in numerous chemical compounds such as glucose, glycerol, and glycine. As mentioned, Davy tried in vain to isolate the metallic element, which at the time was also known as glucium and later more commonly as glucinium. An alternative name and the one accepted today is *beryllium*, which evidently refers to beryl.

The two names and their chemical symbols, glucinium (Gl) and beryllium (Be), coexisted for more than a century. In the 1820s, Davy's glucium was the preferred term in England and France, whereas Berzelius in Sweden insisted on beryllium and also advocated the terms *natrium* and *kalium* for sodium and potassium, respectively. According to a linguist, the origin of 'natrium' is a Hebrew name *neher* which designated soda (sodium carbonate, Na_2CO_3), and the common name sodium 'stems from the Arabic *sūda* "splitting headache", for soda water was often used as a remedy in those ancient days'.[99] When Friedrich Wöhler and Antoine Bussy independently succeeded in isolating the metal beryllium in 1828, they both announced to have discovered glucinium. Only in 1921 did the International Committee on Atomic Weights decide to abandon glucinium and replace it with the present word beryllium, element number 4 in the periodic table.

As far as aluminium is concerned, the new element was known to exist and had a name, but it had not yet been discovered in its elemental form. That only happened in early 1825, when Hans Christian Ørsted in Copenhagen reported how he had isolated the metal by means of a new chemical process involving aluminium chloride $AlCl_3$ as an intermediary product. Ørsted did not refer to either aluminium or aluminum but to 'argillium', a version of the word employed by Morveau and Lavoisier for the corresponding earth or oxide. For the Danish audience he invented the strange name 'leerær', which was never used outside Denmark and only rarely within the country.

With the steady advance of chemical knowledge during the nineteenth century it became accepted that the atomic weight was the defining characteristic of a chemical element. Mendeleev's celebrated periodic system from 1869 was a systematic ordering of all known elements according to their atomic weights. Incidentally, as number 4 in his original table he placed beryllium and not glucinium. There were several vacant places in the early system and Mendeleev famously predicted that these corresponded

---

[98] Jensen (1971), p. 11.
[99] Ellis (1953).

to real if as yet unknown elements. Thus, for the missing element of atomic weight 44 he successfully predicted what he called 'eka-boron' (scandium) and others of the unknown elements were 'eka-aluminium' (gallium) and 'eka-silicon' (germanium). For the symbols he chose Eb, Ea, and Es, respectively.

Why 'eka'? Mendeleev briefly explained that 'the prefix *eka* comes from the Sanskrit word for *one*'.[100] Words in Sanskrit are uncommon in science, but Mendeleev knew about the language and its grammar, which directly inspired his nomenclature and even his thinking about the periodic system.[101] Apart from the prefix *eka-* he also suggested using the prefixes *dvi-* and *tri-*, Sanskrit for two and three. The terminology was meant to be provisional only—when germanium was discovered, eka-silicon disappeared—but for some of the elements it persisted well into the twentieth century. During the 1920s chemists spoke of eka-tantalum ($Z = 91$) and eka-cesium ($Z = 87$), the two elements eventually known as protactinium and francium, respectively.

The recognition and naming of new elements were not formalized at the time when Mendeleev introduced his system and it was merely an unwritten rule that the discoverer had the right to name a new element. With the International Committee on Atomic Weights established in 1899 a more formal system was introduced, but the committee only became an effective instrument after the end of World War I. In the meantime, the definition of an element had changed or was about to change, namely in terms of the *atomic number* and not the older and well-established atomic weight.

The study of radioactive decay series resulted in much confusion with regard to the chemical nature of the decay products and how to place them in the periodic system. What at a first glance appeared to be a host of new 'radioelements' were given provisional names and symbols which served a practical purpose but eventually disappeared and are no longer to be found in the scientific literature. During the first two decades of the new century, chemists and physicists spoke of, for example, ionium, mesothorium, radium C, thorium B, radioactinium, radium emanation, and uranium X. The abbreviated term 'radioelement' for a radioactive element first appeared in English language about 1903, as either 'radio element' or the hyphenated 'radio-element'. The last form was used throughout in Frederick Soddy's monograph *Radio-Activity* from 1904.

The problem of the radioelements led to the recognition that some of the substances identified by their radioactive properties were chemically inseparable from known elements and yet they were not identical to them as they had different atomic weights. A chemical element might not consist of identical atoms, such as traditionally believed since the days of Dalton, but generally be a mixture of lighter and heavier atoms. By late 1913 such atoms were called *isotopes*, a name introduced by the British radiochemist and later Nobel laureate Frederick Soddy in a paper in *Nature* of 4 December 1913:

> The same algebraic sum of the positive and negative charges in the nucleus, when the arithmetical sum is different, gives what I call 'isotopes' or 'isotopic elements', because they occupy the same place in the periodic table. They are chemically

[100] Gordin (2004), p. 36.
[101] Ghosh and Kiparsky (2019).

identical, and save only as regards the relatively few physical properties which depend upon atomic mass directly, physically identical also.[102]

Although Soddy is rightly credited for having introduced isotope in the scientific language, he did not coin the word, which was suggested by Margaret Todd at a dinner party in Glasgow with Soddy and other scientists.[103] Todd, who was a medically trained novelist and proficient in Greek, listened to the scientists' conversation and proposed the term isotope meaning 'same place' (in the periodic table). Soddy never mentioned her contribution to the origin of the new term.

While the noun isotope was a genuine neologism, the adjective 'isotopic' had been introduced nine years earlier by two organic chemists at the University of Leeds, albeit with a meaning quite different from Soddy's. In a report on chlorine and bromine derivatives of the organic compound toluene $C_6H_5CH_3$, Julius Cohen and James Miller discussed what they called 'isotopic dichloro-, chlorobromo-, and dibromo-toluenes', explaining in a footnote: 'We propose, in future, to employ the word "iso-topic" (ἴσος, equal; τόπος, place) in place of the rather awkward expression "similarly substituted".'[104] In all likelihood, Soddy was unacquainted with the word introduced by Cohen and Miller which seems to have been a protologism not used by other organic chemists either then or later.

A little before Soddy's announcement of isotope, the Polish chemist Kasimir Fajans suggested his own name for essentially the same concept but without basing it on the Rutherford–Bohr nuclear atomic model. He called a group of chemically identical elements a *pleiade*, but this term was only used by him and a few other scientists.

'Pleiades' in plural is a well-known term in astronomy, where it refers to a star cluster in the constellation Taurus with a name derived from Greek mythology. Whereas astronomers do not speak of 'a pleiade' (or pleiad), the singular form can and has been used figuratively in the sense of a brilliant group of persons, e.g. 'a pleiad of writers' (*OED*). In any case, whereas Fajan's pleiade was a fiasco, Soddy's isotope and derived words such as isotopy and isotopic were hugely successful. Web of Science gives a total of 351,900 results for isotope, 156,010 for isotopic, and 1,660 for isotopy.

Although as a rule the elements consist of several non-radioactive isotopes, some of them, such as sodium, have only a single stable isotope. These elements are called 'monoisotopic'. Thus, Na-23 is not *an* isotope of sodium, but *the* isotope of sodium. Note that one can speak of isotope in singular only if in conjunction with something else and not as an isolated term. Uranium-235 is an isotope *of* the element uranium, but 'uranium-235 is an isotope' alone is strictly speaking a meaningless sentence as long as the original definition is retained. However, as Soddy pointed out in a paper of

---

[102] Soddy (1913). It was believed at the time that the atomic nucleus consisted of protons and electrons. The 'algebraic sum' takes into account the charge signs of the particles and corresponds to the atomic number; the 'arithmetical sum' disregards the signs and is thus approximately the same as the atomic weight on the H = 1 scale.

[103] Hudson (2019).

[104] Cohen and Miller (1904). Although the pre-Soddy appearance of isotopic is mentioned in *OED*, it has not been previously noticed by historians of science. Julius B. Cohen (1859–1935) was appointed professor of organic chemistry in 1904. The University of Leeds houses the Cohen Geochemistry Laboratory named after him.

1935, the *iso-* in isotope is not always taken seriously: 'Of recent years the word "isotope" has changed its meaning and is now used, for lack of another, to designate any atomic species. ... Such changes, though troublesome, are inevitable for *the language of science is a living rather than a dead one*.'[105]

The individual isotopes of an element are usually designated just with their mass numbers as in uranium-235 or carbon-14, but a few have been granted their own *-ium* names and symbols. The heavy hydrogen isotope discovered in 1932 is called *deuterium* (D = ^{2}H) after the Greek word *deuteros* (δεύτερος) for 'second', and the radioactive mass-3 isotope detected two years later is known as *tritium* (T = ^{3}H). The two names are recognized by IUPAC together with the rarely used *protium* for the much more abundant H = ^{1}H. As stated by IUPAC in 1957: 'All isotopes of an element should have the same name. For hydrogen the isotope names protium, deuterium and tritium, may be retained, but it is undesirable to assign isotopic names instead of numbers to other elements.'[106] Disregarding or unaware of the IUPAC recommendation, in his theory of nucleosynthesis in the early universe from about 1950 (Section 6.5), George Gamow coined the word *tralphium* for the light helium isotope ^{3}He consisting of two protons and one neutron.[107] However, Gamow's name was a protologism that other physicists and chemists either were unaware of or declined to adopt.

The inert radioactive gas radon (Z = 86) has a large number of isotopes and in 1923 IUPAC accepted special names and chemical symbols for three of them, namely radon (Rn), thoron (Tn), and actinon (An) for the isotopes ^{222}Rn, ^{220}Rn, and ^{219}Rn, respectively. The names were officially recognized until 1957, after which 'radon' was elevated from an isotope name to an element name. Despite 'thoron' alias ^{220}Rn or Rn-220 thus no longer being part of the IUPAC nomenclature, the isotope continues to be called so today. According to two chemists: 'Thoron is far easier to say than "radon-two-twenty", perhaps explaining why the annual count of scientific papers mentioning thoron has increased over twenty-fold since thoron was "disallowed" in 1957.'[108]

As a neutron is different from a neuron (Section 3.3), so the term isotropic is different from isotopic. In both cases the words differ only by a single letter, *t* in the first case and *r* in the second. The ending *-tropic* based on the Greek *trōpos* (turning, way, direction) preceded by *iso-* (equal, same) gives the word *isotropic*, which generally means independence of direction in space. Whereas the adjective isotopic corresponds to the noun isotope, there is no 'isotrope' noun form. In mineralogy and materials science an isotropic substance is one with the same physical properties in all directions. According to the generally accepted 'cosmological principle', on a large scale the universe is not only homogeneous but also isotropic: there is no preferred direction in cosmic space. Models of the universe which do not satisfy this criterion are said to be *anisotropic*, such as are substances with direction-dependent properties. William Thomson and other physicists in the Victorian era often preferred *æolotropic* (αἴόλος, changeful) as a synonym for anisotropy.

---

[105] Soddy (1935), emphasis added.

[106] Bassett (1960).

[107] Gamow (1952), p. 72. For the tralphium nucleus he suggested 'tralpha particle' in analogy with alpha particle.

[108] Thornton and Burdette (2013) with details on the naming history of radon.

The internationally renowned Danish linguist Otto Jespersen reflected in a paper of 1929 on the names used in science. 'If you look through a list of chemical elements you will find a curious jumble of words of different kinds', he observed. Moreover: 'The latest fashion is to add the ending -*um* to the place where the element was discovered: this may have originated in a misapprehension of *gallium* as if from *Gallia* France, though it really came from a translation of the name of the discoverer *Lecoq* (1875); but place-names are found in *germanium*, *ytterbium* (from the Swedish town Ytterbo), *hafnium* (from the Latin name of Copenhagen).'[109] The last of the elements mentioned by Jespersen may serve as an example of a relatively new name controversy that occurred after X-ray spectroscopy had been introduced as a powerful technique to determine the atomic number of an element and thereby its position in the periodic table.[110]

In 1922 two French scientists, Georges Urbain and Alexandre Dauvillier, announced that they had discovered the hitherto unknown element 72 as a member of the group of rare earths. The proudly called it *celtium* in honour of the Celtic peoples who in the Roman era lived in the region corresponding to present France. However, scientists at Niels Bohr's institute in Copenhagen were convinced that element 72 was homologous to zirconium and thus not a rare earth, and they consequently disbelieved the French claim. Following up on Bohr's theoretically based prediction, in late 1922 two of his collaborators, Dirk Coster and Georges de Hevesy, succeeded to obtain X-ray evidence in support of their alternative, which was officially reported as a discovery claim on 20 January 1923: 'For the new element we propose the name Hafnium (Hafniae = Copenhagen).'[111] The Copenhageners had originally agreed to call the element 'danium', but due to some misunderstandings in their communication with *Nature*, this name never appeared in print.

By choosing a name for element 72, Hevesy and Coster challenged the 'celtium' of Urbain and Dauvillier, who maintained their priority and right to name the element. Apart from celtium and hafnium, for a short while also 'oceanium' claimed by a British chemist entered the dispute. The celtium–hafnium controversy was not simply about words but also and more so about the political and national issues implied by the words. According to the editor of *Chemical News*, 'We adhere to the original word celtium given to it by Urbain as a representative of the great French nation which was loyal to us throughout the war. We do not accept the name which was given to it by the Danes who only pocketed the spoil after the war.'[112] Although the scientific community outside France soon accepted hafnium, French scientists continued to cling to celtium and its symbol Ct for several years. During the 1920s IUPAC carefully avoided to take a side and instead accepted as an unprecedented compromise two names and two symbols, hafnium (Hf) and celtium (Ct). Even today one can find shadows from the controversy.

---

[109] Jespersen (1929). The village Ytterby (not Ytterbo) in the outskirts of Stockholm has given name to four elements: yttrium, ytterbium, terbium, and erbium. Jespersen's assertion that gallium was named after its discoverer (*coq* = gallus in Latin) is incorrect. There is little doubt that Lecoq de Boisbaudran named the element in honour of France.

[110] Kragh (1980).

[111] Coster and Hevesy (1923).

[112] Quoted in Kragh (1980), p. 294. Denmark stayed neutral during World War I.

Apart from glucinium–beryllium and celtium–hafnium, there are a few other cases in which two different names for the same element have been used internationally over a longer period of time. The high-density element with atomic number 74 (with density the same as gold, namely $19.3\,\mathrm{g/cm^3}$) is today officially called *tungsten* but in many countries it is still known as *wolfram*. Its chemical symbol is W, which refers to the latter name and not to the official one. Despite tungsten being a Swedish name literally meaning 'heavy stone', in Sweden and several other European countries it is called wolfram. This is largely due to Berzelius, who in his influential textbook of 1834 also commented on the proposal of renaming the element to 'scheelium': 'Though some chemists have suggested that it should be named scheelium in honour of Scheele, not only does this name fit poorly with the Swedish language, the immortality of our fellow countryman requires no such additional support; thus I have given precedent to the name wolfram instead.'[113]

The name of the element is 'wolframio' in Spanish, but 'tungsténio' in Portuguese. The discovery of the metal is credited to the Spanish brothers Fausto and Juan José Delhuyar, who in 1783 isolated it from a mineral called wolframite, an old German word which translates into English as 'wolf's froth'. The two brothers proposed 'wolframium' for the new metal. In the tables of elements prepared shortly later by Morveau and Lavoisier it appeared as 'tungstène', a name also adopted (as tungsten) in England and the United States. Only in 1949 did IUPAC address the question of element 74 and then to choose wolfram as the one and only correct name. However, four years later IUPAC was forced to retreat and allow also the alternative tungsten. This situation with two very different accepted names for the same element lasted until 2008, when IUPAC decided to get rid of wolfram and only keep tungsten for the metallic element.

Despite the change of name, the symbol for tungsten remained W, thus adding to the list of elements with symbols that somewhat confusingly do not correspond to their names. The list contains mostly ancient elements, such as Fe for iron (ferrum), Au for gold (aurum), Sn for tin (stannum), and Hg for mercury. The latter symbol stands for *hydrargyrum* (hydr-argyrum), a version of the Greek *hydrargyros* meaning 'water silver'. Antimony has the chemical symbol Sb for *stibium*. The etymology of antimony is uncertain, but according to a source from 1736 the name reflected the fact that the mineral containing the metal, the sulphide Sb_2S_3 called stibnite, rarely was found alone but usually with other minerals: 'Antimony is sometimes found in a particular Oar, but most commonly with other Metals; and hence its Name may have been derived. Antimony being the same as ἀντίμονον [antimonon], an Enemy to Solitude.'[114] The symbol Sb was due to Berzelius, who introduced it in his chemical nomenclature of 1814. A few of the more recent names belong to the same category, notably Na for sodium (natrium) and K for potassium (kalium).

---

[113] Quoted in Jensen (2008). A mineral with the composition $extCaWextO_4$ is called scheelite after Scheele.

[114] Wothers (2019), pp. 37–45. Weeks and Leicester (1968), pp. 95–103. In old English 'oar' was sometimes used as a variant of 'ore' (*OED*).

5.5   Transuranic chemical elements

The twenty-six currently known chemical elements at the end of the periodic table are all artificial, that is, manufactured by means of complex nuclear reactions in the laboratory. Uranium is the heaviest of the elements occurring in nature and for this reason the still heavier radioactive elements are called 'transuranic'. The name was possibly coined by the German physicist and engineer Richard Swinne, who in 1926, many years before the first of these elements were discovered, introduced the noun *Transurane* for elements heavier than uranium.[115]

Except for elements 117 and 118 (with terminations *-ine* and *-on*, respectively), the officially accepted names of the transuranic elements end with the suffix *-ium* in agreement with the nomenclature for metals in the periodic system. Elements 93 and 94, neptunium and plutonium, are the only ones with names derived from classical languages. Ten of the transuranic elements carry names related to a locality (a country, city, or research institution), which is also in agreement with previous naming practice. They are toponyms. What is new is that fourteen of the elements are eponyms named after scientists, something not or almost not to be found in the earlier collection of names. Moreover, a few of the scientists honoured in this way had no share whatsoever in the discovery of the new elements, Copernicus (after whom element 112 is named) being the most striking example.

Elements 93–6 were identified during World War II primarily by a group of Californian physicists and chemists led by Glenn Seaborg, Albert Ghiorso, and Edwin McMillan. Element 93 was introduced without much fanfare in *Physical Review* on 27 May 1940, before the United States had entered the war. The two discoverers, McMillan and Philip Abelson, only suggested the name 'neptunium' in a classified memorandum of March 1942. It was a natural choice given that element 92, uranium, related to the planet Uranus named after the Greek god of the heavens. Uranus was discovered in 1781 and Neptune in 1846 (Section 6.2). Apart from uranium, neptunium, and the ancient mercury, several other elements have been named after celestial bodies. Tellurium (1783) and selenium (1817) are named after the Earth (*tellus*, Latin) and the moon (*selene*, Greek), respectively. Two elements, palladium (1803) and cerium (1804), are named after asteroids, namely Pallas (1802) and Ceres (1801). Finally, helium (1895) is named after the Sun (*helios*, Greek). Thus, the names of these elements are toponyms as well.

Because of the war *plutonium* appeared in a scientific publication only in 1948, but as early as 18 August 1945—just eight days after the destruction of Nagasaki—*Science News Letter* referred to the 'plutonium bomb'.[116] This name too was in the astronomy-element tradition as the planet Pluto identified in 1930 referred to another Greek god, the ruler of the underworld. Of course, when the International Astronomical Union in 2006 decided that Pluto was not a planet after all (but a dwarf planet, see Section 6.2), no one thought of changing the name of the element. Incidentally, it was not the first

---

[115] Much of this section relies on material in Kragh (2018b), where further references can be found.

[116] 'Idea for plutonium bomb credited to Dr. Lawrence', *Science News Letter*, 18 August 1945, p. 102, quoting a memorandum of May 1941 in which Ernest Lawrence suggested that element 94 might be used for a 'super bomb'.

time that the names neptunium and plutonium appeared in the annals of chemistry, albeit in one case for a new metal that turned out to be spurious and in the other as a renaming of an element. The nonexistent neptunium element was proposed in 1877 and in 1817 an English mineralogist insisted on plutonium for what Davy and other chemists called barium.[117]

Seaborg recalled the discussions in his research group concerning the name of element 94, for which they initially considered 'extremium' or 'ultimium' because they thought, in retrospect somewhat naively, that they had reached the upper limit of the periodic system. However, why not name the element in analogy with neptunium and therefore focus on Pluto? Euphonious considerations played a role in their final decision:

> But, and this is a little known story, it seemed to us that one way of using the base Pluto was to name the element 'plutium'. We debated the question of whether the name should be 'plutium' or 'plutonium', the sound of which we liked much better. We finally decided to take the name that sounded better. I think we made a wise choice, and I believe it is also etymologically correct.[118]

Elements 95 and 96 proved frustratingly difficult to separate, for which reason they were at first informally and only in jest referred to as 'pandemonium' and 'delirium' (the first term was the capital of Hell in Milton's *Paradise Lost*). In late 1945 Seaborg appeared on a national radio broadcast where he read a long list of proposed names for the two elements, many of which were rather bizarre (such as 'moonoium' and 'rooseveltium').

The more seriously meant names americium and curium were proposed in March 1946 and officially recognized, together with neptunium and plutonium, by IUPAC three years later. The names for elements 95 and 96 were in part chosen for the reason that they corresponded to the homologous lanthanides called europium and gadolinium: 'Element 95 was given the name "americium" (symbol Am) after the Americas, in analogy to the naming of its rare-earth homologue europium after Europe. For element 96 we suggested the name "curium" (symbol Cm) after Pierre and Marie Curie, in analogy to the naming of its homologue gadolinium after Johann Gadolin.'[119] Although americium was proposed as a toponym, it is also, if only indirectly, an eponym given that America is named after the Italian merchant and navigator Amerigo Vespucci.

While curium is definitely an eponym, if in this case relating to a couple and not to an individual, it is doubtful whether gadolinium belongs to the same category such as Seaborg apparently believed.[120] The Finnish-Swedish chemist Johan Gadolin gave name to the mineral gadolinite discovered in 1787, whereas the name of element 64 discovered in 1886 by Paul-Émile Lecoq le Boisbaudran most likely referred to the mineral and not to the person. Likewise, the name samarium for element 62

[117] Fontani, Costa, and Orna (2015), pp. 48–50.
[118] Seaborg (1994).
[119] Seaborg (1994). Fontani, Costa, and Orna (2015), pp. 376–8.
[120] Rayner-Canham and Zheng (2008).

refs to the mineral samarskite and only indirectly to the Russian mining engineer Vassilii Samarsky-Bykhovets after whom the mineral is named (nor is samarium a toponym referring to the Samaria of the Bible). Thus, the tradition of using eponyms for chemical elements only started in earnest with curium.

One of the difficulties of identifying heavy transuranic elements was that they had very short half-lives. They might have been created in the laboratory but disappeared before the scientists succeeded in detecting them. Or perhaps they had once existed naturally but only for a brief period of time in the earliest phase of the history of the universe. It may have been such speculations that gave rise to the metaphoric expression 'dinosaur element' in the media in the early 1960s. The *New York Times* reported in April 1961: 'Because of the very short half-life, element 103, according to one theory, is a "dinosaur" element which was formed at the birth of the universe but which decayed out of existence in a few weeks.' Another source defined a dinosaur element as 'an element that came into being when the universe originated and which deteriorated almost immediately'.[121] Given that the dinosaurs roamed on Earth for about 165 million years and were preceded by numerous other species, the metaphor was decidedly poor.

With the discovery of more transuranic and 'superheavy' elements (atomic number $Z > 103$), and with the simultaneous tough competition between American and Russian research groups, the naming of new elements resulted in a series of controversies. The accepted name of an element may directly or indirectly refer to the discoverer and thus suggest which scientists are to be credited with the discovery. Consider a scientist X who proposes the name A for a new element he claims to have found, while scientist Y independently finds what he names B and believes is the same element. In this case a dispute about the name reflects a controversy about discovery.

The phenomenon of priority questions and terminological preferences going hand in hand is nicely illustrated by element 106, the first element ever to be named after a living scientist. In short, in 1994 a group of Californian scientists confirmed earlier experiments indicating the synthesis of atomic nuclei belonging to element 106. They wanted to name the new element *seaborgium* in honour of the 82-year-old veteran in transuranic research Glenn Seaborg. The suggestion to name an element after Seaborg seems to be much older, though, for 'seaborgium' appeared in print as early as 1953: 'More elements will be produced by the irradiative techniques of the cyclotron; elements no. 99 to no. 103 have already been predicted. The name seaborgium has already been suggested for the next element, in recognition of Seaborg's achievements.'[122] However, the nomenclature commission under IUPAC was in favour of 'rutherfordium'—obviously from Ernest Rutherford—and resisted the name suggested by the Americans. IUPAC insisted that although discoverers had a right to *suggest* a name, it was IUPAC alone which made the decision.

Besides, could an element be named after a living person? IUPAC's answer was no, one of the arguments being that there were no precedents for such names. On the other hand, supporters of seaborgium objected that there were in fact a few

---

[121] Algeo (1991), p. 160. The term 'dinosaur element' did not appear in scientific publications and also not in *OED*. Element 103 = lawrencium (Lr).

[122] Ellis (1953). The same suggestion appeared in the *New York Times* of 28 February 1954 (*OED*).

precedents of this kind such as element 99 called einsteinium. However, since this element and its suggested name was only announced in public on 20 June 1955, two months after Einstein's death, the argument was ineffective. It did not help that einsteinium had circulated informally while Einstein was still alive. IUPAC could have maintained its position, but after three years of sustained pressure it 'decided to modify its decision that the name of a living scientist should not be used as the basis for an element name'.[123] And so it happened that Seaborg could enjoy having an element officially named after him for the last two years of his life (since 1983 a minor planet also carried his name). As to Rutherford and rutherfordium, in 1997 IUPAC decided to assign the name to element 104. The new IUPAC rules implied that in principle the discoverer of a new element could propose to name it after himself (or herself), but this has never happened.

The still living Russian nuclear physicist Yuri Oganessian, born in 1933 and a highly respected veteran in the creation of superheavy elements, similarly enjoyed to have an element named after him and that for a longer time than Seaborg. Element 118 called *oganesson* was officially sanctioned by IUPAC in 2016. When the suffix of the name of this element is *-on* and not *-ium* ('oganessium') it is to underline that it belongs to group 18 of the periodic table, as do xenon, radon, and other inert gases. The only exception to the *-on* suffix is helium, the name of which has been retained for historical and practical reasons. It would seem too artificial to rename the light noble gas 'helion', although this was what a few British scientists proposed in the 1920s.[124] It is unknown what the upper limit to the periodic table is, but most chemists and physicists believe that there are elements yet to be discovered—that is, manufactured—beyond oganesson.

With regard to helium it is often stated that when Norman Lockyer in 1868 discovered an unidentified line in the solar spectrum, he named the substance responsible for it *helium*, the solar element. However, it took a long time before he referred to the name in print. Referring to his collaboration with the chemist Edward Frankland, he wrote in 1896: 'We had to do with an element which we could not get in our laboratories, and therefore I took upon myself the responsibility of coining the word helium, in the first instance for laboratory use.'[125] The name first appeared in public in 1871, namely in an address to the British Association in which William Thomson ascribed it to Frankland and Lockyer.

According to a IUPAC report of 2002, for reasons of 'linguistic consistency' all new elements should end in *-ium*, but the recommendation was changed in a later report of 2016, which stated

In keeping with tradition, newly discovered elements can be named after: > a mythological concept or character (including an astronomical object), > a mineral, or similar substance, > a place, or geographical region, > a property of the element, or > a scientist. ... The names of all elements should have an ending that reflects and maintains historical and chemical consistency. This would be, in general, '-ium' for

---

[123] IUPAC (1997).
[124] Lodge (1924), p. 31. Friend (1927). Jensen (2004).
[125] Lockyer (1896). For historical details, see Kragh (2009a).

elements belonging to Groups 1–16, including lanthanoids and actinoids, '-ine' for elements of Group 17, and '-on' for elements of Group 18.[126]

Thus, not only did element 118 become 'oganesson', element 117 also became 'tennessine' (after the state of Tennessee) in agreement with fluorine, chlorine, bromine, iodine, and astatine.

Another of the controversial transuranic elements was number 102 in the periodic system which in 1957 was announced by a team of scientists working at the Nobel Institute of Physics in Stockholm. As a name for the allegedly new element they suggested 'nobelium' in recognition of Alfred Nobel and his services to science. However, the discovery claim was premature as the Swedish experiments could not be reproduced by either American or Russian specialists. The problem of who had really discovered element 102, the Americans or the Russians, soon evolved into a major priority controversy.

Remarkably, although the Swedish claim was known to be wrong, IUPAC welcomed the name 'nobelium', which is still what the element is called. Nobelium had come into common usage and entered textbooks as well as periodic tables, and for these and other reasons the name of the Swedish philanthropist survived despite being originally based on erroneous data. For a while Russian and German scientists advocated 'joliotium' in honour of the French physicist and Nobel laureate Frédéric Joliot-Curie, but their attempt to replace 'nobelium' with 'joliotium' failed. The latter name was also suggested for elements 103 and 105, and for element 105 it was even recommended by IUPAC in 1994. However, three years later 'joliotium' disappeared for good when 'dubnium', a toponym for the Russian laboratory in Dubna outside Moscow, was finally adopted for the element.

To avoid unnecessary disputes about names for elements not yet firmly established, in 1978 IUPAC suggested a systematic if provisional nomenclature system based directly on the atomic number Z.[127] The new names and their corresponding symbols were primarily based on Latin numbers, such as 0 = nil, 1 = un, 2 = bi, etc. For example, the systematic name and symbol for $Z = 105$ would be unnilpentium (un-nil-pentium, Unp) and for $Z = 112$ ununbium (un-un-bium, Uub). In some cases, the names included the Greek prefixes *penta-* and *hexa-*, as in $Z = 106$ unnilhexium (un-nil-hexium, Unh). In contrast to the symbols of the known elements, the systematic symbols consisted of three letters. But although systematic and neutral, the unwieldly names were rarely used by the scientists who preferred to refer directly to the atomic numbers. After all, it was easier to say 'element 108' or just '108' than 'unniloctium'. A search in Google Scholar shows that in the period 1980–95 not a single research paper referred to unnilhexium, the systematic name for what later became seaborgium. More than 1,000 papers referred to 'element 106' and about 70 to 'seaborgium'.

Nor did the terminology please the linguists. One of them, the French specialist in onomastics (the study of names) Henri Diament, characterized the IUPAC system as 'a masterpiece of diplomacy but also of blandness and lack of imagination,

---

[126] Öhrström and Reedijk (2016).
[127] Chatt (1979).

as well as an etymological hodge-podge'. He concluded his critical review: 'What emerges even now is the following: Onomastically speaking, the names are (a) under-differentiated, (b) their symbols are also underdifferentiated, with duplication, even triplication of letters, (c) the names are unwieldly, and (d) the symbols expose scientists to ridicule.'[128] Diament illustrated his last point by referring to the hypothetical element 131, the symbol of which would be Utu. In America, he said, this 'will evoke a pun, "you too!", or else the ill-fated U-2 spy plane of the CIA'.

Elements 110–12 were discovered by a research group in Darmstadt, Germany, and recognized by IUPAC in the years from 2003 to 2009. Atomic nuclei belonging to element 113 were synthesized in 2004 by a Japanese research team and approved twelve years later. With the approval its name changed from ununtrium (Uut) to 'nihonium' derived from the country where it was discovered, 'Nihon' literally meaning 'the land of the rising Sun'. It was the first time that a chemical element was discovered in Asia and by scientists from this continent (indium, element 49, was discovered by a German chemist in 1864 and its name relates to India only indirectly[129]). An element named after Japan had been proposed about a century earlier, but when Matasaka Ogawa announced the discovery of 'nipponium' in 1908, he was less fortunate. It turned out to be spurious.[130]

The names for the elements discovered in Darmstadt referred to the German city and two of the celebrated figures in the history of science, Röntgen and Copernicus. As to 'darmstadtium' or the former ununnilium, the German group originally considered naming it 'wixhausium' after the suburb Wixhausen where the laboratory was located. Whatever can be said of these names, they are not euphonious. While the reference to Röntgen in 'roentgenium' is perhaps understandable—X-rays have played a crucial role in element identification—the eponymous 'copernicium' is more puzzling. Why refer to the famous sixteenth-century astronomer who had no interest in chemistry and no idea of what a chemical element is? According to a press release from the Darmstadt laboratory, the eponym was chosen for very general reasons:

> In honor of scientist and astronomer Nicolaus Copernicus (1473–1543), the discovering team around Professor Sigurd Hofmann suggested the name 'copernicium' with the element symbol 'Cp' for the new element 112, discovered at the GSI Helmholtzzentrum für Schwerionenforschung (Center for Heavy Ion Research) in Darmstadt. It was Copernicus who discovered that the Earth orbits the Sun, thus paving the way for our modern view of the world.[131]

Whereas 'copernicium' was accepted by IUPAC, the proposed symbol Cp was not. The reason was that Cp had earlier been used for the element 'cassiopeium' (now lutetium, $Z = 71$) and could also refer to an organic compound called cyclopentadiene. Consequently, the symbol was changed to Cn. Incidentally, in 2015 the

[128] Diament (1991).
[129] Gutman (1985).
[130] Fontani, Costa, and Orna (2015), pp. 219–23.
[131] Press release of 14 July 2009. Available at https://web.archive.org/web/20090718113516/https://www.gsi.de/portrait/Pressemeldungen/14072009_e.html

International Astronomical Union approved the name Copernicus for the star 55 Cancri A, one of the very few star eponyms recognized by the astronomical community.

As there are several named imaginary elements in the history of chemistry, so there are molecules and chemical compounds that were once discussed but turned out not to exist. One of the spurious molecules was the *antozone* which the German-Swiss chemist Christian Schönbein thought to have detected in the late 1850s.[132] One of the leading chemists of the century, in 1840 Schönbein had discovered a new colourless gas for which he coined the term *ozone* from the ancient Greek verb *oūzo* meaning 'to smell'. He believed that the molecule consisted of three oxygen atoms (O_3) but without suggesting a structural formula. Further work on ozone made him to suggest in 1858 that there were two species of the molecule, one negatively charged for which he kept the name ozone and another positively charged which he called *antozone*, an abbreviation of anti-ozone. Schönbein proposed the name to account for the formation of ordinary oxygen from ozone ($2O_3 \rightarrow 3O_2$), which he interpreted as ozone + antozone → oxygen or

$$O_2O^- + O_2O^+ \rightarrow 3O_2.$$

Compounds formed by the action of ozone he called *ozonides* and those derived from antozone *antozonides*. Schönbein was convinced that he had made an important discovery, such as he enthusiastically reported in a letter to Faraday:

> Oxigen [*sic*], as you well know, is my hero as well as my foe and being not only strong but inexhaustible in strategies and full of tricks, I was obliged to call up all my forces to lay hold of him and make the subtle Being my prisoner. Now to drop the metaphor, I will tell you, that I have been working very hard these many months to get the 'Antozone' or –O in its insulated state and I flatter myself to have succeeded in that undertaking, at least to a certain extent. ... Now, what do you say to the extraordinary fact, that the antipode of ozone has these many thousand years been ready formed and incarcerated, only waiting for somebody to recognize and let it loose out of its prison.[133]

Although a few chemists believed in antozone and studied its properties, in general Schönbein's discovery claim was met with resistance or indifference. In a paper of 1879, an American chemist summarized what by then was undisputed, namely that antozone does not exist: 'By far the most important fact in the long and perplexing history of Antozone, is the recent discovery that there is no Antozone.'[134]

---

[132] Rubin (2009).
[133] Schönbein to Faraday, 11 December 1860, in Faraday (2011). Available at https://epsilon.ac.uk/view/faraday/letters/Faraday3928.
[134] Cited in Rubin (2009).

6
Heavenly sciences

For a very long time astronomy was almost the same as planetary astronomy. As the names of the chemical elements each has a story to tell, so it is the case with the much smaller number of solar planets varying from six to nine through history. In addition to the currently recognized planets there are the asteroids discovered in the early nineteenth century. What these small planet-like objects were and how they should be named and categorized gave rise to considerable controversy. The exoplanets discovered nearly two centuries later also belong to this chapter. And so do the strange celestial bodies predicted and discovered in the twentieth century such as quasars, pulsars, and neutron stars. The strangest of all are the black holes which for long were hypothetical but now have entered the cosmic zoo as real and abundant objects. The non-academic but highly successful name 'black hole' dates from the early 1960s and not from 1967 as usually stated.

Section 6.3 is about the words related to the new and surprising astronomical discipline called either astrobiology or exobiology. The discipline is surprising because biological species are not known to exist elsewhere than on Earth. For this reason, astrobiology has been called a science without a subject. By looking at ideas of extraterrestrial life in the universe in a broad historical perspective one will encounter words and concepts that are no longer found in glossaries of astronomy or any other science. More recently the fascination of astrobiology has resulted in speculative attempts to create an artificial 'astrolanguage' which supposedly will be understandable to all intelligent beings in the universe.

The last two sections of this chapter deal with terms used in cosmology with an emphasis on the modern era when the universe as a whole became recognized as a worthy if exceptional field of scientific study. 'Cosmology' as a scientific term belongs essentially to the twentieth century. To describe in words what a closed universe is, or what it means that the universe is expanding, physicists and astronomers were forced to invent metaphors. Many of these metaphors are still with us and used not only for popular and educational contexts but also in scientific presentations. The most successful of the cosmological neologisms is undoubtedly the 'Big Bang' originally introduced in 1949 as a nickname for the origin of the universe in the far past.

6.1 From archaeoastronomy to black holes

With its origin in the Mesopotamian and Egyptian civilizations, and subsequently in the more advanced form known from ancient Greece, astronomy is arguably the oldest natural science. And yet its roots stretch much farther back in time, namely to the Neolithic ages where our distant forefathers studied the sky not for scientific reasons but for social, cultural, and religious reasons. This kind of prescientific astronomy is known as *archaeoastronomy*, an apt name that first appeared at about 1970 in the context of anthropological studies. *Archaios* (ἀρχαῖος) means 'ancient' in Greek and is known from *archaeology*, the scientific study of past peoples and cultures through their material remains. It is also known from the adjective *archaic* which often appear as a synonym of 'outdated' or 'antiquated'. An archaic word is one which was common in the past but rarely used in modern language. It is archaic to believe that the Earth is in the centre of the universe.

The word archaeoastronomy may have been suggested by the astronomer Gerald Hawkins, while another of the pioneers, the engineer Alexander Thom, preferred 'megalithic astronomy'. A megalith is a very large stone, *megas* (μέγας) meaning large or mighty in Greek and *lithos* (λίθος) meaning stone. One more word is 'ethnoastronomy' defined as 'a closely allied research field which merges astronomy, textual scholarship, ethnology, and the interpretation of ancient iconography'.[1] Although archaeoastronomy is a new branch of scholarship, suggestions that Stonehenge and other megalithic monuments served astronomical purposes, at least in part, go back to the eighteenth century. Today, the field of 'Archaeoastronomy and Astronomy in Culture' is recognized by IAU, the International Astronomical Union.

For a long time, *astrology* was an integrated and often inseparable part of astronomy, a close association which only ended in the mid-seventeenth century when astrology was declared nonscientific and finally separated off from astronomy. The two words mean basically the same, namely the learning or science about the stars or heavenly bodies generally. In English, 'astronomy' is earlier than 'astrology' and originally included the meanings now associated with both words.

The term *astroscopy* was used until the mid-eighteenth century to designate observations of the stars primarily for astrological purposes. The Greek *skopos* (σκοπός) means 'observer' or 'one who watches'. In a scientific context the suffix *-scope* typically refers to an instrument used for observing or showing something but not measuring it, for example microscope, spectroscope, thermoscope, gyroscope, and stethoscope. Helmholtz's celebrated invention in 1851 of what he first called an *Augenspiegel* (eye mirror) soon became known as an ophthalmoscope, *ophthalmos* meaning eye in Greek. For the art of using an instrument of this type, the suffix is *-scopy* (microscopy, spectroscopy, etc.). In contrast to the obsolete astroscopy, the term *astrometry* refers to precision measurements of the positions of the stars and other celestial bodies. Although astrometry has always been practised by astronomers, the term is of relatively recent origin as it may first have appeared in about 1810: 'Oryctometry [is] ... the art of measuring fossils... . No one has ever said that astronomy is astrometry or uranometry, although the heavens and the stars are measured.'[2]

The word 'uranology', corresponding to 'uranometry', is older and was used well into the nineteenth century, often as a synonym of astronomy. In Greek mythology Uranus was the personification of the sky and the heavens. According to Benjamin Martin's *Philosophical Grammar* of 1735, the term uranology signified 'a *Discourse* or *Treatise* of the *Heavens*, or *heavenly Regions*, and Bodies therein', which comes pretty close to the modern standard definition of astronomy. Martin further explained

---

[1] Baity (1973).
[2] Chenevix (1811). See Section 1.6 for 'oryctometry'.

that uranology comprised a number of branches such as heliography, selenography, planetography, and cometography.

Whether with roots in the astronomical or astrological tradition, many of the words relating to astronomy have entered common language in the form of colloquial phrases. For example, we may speak of an artist who 'rose to meteoric fame and at the zenith of his career was considered a shining star in the firmament of modern art'. An actor can be a 'movie star' whose performance in a movie is a 'stellar attraction' and earns her 'an astronomical sum of money'. She always stays at a 'five-starred hotel'. Strangely, the adjective 'astronomic(al)' can also refer to something which is very small or imperceptible. The chance that an event happens can be judged to be 'astronomically small', that is, close to zero. The *asterisk* (∗) used in academic writing and elsewhere is derived from the Greek word for star (*astēr*), which also appears in the family of flowering plants called aster. Moreover, the common word 'disaster' (dis-aster) refers in its origin to the stars, in this case to the astrological claim that certain constellations are signs of misfortune. The Latin prefix *dis-* means 'apart' but can also indicate negation or deprivation as in words such as disabled, distrust, and disgrace.

In the classical era of astronomy, two works shine above all others, namely Ptolemy's *Almagest* from about 150 AD and Copernicus' *De Revolutionibus* from 1543. The original title of Ptolemy's famous work was *Syntaxis Mathematica* (Mathematical Compilation) and in the Arab world it became known as *Megale Syntaxis* and later *al-majisti*, meaning 'the greatest'. In Medieval Latin it was rendered as *Almagestum*. After Ptolemy's work was translated from Greek into Latin by Gerard of Cremona in 1175, it exerted an enormous influence on medieval and Renaissance astronomers. Nicolaus Copernicus admired the *Almagest* but thought that he could improve upon Ptolemy's system by replacing the central Earth with the Sun and thus turning the Earth into a planet. He lyrically referred to the Sun with a number of metaphors:

> For who would place this lamp of a very beautiful temple in another or better position than this wherefrom it can illuminate everything at the same time? As a matter of fact, not unhappily do some call it the lantern; others, the mind and still others, the pilot of the world. [Hermes] Trismegistus calls it a 'visible god': Sophocles' Electra, 'that which gazes upon all things'. And so the Sun, as if resting on a kingly throne, governs the family of stars which wheel around.... The Earth moreover is fertilized by the Sun and conceives offspring every year.[3]

The title of Copernicus' great work of 1543 was *De Revolutionibus Orbium Coelestium Libri Sex* or in English 'Six Books on the Revolutions of the Heavenly Spheres'. The word 'revolution' in the title is from Latin, referring to something that moves repeatedly in circles (*volvo*, to roll or turn round). It should not be confounded with the current meaning of the term, namely a drastic and often violent change of the old political (or scientific or cultural) order into a new one. This meaning, so different from the original, was scarcely known at the time and became common only about two centuries later. Nonetheless and despite Copernicus' intention,

---

[3] Copernicus (1995), pp. 24–6. For the mythical Trismegistus, see Section 5.2.

De Revolutionibus is frequently considered the book which ushered in the so-called scientific revolution. Bernard Cohen, a pioneer historian of science, writes:

> Revolution means to return again, to go through a cyclical succession, as in the seasons of the year, or to ebb and flow, as in the motion of the tides.... It is the historian's task to find out how and when an innocent scientific term that implies permanence and recurrence became transformed into an expression for radical change in political and socioeconomic affairs, and then to discover the way in which this altered concept was applied to science itself.[4]

Copernicus, in some ways a conservative, did not consider his theory to be a revolutionary break with the past, and yet the term 'Copernican revolution' has long been adopted for the new world view that took roots in the early seventeenth century. The same term is used in the more general meaning of something radically new, the addition of Copernicus' name being merely ornamental. 'Democracy needs a Copernican revolution to survive climate change', one home page states.[5] Philosophers and historian of ideas have long associated the so-called Copernican revolution with Immanuel Kant, who supposedly referred to his own work with this phrase. Thus, in his *Conjectures and Refutations* published in 1962, Karl Popper wrote that 'Kant's proposed solution ... consisted of what he proudly called his "Copernican Revolution" of the problem of knowledge.'[6] However, as Cohen has convincingly argued, Kant did not, in fact, speak of his ideas as constituting a 'Copernican revolution'.

The 'clockwork universe' is a powerful and much-used metaphor that supposedly encapsulates the essence of the dramatic transition from the animistic medieval conception of the universe to a new one governed by the stern laws of mechanics. Indeed, in 1609 Johannes Kepler triumphantly wrote in his *Astronomia Nova* that now 'the celestial machine is to be likened not to a divine organism but rather to a clockwork.'[7] The clockwork metaphor was popular in the seventeenth century when it was employed by Leibniz and many others, but the great Newton—with whom it is often associated—never mentioned the image of the clock. In fact, he much disliked it. His universe was mechanical but without running like a perfect clock and not free of vital principles or spirits.

Perhaps surprisingly, the clock metaphor goes back to the Middle Ages, where it can be found in the writings of the scholastic philosopher Nicole Oresme. In his *Le Livre du Ciel et du Monde* from 1377, Oresme wrote that God had created the heavens 'much like that of man making a clock and letting it run and continue its motion by itself'.[8] Much later, in a work of 1752, the great French mathematician and physicist Jean le Rond d'Alembert used the clock metaphor to emphasize that even the best scientific theories are uncertain and hypothetical. Nature was indeed a 'vast machine',

---

[4] Cohen (1985), p. 6.
[5] https://www.thenewfederalist.eu/democracy-needs-a-copernican-revolution-to-survive-climate-change?lang=fr
[6] Popper (2010), p. 126. Cohen (1985), pp. 237–53.
[7] Cited in Holton (1988), p. 56.
[8] Cited in Grant (2007), p. 284.

but one we can only know about 'through a veil which hides the workings of its more delicate parts from our view'.[9]

Astronomy, the quintessential observational science, has for most of its history been conceived as closely connected with mathematics. Indeed, at the time of Kepler it was a mathematical science with astronomy professors sometimes occupying chairs in 'higher mathematics' in contrast to the 'lower mathematics' taught by the mathematicians. According to Harris' *Lexicon Technicum* published in 1704:

> ASTRONOMY, is a Mathematical Science, teaching the Knowledge of the Stars or Heavenly Bodies, and their Magnitudes, Distances, Eclipses, Order, and Motions: By some 'tis taken in so large a Sense, as to contain also the Doctrine of the Mundane System, the Laws of the Planetary Motions, &c. which others reckon as a part of Physicks or Natural Philosophy.

The links to mathematics were further strengthened with the 'celestial mechanics' developed on the basis of Newton's law of gravitation. The term for this kind of mathematical astronomy may have been coined by one of its greatest experts, the French physicist Pierre-Simon Laplace, who used it in the title of his classic *Traité de Mécanique Céleste* published in five volumes between 1799 and 1825.

Laplace was also the creator of what became generally known as *Laplace's demon*, an omniscient being who could precisely account for everything in the past and future of the universe on the basis of the deterministic laws of mechanics. However, when Laplace introduced his thought experiment in 1814 he did not refer to 'demon' but to 'intelligence':

> Given for one instant an intelligence which could comprehend all the forces by which nature is animated and the respective situation of the beings who compose it—an intelligence sufficiently vast to submit these data to analysis—it would embrace in the same formula the movements of the greatest bodies of the universe and those of the lightest atom; for it, nothing would be uncertain and the future, as the past, would be present to its eyes.[10]

The demonization (so to speak) of his intelligence only began in the 1930s. About fifty years after Laplace had introduced his so-called demon, Maxwell introduced another demon in a discussion of reversibility in thermodynamics. But as mentioned in Section 4.1, Maxwell did not refer to 'demon' any more than Laplace did.

Still in the early part of the nineteenth century the consensus view among astronomers was that their science, whether primarily mathematical or observational, was distinctly different from physics. The German astronomer Wilhelm Bessel is known not only for his discovery of a stellar parallax in 1828 but also for the differential equation and class of functions named after him. As he emphasized in a popular

[9] Cited in Pulaczewska (1999), p. 167.
[10] Laplace (1902), p. 4. Available at https://www.gutenberg.org/ebooks/58881. See also Canales (2020), pp. 29–48.

lecture of 1832, the business of astronomy was solely about precise measurements and calculations of celestial bodies:

> Everything else that can be learned about the heavenly bodies, for example their appearance and the constitution of their surfaces, is certainly not unworthy of attention; but is not properly of astronomical interest. Whether the lunar mountains have this or that shape is no more interesting for the astronomer than is such knowledge of terrestrial mountains for the non-astronomer.[11]

Nonetheless, at about that time the first germs of what came to be known as *astrophysics* saw the light of day. One of these germs was the announcement in 1842 by the Austrian physicist Christian Doppler of the effect named after him, namely a shift in the wavelength of light $\Delta\lambda$ from a source moving with a radial velocity v relative to the observer: $\Delta\lambda/\lambda = v/c$ with c denoting the speed of light. The 'Doppler effect' came to play a crucial role in the development of astrophysics and cosmology, but initially it was controversial and in France known as the 'Doppler–Fizeau effect' because the French physicist Hippolyte Fizeau had independently discovered it. The double eponym was adopted in France, where it is still occasionally used, but not elsewhere in the world.

Doppler and his contemporaries did not refer to astrophysics, a term which in its modern meaning had not yet been coined. Another and apparently equivalent term, namely 'physical astronomy', was used in the late eighteenth century by Laplace and others, but in the sense of celestial mechanics. 'Astrophysics' made its entrance in a treatise on photometry of 1865 written by the German astronomer Friedrich Zöllner, who was also the first to occupy a chair in astrophysics established at the University of Leipzig in 1872. The word first appeared in English in 1870, in a review in *Nature* of a book by Zöllner on the physical constitution of the Sun. Two decades later astrophysics had grown into a large and exciting research field. The term first entered the name of a journal in 1893, the *Astronomy and Astro-Physics* published by the Goodsell Observatory in Minnesota. When sold to the University of Chicago two years later it was renamed the *Astrophysical Journal*. This journal, subtitled 'An International Review of Spectroscopy and Astronomical Physics' and edited by George Hale and James Keeler, quickly became the world's preeminent astrophysics journal.

From the perspective of the twenty-first century it may be tempting to identify modern astronomy with astrophysics, as if the two terms are synonymous. Perhaps astronomy has largely become absorbed by astrophysics today, but if so it is a recent phenomenon and one which definitely disagrees with the broader historical development given that physics is after all a latecomer to the astronomical sciences.

Twentieth-century astrophysics has resulted in the discovery of several new and highly surprising objects in the universe. The *supernovae* have been mentioned in Section 2.5 and the *quasars* in Section 4.4. The suffix *-ar* in quasar refers to 'stellar' and the same is the case with the *pulsars* discovered a few years later. The latter name, a contraction of 'pulsating star', was possibly introduced by Anthony Hewish, discoverer or co-discoverer of pulsars. However, it seems that the coining of the new

---

[11] Quoted in Hufbauer (1991), p. 43.

acronym was originally suggested by Anthony Michaelis, a science correspondent of *The Daily Telegraph* and then accepted by Hewish in an interview. According to Michaelis' article of 5 March 1968: 'An entirely novel kind of star ... came to light on Aug. 6 last year and, at first, ... the star was referred to by astronomers as LGM (Little Green Men). Now it is thought to be a novel type between a white dwarf and a neutron [star]. The name Pulsar (Pulsating Star) is likely to be given to it.'[12]

During 1968 there appeared about two dozen papers with pulsar in their title. In contrast to quasar, no one objected to the name. The names of both kinds of these astronomical objects were neologisms and they also have in common that they have been used commercially and in other social contexts. Pulsar, singular and with capital P, is an extremely precise digital wristwatch manufactured since 1970. There is little doubt that the name was borrowed from the heavenly pulsars, which are the most accurate of all celestial timekeepers.

In addition to quasar and pulsar, the astrophysical *-ar* family also includes so-called *blazars*, a kind of quasars with an origin in galactic nuclei. This kind of variable radio object was originally thought to be stars located within the Milky Way and named BL Lacertae or BL Lac for short (Lacerta, Greek for lizard, is a small Milky Way constellation). By the early 1970s they were recognized to be extra-galactic objects belonging to a new class of active galaxies. The present name dates from 1978: 'In a memorable banquet speech at the Pittsburgh meeting on BL Lac objects ... Ed Spiegel suggested the name "blazar" for this class of object. A combination of BL Lac object and quasar, with a strong feeling of the characteristic violent optical flaring, blazar seems an excellent name, one which we will adopt.'[13] Yet another *-ar* object is the *magnetar*, a rare type of neutron star proposed in 1992. According to the two astrophysicists introducing the concept and name, 'We refer to such a highly magnetized neutron star as a *magnetar*.'[14]

However, by far the most popular of the new creatures of the universe, and also the strangest, are the *black holes* which long remained hypothetical but are now known to be real and abundant. In the 1950s physicists realized that the space-time singularity predicted by general relativity for the gravitational collapse of a massive star was a horizon surrounding the centre of the imploding star. With the recognition of the horizon a better name was needed for what was traditionally called a 'frozen star', a 'dark star', a 'gravitationally completely collapsed object', or a 'collapsed star'. From about 1970 the latter term was contracted to 'collapsar', thus adding one more *-ar* word to the astrophysicists' vocabulary.[15] The leading American physicist John Wheeler responded to the need when he introduced the name 'black hole' in a talk given at the end of 1967. The name appeared in print in the spring of 1968:

The core like the Cheshire cat fades from view. One leaves behind only its grin, the other, only its gravitational attraction. Gravitational attraction, yes; light, no. No

---

[12] A. Michaelis, 'Space "Signals" May Be from Intelligent Being', *The Daily Telegraph*, 5 March 1968. The first pulsar was observed in 1967 by Jocelyn Bell (later Bell Burnell), who at the time was a graduate student with Hewish as her supervisor.
[13] Angel and Stockman (1980).
[14] Duncan and Thompson (1992).
[15] Truran and Cameron (1970), who wrote of 'a black hole or "collapsar"'.

more than light do any particles emerge. Moreover, light and particles incident from outside emerge and go down the black hole only to add to its mass and increase its gravitational attraction.[16]

Wheeler did not italicize the term black hole or put it in quotation marks. He just used it without paying attention to it, as if the word was well known to his audience—which it probably was not.

Although Wheeler is generally credited for having placed black hole in the scientific lexicon, he did not coin the term which in a similar astrophysical context can be found earlier. Robert Dicke, another specialist in the theory of general relativity, used it informally in lectures from about 1960 and in January 1964 it appeared in a report in *Science News Letter*: 'As mass is added to a degenerate star a sudden collapse will take place and the intense gravitational field of the star will close in on itself. Such a star then forms a "black hole" in the universe.'[17] Note the inverted commas. A few days later the term also appeared in an article in *Life Magazine* mentioning that the energy of quasars might be due to an 'invisible black hole in the universe.'[18]

Despite the term thus being known in parts of the physics and astronomy communities prior to Wheeler, it was only with him that it caught on. Dicke (and possibly also Wheeler through conversations with Dicke) was aware of the origin of the term in the 'black hole of Calcutta', a reference to a terrible and well-known incident in British–Indian history. In brief, in 1756 a large number of captive British soldiers suffocated to death when crammed into a small dungeon in Calcutta. The 'black hole in Calcutta' was used as a phrase not only for the historical incident but also, for example, colloquially for a very crowded space. Besides, without referring to the Calcutta tragedy astronomers in the nineteenth century sometimes referred to regions in the sky seemingly devoid of stars as black holes.

The name promoted by Wheeler was quickly adopted by physicists and astronomers as well as the general public. As Wheeler stated in his memoirs:

> The advent of the term black hole in 1967 was terminologically trivial but psychologically powerful. After the name was introduced, more and more astronomers and astrophysicists came to appreciate that black holes might not be a figment of the imagination but astronomical objects worth spending time and money to seek.[19]

For the period 1985–2023, Web of Science lists 4,234 papers with 'black hole(s)' in the title and 14,034 with the term in the topic. *OED* reports that in 2010 the term occurred with a frequency of 2.5 per million words in written English, which makes it more common than, for example, Big Bang and quasar and only slightly less common than spacecraft. Like so many other popular science words it is also used figuratively, as in phrases such as 'My handbag is a black hole—I can't find things in it'

---

[16] Wheeler (1968).
[17] Ewing (1964).
[18] Rosenfeld (1964). For details on the naming history, see Bartusiak (2015), pp. 137–40 and Herdeiro and Lemos (2019).
[19] Wheeler (1990), p. 211.

and 'the company division has become an economic black hole'. In 1979 Disney Pictures released a science fiction movie called *The Black Hole* which at the time was the most expensive movie ever produced by the company.

In their endeavours to make black holes comprehensible to laypeople, scientists and science reporters inevitably make use of metaphors. There is no other way. One of the most popular metaphors is to portray black holes as hungry cosmic beasts with an endless appetite for innocent stars. When a team of scientists in late 2019 observed a huge flare in the centre of a galaxy in the Eridanus Cluster, they interpreted it as the result of a star being swallowed by a black hole. 'This black hole was a messy eater', said one of the scientists. The violent cosmic event appeared in the *New York Times* under the headline 'A black hole's lunch: Stellar spaghetti. Astronomers observed a star become a "feast" for a cosmic monster.'[20]

By their very nature black holes are not discovered in the ordinary sense but only identified by indirect methods, which is one reason why there is no systematic IAU nomenclature for them. In some cases, when they are identified in galactic cores, they are named after the galaxy in question, an example being the supermassive black hole M87* found in the giant galaxy M87 alias Virgo A. Black hole M87* is also known as NGC 4486 or 3C 274, NGC and 3C referring to two astronomical catalogues: New General Catalogue of Nebulae and Clusters of Stars (NGC) and Third Cambridge Catalogue of Radio Sources (3C). In other cases, black holes are named after the instruments or collaborations that first detected the objects. Thus, the technical name XTE J1118+480 refers to a black hole identified by the Rossi X-Ray Timing Explorer (XTE) satellite at the celestial coordinates 1118+480.

As one commentator pointed out, these and similar designations 'don't exactly spark the imagination like the names of planets, asteroids, comets and other cosmic objects that recall gods or other figures from ancient mythologies.'[21] True enough, but this is a problem (if it is a problem) that black holes have in common with many other and less exotic astronomical objects. Astronomers call the Jupiter-like exoplanet found in 1999 in revolution about a foreign sun HD-209458-b, another useful but unimaginative identifier (see also Section 6.3).

At about the time that black holes made scientific headlines, another important astrophysical phenomenon based on general relativity theory began to attract attention. In 1936 Einstein published a short note in *Science* in which he argued that massive stars might have the effect of magnifying and focusing the light from a background source, a phenomenon known as *gravitational lensing*.[22] Einstein did not use this term, but he did apply the lens metaphor when writing about a 'lens-like effect of a star'. He actually got this insight as early as 1912, but without publishing it. A few other physicists discussed similar ideas before Einstein's paper, amongst them the British physicist Oliver Lodge who in 1919 referred to a possible cosmic lens effect. However, scientific metaphors can sometimes be misleading and Lodge thought that the lens metaphor was because no focal length could be ascribed to the astronomical

---

[20] Dennis Overbye, *New York Times*, 17 October 2020. https://www.nytimes.com/2020/10/17/science/astronomy-black-hole-at1910qix.html
[21] https://www.livescience.com/65223-black-hole-names.html
[22] Renn and Sauer (2003).

phenomenon. The term 'gravitational lens' was coined by Fritz Zwicky in a paper of 1937 in which he made the important suggestion that galaxies would be better lenses than stars.[23] It took more than three decades before this term and the equivalent 'gravitational lensing' became common among astrophysicists. The first gravitational lens, a huge galaxy named YGKOW G1 and located some four billion light years away from the Earth, was discovered in 1979.

Given Einstein's unique status in modern science, relatively few eponyms are named after him. One of them is the now forgotten einstein unit for light intensity mentioned in Section 3.4, and another is a minor planet named after him in 1973. Moreover, the 'Einstein ring' is a name for a particular kind of gravitational lenses where light from the discrete background source appears as a ring. The eponymous term was suggested in a 1984 paper in *Astrophysical Journal*. This class of gravitational lenses are sometimes referred to as 'Einstein–Chwolson rings' because the Russian physicist Orest Chwolson in a note of 1924 was the first to publish the idea of a ring-shaped image produced gravitationally. However, the double eponym is rare and of later date. There are numerous scientific papers with 'Einstein ring' in the title, but none with 'Einstein–Chwolson ring'.

Physicists and philosophers, and also a large part of the educated public, know about the EPR paradox, a famous thought experiment concerning the completeness of quantum mechanics proposed in 1935. They all know that the E stands for Einstein, but not all are aware that the P and R refer to his two collaborators Boris Podolsky and Nathan Rosen (it was actually the lesser known Podolsky who authored the article in *Physical Review*). The paradox or argument is usually cited as the acronym EPR with only one tenth of the scientific references using Einstein–Podolsky–Rosen. Einstein also belongs to the very few scientists whose name is used colloquially to express a person's intellectual capacity or lack thereof. Sometimes 'Einstein' appears as a synonym of 'genius'. Sentences such as 'he is another Einstein', 'to be sure, he is not exactly an Einstein', or the plural form 'there are no Einsteins around here' will be understood even by people who have never heard of the theory of relativity. This kind of usage was around even when Einstein was still alive and in the United States it can be found as early as in the 1920s.

6.2 Celestial names

With the adoption of Copernicus' heliocentric system in the early part of the seventeenth century the number of planets was reduced from seven to six: Sun, Mercury, Venus, Earth, Mars, Jupiter, Saturn. Of these, the Earth is the only one without an internationally recognized name. It is *Erde* in German, *Tierra* in Spanish, *Jorden* in Swedish, and *Aarde* in Dutch. The other planetary names refer to Roman gods or goddesses, but Earth (or earth) does not. The Latin equivalent would be *Tellus* or *Terra*, names which are, however, only rarely used although derivations of them are common. One of the derivations is the chemical element tellurium, fittingly a close homologue to selenium (the Moon), and others are the adjectives 'terrestrial' and

---

[23] Zwicky (1937).

'telluric', both of which mean Earth-like or belonging to the Earth. To illustrate the cultural naming history of the ancient planets, I single out the planet of love also known as Venus.

Planet Venus is named after the Roman goddess of beauty, love, and fertility, the equivalent of Aphrodite in Greek mythology. According to legend, she emerged from the foam of the sea onto the Island of Cythera (Cyprus), for which reason Venus was sometimes called the Cytherean planet. The adjectives 'Venerean' or 'Aphrosidian' have been proposed, but both names have unfortunate connotations as indicated by words such as 'venereal disease' and 'aphrodisiac'. The legend is also responsible for the old association of the planet Venus and the element copper, a metal which was abundant on Cyprus, *Kypros* in Greek, and named after the island (or perhaps it was the other way around). The alchemical symbol for copper is ♀, the same as the astrological symbol for the planet and also, and today better known, for the female sex.

Venus may have been the first celestial object clearly recognized as a planet. It is the only one mentioned in the Homeric writings, where it appears in the *Iliad* announcing the close of day and darkening of night. However, Homer and his contemporaries seem to have regarded the planet as two distinct objects, without realizing that the morning star (Phosphoros or Lucifer) is the very same as the evening star (Hesperos or Vesper). Nor is this insight to be found in the Bible, where there is a rare reference to Venus in Isaiah 14:2. In the King James version, it reads: 'How are thou fallen from heaven, O Lucifer, son of the morning! how art thou cut down to the ground, which didst weaken the nations!'

The first newcomers to the solar system—a term which is meaningful only within the framework of heliocentric cosmology—were not new planets but the moons that Galileo sensationally discovered in 1610 revolving around Jupiter. The four moons are now called Io, Europa, Ganymede, and Callisto, all from Greek mythology, but Galileo called them the 'Medicean stars' (*Medicea Sidera*) in honour of his patron, the powerful Medici family. In the dedication of *Sidereus Nuncius* (The Starry Messenger) to Cosimo II, Galileo explained at length his choice of eponym:

> Scarcely have the immortal graces of your soul begun to shine forth on earth than bright stars offer themselves in the heavens which, like tongues, will speak of and celebrate your most excellent virtues for all time.... And hence, since under Your auspices, Most Serene Cosimo, I discovered these stars unknown to all previous astronomers, I decided by the highest right to adorn them with the very august name of Your family. For since I first discovered them, who will deny me the right if I also assign them a name and call them the Medicean Stars?[24]

At the time a 'star' often meant just a luminous celestial body and Galileo referred to neither 'moon' nor 'satellite' in his *Sidereus Nuncius* published in March 1610. Nor did he describe his optical tube as a telescope, a word which had not yet been coined. Galileo wrote of a 'spyglass' and also used the Latin word formation *perspicillum* based on the verb *perspicere* meaning 'to see through'. When something is clear and

---

[24] Galilei (1989), pp. 31–33.

understandable, it is perspicuous. Telescope in the Italian version *telescopio* was suggested in the summer of 1611 and the Latin *telescopium* appeared in Kepler's writings two years later. The word combines the Greek terms *tēle* ('far') and *scopeo* ('I see').

To distinguish the stars on the firmament from those which appeared to move slowly relative to them (planets, comets, asteroids) astronomers spoke of the first group as 'fixed stars'. According to a dictionary of 1720,

> The Stars of the several Constellations, which tho' carry'd about daily from East to West by the *Primum Mobile*, and back again by the slow motion of the Firmament; yet because they do not move of themselves, but always keep the same Place, they are justly counted, in respect of the others, fixed and unmoveable.[25]

The term was in common use for centuries and retained long after it was understood that it is a misnomer. As first demonstrated by the Göttingen astronomer Tobias Mayer in a treatise of 1760, eighty of the brightest stars exhibited small but perceptible proper motions. After Mayer's important work, the proper motions of the stars became a reality and 'fixed stars' merely a name of historical convention. Nonetheless, astronomers continued to use the archaic term which is well and alive even today.

About sixty years after Galileo's discovery, Jean Dominique Cassini in Paris discovered two more moons around Saturn, and he too suggested an eponymous term. He called the moons, which are now Iapetus and Rhea, *Sidera Lodaica* after the king of France, Louis XIV. The first of Saturn's moons had been observed in 1655 by Christiaan Huygens, who simply named it *Saturni Luna* (Saturn's moon). Its present name Titan was introduced by John Herschel in 1847. The name satellite for a planetary moon was introduced by Kepler, who in a tract from 1611 wrote of *quatuor Jovis satellitibus*, that is, the four satellites of Jupiter. He derived the term from the Latin *satelles* meaning attendant. The word was already used in English and other languages for a guard or attendant of an important person, so Kepler merely extended the meaning of it from the social sphere to an astronomical context.

There are many later connotations of the word, which is used in phrases such as 'a satellite city to London' and 'a satellite nation of the Soviet Union'. Today the word is usually associated with artificial and not natural satellites. *Lexicon Technicum* from 1704 said about the term that it designated

> ...those Planets who are continually, as it were, waiting upon, or revolving about other Planets; as the *Moon* may be called the *Satellite* of the *Earth*; and the rest of the Planets, *Satellites* of the *Sun*. But the word is chiefly used for the new discovered small *Planets*, which make their Revolution about *Saturn* and *Jupiter*.

When the world's first artificial satellite was sensationally launched into orbit on 4 October 1957, it was called *Sputnik* in Western media. However, the Russians did not use that name, the reason being that 'sputnik' in Russian is a general term for satellites

---

[25] Phillips (1720).

of any kind, artificial or natural.[26] It has the same meaning as the Latin *satelles*, a companion or fellow traveller.

Apart from satellite Kepler coined another and much more important term for the paths of the planets in their motions around the Sun. We are today so used to the concept and word *orbit* that it is hard to imagine that in its technical meaning it came into existence only in the early seventeenth century.[27] Before that time, it is strictly speaking anachronistic to speak of orbits or orbital motions of the heavenly bodies. Copernicus and contemporary astronomers described planetary models in terms of 'orbs' in the sense that a planet was attached to a particular *orb* and its apparent motion was due to the motions of its orb or spherical shell. In his *Astronomia Nova* from 1609, Kepler introduced orbit for the path of a planet as determined by the action of physical causes. In the later textbook *Epitome Astronomiae Copernicanae*, Kepler defined the term: 'What is understood by the name "orbit"? Properly speaking, it is that line [curve] which the planet describes around the Sun by means of the centre of its body.'

The adjective 'orbital' only became commonly used in the 1930s and then mostly for atomic and molecular electrons and not for planets. The verb orbit—to travel around in an orbit, as in 'the spacecraft orbited the planet'—is even more recent (*OED*). According to the so-called molecular orbital or MO theory dating from the early 1930s, the chemical bond could be explained in terms of 'orbitals' (not orbits) which here appear as a noun and not an adjective. In this sense the word was coined in 1932 by Robert Mulliken, who in his later Nobel Prize lecture explained that 'an *orbital* means ... something as much like an orbit as is possible in quantum mechanics.'[28]

It was by careful observations of the Jupiter moons that Ole Rømer, a young Danish astronomer working at Cassini's observatory in Paris, concluded that light propagates with a finite velocity during its journey from Jupiter to the Earth. The conclusion was turned into a proof when James Bradley in England discovered in 1728 what is now called the 'aberration of light', a most important phenomenon in astronomy and physics. Only with this discovery did the velocity of light become a constant of nature. The term aberration of light did not appear in Bradley's discovery report to the *Philosophical Transactions* but only in some of his later works. 'Aberration' was at the time well known in the meaning of deviation or departure from the norm and typically used in a social or moral sense ('His strange behaviour was dismissed as an aberration'). It was also used in other scientific contexts, especially in optics where it denoted a defect in an optical system such as a lens where the rays of light did not focus properly.

The first planet ever to be discovered (and not merely observed) was Uranus, which is credited to German-born William Herschel in 1781 after he had first, for almost a year or so, mistaken it for a comet. While moons could be named after minor figures in Roman mythology, the name of the new planet was a more serious matter and one

---

[26] https://commons.trincoll.edu/eclectic/where-did-sputnik-get-its-name/
[27] Goldstein and Hon (2005).
[28] https://www.nobelprize.org/uploads/2018/06/mulliken-lecture.pdf. In the old Bohr–Sommerfeld model, electrons moved in definite orbits around the nucleus, but according to quantum mechanics this is not possible.

for which there was no precedent. Following in the footsteps of Galileo and Cassini, in 1782 Herschel suggested to call the planet *Georgium Sidus* (George's star) after the king of England, George III. As he explained: 'The name of *Georgium Sidus* presents itself to me ... as a native of the country from whence the Illustrious Family was called to the British throne ... [and] as a person now more immediately under the protection of this excellent Monarch.'[29] Understandably, the idea to name the planet after the British king was not welcomed on the Continent. Johann Bode in Germany proposed *Uranus*, a fit name given that the god Uranus was the father of Saturn, who was again the father of Jupiter.

Bode's name was generally adopted except in Britain where *Georgian Planet* (not star) was the favoured term until about 1830 after which also British astronomers accepted 'Uranus'. Before they switched over to Uranus some twenty years earlier, French astronomers mostly used the eponymous 'Herschel' label. Eight years after Herschel had discovered the new planet, the German chemist Martin Heinrich Klaproth announced that he had found a new metallic element, which he called uranium. As Klaproth pointed out, he suggested the name in analogy with the planet and Bode's name for it. In an address to the Prussian Academy of Science in Berlin, he said: 'A few years ago we thrilled to hear of the discovery of the final planet by Sir William Herschel. He called this new member of our solar system Uranus. I propose to borrow from the honour of that great discovery and call this new element—Uranium.'[30]

At about the same time that Herschel announced George's star he observed a nebulous body that superficially looked like the planet Saturn and which he therefore referred to as the Saturn Nebula. Within a few years he found some other planet-looking nebulae, wondering what they were and if they made up a new class of celestial bodies. In a letter of 17 March 1785 to the French astronomer Jerôme Lalande, he wrote that he had found 'the locations of half-a-dozen *planetary nebulae*, as I call them'. He had a name for the objects, but nothing more: 'These are celestial bodies of which as yet we have no clear idea and are perhaps of a type quite different from those that we are familiar with in the heavens.'[31] The nature of the planetary nebulae remained a mystery well into the twentieth century. Although it was realized early on that the new class of nebulae do not relate to planets at all, Herschel's misleading name has been retained to this date. It is a prime example of a scientific misnomer.

As it turned out more than two decades after the death of William Herschel, Uranus was not the most distant of the planets revolving round the Sun. The naming of the new planet Neptune was controversial because it was mixed up with a major discovery controversy between French and British astronomers concerning its prediction. Should it be credited to Urbain Le Verrier in France or to John Adams in England? The planet was actually discovered in 1846, on the basis of the theoretical predictions, by a third party, namely the German astronomer Johann Galle, who

---

[29] Quoted in Leverington (2003), p. 152. George III was the third British monarch of the House of Hanover, the same part of Germany in which Herschel was born and lived before he emigrated to Britain.
[30] Quoted in Bickel (1979), p. 21, with no reference to the original source.
[31] Translated from French in Hoskin (2014).

first proposed calling it *Janus*. Le Verrier came up with *Neptune*, while his compatriot François Arago suggested *Le Verrier* in analogy with the former 'Herschel' for Uranus. For a while a flattered Le Verrier endorsed Arago's eponym. To the Holsteinian astronomer Heinrich Christian Schumacher he wrote: 'I have been valiantly defended by M. Arago. In another epoch I would perhaps have fended off the honour which he wanted me to have in giving my name to the planet; but the singular pretensions of the British have decided me to accept his friendly gesture.'[32]

The Cambridge astronomer James Challis proposed the name *Oceanus* and John Herschel suggested *Minerva* as a compromise. Most Britons were, however, in favour of *Neptune* whereas they flatly rejected *Le Verrier*. This was not only for nationalistic reasons but also because they felt that the eponym would set a problematic precedent by breaking with the mythological tradition. As the president for the Royal Astronomical Society, William H. Smyth, wrote in a letter: 'Mythology is neutral ground... . Just think how awkward it would be if the next planet should be discovered by a German, by a Bugge, a Funk, or your hirsute friend Boguslawski.'[33] By the end of 1847, a plethora of names had been proposed, apart from those already mentioned such as Minerva, Hyperion, Chronon, and Demorgogon. The latter term refers to a demon or deity associated with the underworld in Greek mythology. In the end it was the astronomers who informally made the decision, and since most of them used Neptune in their publications, latest by 1848 the question was settled.

Neptune is the outermost planet in the solar system, but in the period from 1930 to 2006 the position was occupied by small Pluto discovered by Clyde Tombaugh at the Lowell Observatory, Arizona, founded by Percival Lowell in 1894. What originally had been a search for 'Planet X' deserved a less anonymous name. More than 1,000 names were proposed, some by the observatory staff but most from the public, including Atlas, Cronus, Odin, Prometheus, and Zeus, among others. Constance Lowell, the widow of Percival Lowell, presumptuously suggested to call the new planet by her own first name. At the time it was taken for granted that the privilege of naming the planet belonged to the observatory or the discoverer and not to IAU, the International Astronomical Union.

After having considered the many name proposals, Tombaugh and Melvin Slipher, the director of the observatory, settled on Pluto, which was announced on 1 May 1930.[34] Remarkably, the suggestion of naming the planet after the Roman god of the underworld came from an 11-year-old English schoolgirl by the name Venetia Burney. Pluto is still Pluto, but it is no longer a planet. After the IAU General Assembly in 2006 decided to change the definition of a planet, it was relegated to a new category called 'dwarf planets' and Neptune reinstated in its former role as the most distant of the planets. The official name for Pluto is now 134340 Pluto, where the number in front indicates the chronological order of when it was admitted as a member of the minor planet family.

---

[32] Quoted in Kollerstrom (2009), where the complex naming history is fully described.
[33] Standage (2000), p. 147. P. H. Ludwig von Boguslawski was a German astronomy professor at the University of Breslau.
[34] Leverington (2003), pp. 273–4. https://earthsky.org/space/this-date-in-science-pluto-gets-its-name/

As there are names for nonexistent chemical elements and compounds—and there are many of them—so there are names for hypothetical planets and moons that turned out to be imaginary. The best known of this group of chimeras is probably *Vulcan*, a very small intramercurial planet that Le Verrier in late 1859 assumed to exist for theoretical reasons, namely to explain the precise motion of Mercury in its orbit around the Sun. Shortly later, when the French medical doctor and amateur astronomer Edmond Lescarbault observed a small object passing the face of the Sun, Le Verrier jumped to the conclusion that the hypothetical planet had been found. At first neither Le Verrier nor Lescarbault provided the intremercurial planet with a name, but within a month of its announcement it got one. An article in the periodical *Cosmos* from 3 February 1860 explained:

> It will be more useful to make another observation of the planet than to argue about a name.... And yet names have been proposed by a great many of our correspondents, several of whom, at their own initiative and without our encouragement, favour Vulcan—the only name, they say, which could possibly be considered for an intramercurial planet. Indeed, most probably it will be called Vulcan.[35]

In Roman mythology, Vulcan was the god of fire and the forge. Alas, although a few other astronomers claimed to have confirmed the existence of the new planet, most were sceptical and after about two decades or so Vulcan disappeared from astronomical research publications. The anomalous motion of Mercury remained a mystery until 1915, when Einstein explained it on the basis of his theory of general relativity.

At about the time that Vulcan was realized to be a mistake, another spurious planet was proposed by the Belgian astronomer Jean-Charles Houzeau, who in 1884 claimed to have discovered a small planet a little farther away than Venus. He proposed to call it *Neith* after an Egyptian goddess: 'I will give the supposed star ... a name. I choose Neith, the name of the mysterious goddess of Saïs, whose veil no mortal raised.'[36] However, planet Neith was much more short-lived than planet Vulcan as the discovery claim was shut down immediately after it had been announced.

The suffix *-oid* used in the formation of adjectives or nouns usually refers to something resembling the preceding word element. If a movement is accused of being fascistoid, it means that it has traits in common with fascism. Gorillas and chimpanzees are sometimes called anthropoids because of their similarity to humans. Opioids are a class of synthetic drugs that mimic natural opium. Other science words with this ending include colloid, alkaloid, thyroid, hyperboloid, meteoroid, and metalloid (for the latter, see Section 5.4). A prominent example from the history of astronomy is the *asteroids* first discovered in the early years of the nineteenth century. Despite their name, they are entirely different from stars. Most of them are small celestial bodies orbiting the Sun in the space between Mars and Jupiter, the so-called asteroid belt.

---

[35] Baum and Sheehan (1997), p. 156.
[36] Cited in [Anon.] (1884). See also Kragh (2008b), pp. 119–24. Sais is an ancient Egyptian city in the Nile River delta.

The first asteroid was found serendipitously in early 1801 by the Sicilian astronomer Giuseppe Piazzi, who named it *Ceres Ferdinandea* after the goddess of Sicily (the Roman goddess of agriculture and fertility) and the king Ferdinand of Bourbon. European astronomers accepted the first part of the name, but not the second. In letters to William Herschel of 1801, Piazzi described the object as a 'planetoid' or 'a star which ... much resembles a planet'.[37] Over the next few years three more asteroids were discovered and again named after Roman mythology: Pallas, Juno, and Vesta.

As the number of asteroids rapidly increased after about 1850, astronomers felt a need for a new and more rational nomenclature. Thus, Le Verrier proposed to discontinue the traditional system based on Roman divinities and replace it with the number in order of discovery followed by the name of the discoverer. Hence Juno would be 3 Harding and Vesta renamed 4 Olbers. However, the suggestion to skip the classical names was controversial and strongly resisted by John Russell Hind, a British astronomer and discoverer of several asteroids. Without using the term 'asteroid', he defended the existing nomenclature:

It has been long understood amongst astronomers that names for the small planets shall be selected from those of classical antiquity. The advantage of adhering to this neutral foundation is clear enough, and I think it is to be regretted that in two or three cases an inclination to depart from it has been evinced. *Angelina* and *Maximiliana* ought, in my opinion, to be rejected, as inconsistent with the strict interpretation of the rule.[38]

However, it soon turned out that even though there are a great many figures in ancient mythology, there are not enough of them to keep up with the ever-increasing number of asteroids. Today there are hundreds of thousands of them, the large majority unnamed and only provided with a catalogue number (thus Ceres = 1 Ceres, Pallas = 2 Pallas, ..., 171 Ophelia; but just 97,136). Many of the asteroids included in the category of 'minor planets' have not even received a number.[39] In 2006, when the IAU chose to classify Pluto as a dwarf planet, Ceres was placed in the same category.

Asteroid 11,059 identified in 1991 is called *Nulliusinverba*, a name that historians will recognize to be a subtraction of the old motto chosen by the Royal Society (Section 1.3). It was so named on the occasion of the 350th anniversary of the Royal Society. Another asteroid with a remarkable name is 20,461 Dioretsa discovered in 1999. This celestial body moves in a highly eccentric and retrograde orbit around the Sun, meaning that it revolves in the opposite direction of the Earth and the other major planets. As the first case of a retrograde asteroid it was named as an anadrome, namely asteroid spelled backwards.

---

[37] Cunningham and Orchiston (2011).
[38] Hind (1861). The asteroids called Angelina and Maximilian were discovered and named so in 1861 by the German astronomer Ernst Tempel.
[39] See Schmadel (2012), a rich source of information concerning the naming histories of minor planets.

A large fraction of the names of asteroids and minor planets are either eponyms or toponyms with numerous scientists (and also many nonscientists) being honoured in this way. Number 1,249, called Rutherfordia, was discovered in 1932. One might believe that the name is an eponym after the physicist Ernest Rutherford, but in fact it is a toponym named after the city of Rutherford in New Jersey. A few of the eponymous minor planets have been named after scientists while they were still alive, such as was the case with 1,024 Hale (astronomer George E. Hale) and 1,069 Planckia (physicist Max Planck) discovered in 1923 and 1927, respectively. Minor planets have been named after most of the founders of modern physics, examples being Einstein (2,001; in 1973), Dirac (5,997; in 1983), Bohr (3,948; in 1985), Sommerfeld (32,809; in 1990), Heisenberg (13,149; in 1995), Schrödinger (13,092; in 1991), Wolfgangpauli (13,093; in 1992), and Feynman (7,495; in 1995). Finally, an example of a nonscientific eponym: 3,768 Monroe named after Marilyn Monroe.

The origin of the word asteroid is credited William Herschel, who, in contrast to most other contemporary astronomers, did not consider the new objects to be small planets. As soon as he heard of Ceres and Pallas he started his own elaborate observation programme, the results of which he reported in a paper read to the Royal Society on 6 May 1802. Concerning these objects, 'we ought to distinguish them by a new name, denoting a species of celestial bodies hitherto unknown to us', he wrote. Herschel argued that Ceres and Pallas differed from planets and comets by looking like small stars when observed in the telescope.

> They resemble small stars so much as hardly to be distinguished from them, even by very good telescopes.... . From this, their asteroidical appearance, if I may use that expression, therefore, I shall take my name, and call them *Asteroids*; reserving to myself, however, the liberty of changing that name, if another, more expressive of their nature, should occur.[40]

He went on, offering a definition that he thought was sufficiently robust to encompass future discoveries: 'Asteroids are celestial bodies, which move in orbits of little or of considerable excentricity [*sic*] round the sun, the plane of which may be inclined to the ecliptic in any angle whatsoever.' Perhaps uncertain about the proposed name, Herschel addressed Joseph Banks, president of the Royal Society, who asked Stephen Weston, a learned antiquarian and philologist, to comment on the matter. In a letter to Banks of 10 June, Herschel wrote:

> The names you have done me the favour to send I have carefully examined, and beg leave to give you my remarks on them.... . If Mr. Weston were to have a definition of the thing we want a name for, he might possibly find a better one than that of asteroids, which is not exactly the thing we want.... . Will you do me the favour to consult him once more upon the subject, and mention to him that the bodies to be named are neither fixed stars, planets, nor comets, but have a great resemblance to

---

[40] Herschel (1802).

all the three? With this view before him he will probably succeed in an appropriate appellation.[41]

Herschel was probably aware that in ancient Greek there were two forms to denote a star and that the form *astēr* was more common than *astron*.[42] Moreover, he may have known that the term *asteroidēs* was sometimes used for a star-like object, which makes his choice of name understandable if also less of a genuine neologism. He could have suggested the alternative form 'astroid' but did not. This term was actually proposed in 1838 by the Austrian astronomer Joseph Johann von Littrow, but for a star-like geometric curve ($x^{2/3} + y^{2/3} = a^{2/3}$, to be exact) and not for a celestial object. It is a well-known mathematical term but not one used in the context of astronomy.

Although Herschel liked his neologism and used it frequently, he did not insist that it was the one and only proper term. As he phrased it in his letter to Banks, it was 'not exactly the thing we want'. In a paper of 1803, Herschel wrote: 'It is not the least material whether we call them asteroids, as I have proposed; or planetoids, as an eminent astronomer [Piazzi] in a letter to me, suggested; or whether we admit them at once into the class of our seven large planets.'[43] Some astronomers liked asteroid, others disliked it, and others again just did not care. To the first group belonged the German astronomer Heinrich Wilhelm Olbers, who in 1802 had discovered Pallas and five years later Vesta, while Piazzi belonged to the second group. 'Whatever the name given to this new star doesn't really matter', the latter wrote to an Italian colleague in 1802. 'Are they moving stars? You can call them planetoids or cometoids, but not asteroids... . If we call Ceres an asteroid, so we must call Uranus an asteroid.'[44]

Asteroid was taken up rather quickly, but not instantly and not without alternative names being used as frequently. Among those who ignored the word was John Herschel, the son of William. By the mid-nineteenth century the number of asteroids had increased to about forty and they were variously referred to as asteroids, planetoids, smaller planets, or minor planets. The latter term was first used in the British Nautical Almanac in 1835. According to the amateur astronomer George Chambers' widely read *Descriptive Astronomy* published in 1867: 'The old name of *asteroids*, proposed by Sir William Herschel, has nearly fallen into disuse. Nothing could be more inappropriate than such a designation; *planetoids* would have been better. However, *minor planets* is preferable to either.'[45] The IAU established in 1919 agreed with Chambers' preference of minor planets over asteroids, a name which is today very popular but without being endorsed or even formally defined by the IAU. What is currently known as minor planets is not a synonym for asteroids but a broader group that also includes other nonplanetary objects in revolution around the Sun. The official name is small solar system bodies.

Since the late 1970s a new and exciting branch of astrophysics has revolutionized the study of the internal parts of the Sun and the stars generally. As indicated by the name *helioseismology* it is possible to examine the dynamics of the Sun by

---

[41] Cunningham and Orchiston (2011).
[42] On Greek and ancient English words for 'star', see Gough (1996).
[43] Hughes and Marsden (2007).
[44] Hughes and Marsden (2007).
[45] Hughes and Marsden (2007).

means of observational and theoretical techniques somewhat similar to those used in traditional seismology or what astronomers prefer to call *geoseismology*. Given that the Greek *seismos* (σεισμός) typically refers to an earthquake, 'helioseismology' is oxymoronic and 'geoseismology' redundant. The first term is often credited to the British astrophysicist Douglas Gough, one of the fathers of the field, but it may first have appeared in a 1979 paper written by three Russian astronomers who used it in inverted commas and only once.[46]

Gough recalled that in 1976 he was uncertain of what to call the new science, which he wanted to give a correct Greek name. He first coined the word *heliology* but after some time,

> I consulted Liddell and Scott [*Oxford Greek Lexicon*, 1889] to learn that I could have adopted my favoured term helioseismology. So I used the term thereafter, although not at first in print, because ... American grant-awarding politicians would not understand such a term.[47]

It took but a few years until helioseismology was extended to other stars than the Sun and thus became *asteroseismology*, a field which since then has developed explosively. The term may have been invented by Gough, who in an article of 1983 prophesied that 'asteroseismology will soon be a reality'.[48]

But why astero- and not astroseismology? The question was raised by the American astronomer Virginia Trimble: 'My only idiosyncratic thought is to urge that the words "astronomy" and "astrology" already tell us that the correct combining form of "aster" is "astr" or "astro" not "astero"... . I hope it will persuade workers in the field to pronounce their subject in a way that makes clear its connections with stars, rather than stereo.'[49] However, Gough defended his neologism in erudite philological details that brings to mind Whewell's nineteenth-century intervention in scientific terminology. As he pointed out, in ancient Greece *astēr* was more commonly used for star than *astron* and the prefix *astero-* is etymologically unrelated to Trimble's 'stereo'. With regard to compound words involving the term seismology, he wrote:

> Surely, one should always prefer thoroughbreds, such as geoseismology, probably the oldest of the seismological disciplines, selenoseismology, planetary seismology, diskoseismology, and the subdisciplines epichorioseismology and telechronoseismology of helioseismology, and leuconanoseismology and erythrogigantoseismology of asteroseismology![50]

Until September 2023, the term asteroseismology has appeared in about 4,100 scientific papers and helioseismology in about 2,700 papers. A few papers still use the spelling astroseismology.

[46] Severnyi, Kotov, and Tsap (1979).
[47] Gough (2022), p. 95. Christensen-Dalsgaard and Gough (1976).
[48] Gough (1983).
[49] Trimble (1995).
[50] Gough (1996).

In the autumn of 2017 the nomenclature committees of IAU were faced with a problem caused by the discovery of a strange object which did not fit into existing categories of classification. The small cigar- or pancake-shaped object was observed from a Hawaiian telescope on 19 October 2017 when passing through the solar system. However, in contrast to other small objects belonging to the class of minor planets it turned out to be a one-time visitor from another stellar system (see also Section 6.3). How to classify the new object? What to call it? According to a message from IAU's Minor Planet Center of 6 November:

> The discovery of A/2017 U1 has presented a slight nomenclature problem... . A solution has been proposed that solves the problem. A new series of small-body designations for interstellar objects will be introduced: the I numbers... . Accordingly, the object A/2017 U1 receives the permanent designation 1I and the name 'Oumuamua. The name, which was chosen by the Pan-STARRS team, is of Hawaiian origin and reflects the way this object is like a scout or messenger sent from the distant past to reach out to us.[51]

For a while there was only one member of the I (interstellar) category, but two years later 1I/Oumuamua was followed by 2I/Borisov, a stray extrasolar comet named after its discoverer, the Ukrainian-Russian amateur astronomer Gennadiy Borisov. At the time of writing no more interstellar objects have been confirmed. So far, there is no 3I.

6.3 Extraterrestrial life

The belief that there is life outside the Earth is at the same time very new and very old. As astrophysics and astrochemistry were once thought to be labels devoid of scientific content, so, until fairly recently, was it the case with 'astrobiology'. After all, literally the term means the science of living organisms on or related to the stars and there still is no shred of evidence for stellar biology in this sense. Nonetheless, today astrobiology is a recognized and thriving interdisciplinary branch of science principally but not only exploring the conditions for life outside the terrestrial domain.

The hypothesis of extraterrestrial intelligent life is primarily a philosophical and sometimes religious speculation, and as such it is not part of modern astrobiology or is only peripherally related to the field. On the other hand, under the name *pluralism* the hypothesis has an old and glorious history going way back before the word 'biology' was coined. The term pluralism refers in this context to suggestions about a plurality of 'worlds' inhabited by higher beings different from humans. The worlds might be anything beyond the Earth, but most often the pluralists had other solar systems in mind. John Kersey, the compiler responsible for the 1720 edition of Phillips' *New World of Words*, included an entry on 'interstellar' in which he explained:

---

[51] https://minorplanetcenter.net//mpec/K17/K17V17.html. The first apostrophe-looking diacritic is usually left out in the name. Pan-STARRS is an acronym for Panoramic Survey Telescope and Rapid Response System.

INTERSTELLAR ... a Word that some Authors make use of to express those Parts of the Universe which are suppos'd to be Planetary Systems, having each a fixed Star for the Center of their Motion, as the Sun is of ours; so that if it be true that every such Star may thus be a Sun to some Habitable Orbs moving round it, the *Interstellar World* will be infinitely the greater Part of the Universe.[52]

Bernard Fontenelle, perpetual secretary of the Royal Academy of Science in Paris, wrote in 1686 a classic work titled *Entretiens sur la Pluralité des Mondes* (Conversations on the Plurality of Worlds) and during the next one and a half centuries a wealth of other books appeared in the same pluralist tradition. Christiaan Huygens' *Kosmotheoros* published posthumously in 1698 was an early example and one of the last was *More Worlds than One* published in 1854 by David Brewster, the eminent Scottish physicist best known for his important contributions to optics.[53]

The word pluralism and its derivations are commonly known today, but in meanings quite different from that conceived by eighteenth-century astronomers. Philosophers sometimes use the term as an antonym to monism, and in the current cultural and political debate it typically refers to the diversity of ethnic, racial, and sexual groups within society. One of the *OED* definitions of pluralism is 'toleration or acceptance of the coexistence of different views, values, cultures, etc'. William Herschel was a time-typical astronomical pluralist who was convinced that higher life forms were to be found all over in God's universe. Not only were there divinely created beings on the Moon and the planets, a species of them were also living on the hot and apparently inhabitable Sun. In a paper of 1795 he explained, after having referred to 'the inhabitants of the satellites of Jupiter, Saturn, and the Georgian planet [Uranus]', that the Sun

is most probably also inhabited, like the rest of the planets, by beings whose organs are adapted to the peculiar circumstances of that vast globe.... . We need not hesitate to admit that the sun is richly stored with inhabitants.... . If the stars are suns, and suns are inhabitable, we see at once what an extensive field for animation opens itself to our view.[54]

Perhaps realizing that the hypothesis of solar inhabitants might appear extravagant even to fellow pluralists, Herschel emphasized that his arguments for 'Solarians' were strictly scientifically based. 'I think myself authorized, *upon astronomical principles*, to propose the sun as an inhabitable world', he wrote.

The pluralist literature in the period usually referred to just 'inhabitants of Mars' but on occasions also to 'Martians' and similarly for the inhabitants of the other globes. While 'Martian' is currently used in scientific language only as an adjective as in 'Martian atmosphere', to the pluralists—and also in H. G. Wells' science

---

[52] Phillips (1720), which was largely copied from the same entry in Harris (1704). The term 'interstellar' was first employed by Francis Bacon in 1626 but rarely used until the late nineteenth century (*OED*).

[53] The literature on pluralism and extraterrestrial intelligent beings is extensive. See Crowe (1999) and Dick (1996).

[54] Herschel (1795). See also Crowe (2011).

fiction classic *The War of the Worlds* from 1898—it was a noun. The adjective relating to the Sun is 'solar', whereas a solar being was a 'Solarian'. Similar nouns were applied to the other planetary inhabitants: Mercurian, Venusian, Jovian, Saturnian, Uranian, and Neptunian. The creatures living on the Moon—and few doubted their existence—were called either Lunarians or Selenites, and those on the large asteroid Ceres were Cererians. These names are so-called *demonyms*, that is, words referring to a group of people in relation to a particular place which can be anything from a village to a galaxy. 'European', 'Texan', and 'Mancunian' (Manchester) are better known examples. Like eponyms, the people to whom demonyms refer can be either real or imagined.

The Swedish physical chemist Svante Arrhenius was not a pluralist in the sense of Herschel, but he did believe that life was abundant throughout the infinite universe and that it would never cease. He was the leading advocate of the theory of *panspermia*, according to which primitive life forms were propelled through space by stellar radiation pressure. The general idea was far from new as speculations of this kind can be traced back to ancient Greece, where it was introduced by Anaxagoras in the fifth century BC. In 1865 it was revived within the context of Darwin's evolution theory by Hermann Richter, a German medical doctor. Richter did not use the term panspermia for the germs travelling through space, but wrote of *cosmozoa*, where the plural suffix *-zoa* refers to an animal or organism as in *protozoa* and *spermatozoa*. The hypothesis was widely and critically discussed by scientific authorities such as William Thomson in England and Hermann Helmholtz in Germany.[55] Those who favoured the controversial idea were sometimes nicknamed 'panspermists' or 'panspermicists'. Although Arrhenius did not coin the word panspermia, which was first used at about 1840, it was only with his influential book *Worlds in the Making* from 1908 that it became widely known.

Although still a fringe theory, a minority of astrobiologists take the panspermic theory seriously and have developed it in various directions. The primitive microbes possibly propelled through space by radiation pressure—Arrhenius' theory—are called *radiopanspermia* and those transferred by means of rocks for *lithopanspermia*. The Greek word *lithos* meaning stone is known from e.g. lithosphere, monolith, megalith, and lithium. More generally much work has been done by astrobiologists on the surprising ability of bacteria and other microscopic organisms to survive the harsh conditions of interplanetary space or elsewhere. Such organisms are called *extremophiles* and those which thrive at extreme temperatures for *thermophiles*. The space-travelling panspermia are only of marginal interest to modern astrobiologists, though. Organisms which grow in an acidic environment (e.g. pH < 3) are called *acidophiles*. The word ending *-phile* is from Greek (φίλος) and means 'loving' or 'friendly'. A bibliophile loves books, a paedophile loves young children, and a thermophile organism loves heat.

Although speculations related to *astrobiology* are old, the term itself belongs to the twentieth century. It may have been introduced in the French popular science magazine *La Nature* in 1935, where the author, a Polish scientist by the name Ary Sternfeld, wrote about the recent progress in astronomy that it 'has led to the birth of a new

---

[55] Kamminga (1982).

science whose main objective is to assess the habitability of the other worlds, this science is called astrobiology.'[56] Probably without knowing of Sternfeld's neologism, in 1941 Laurence Lafleur, a young American philosopher, wrote an article titled 'Astrobiology' in an astronomical journal. Admitting that the subject might seem to be just 'another bit of pseudo-scientific romancing', he argued competently that it belonged to the domain of respectable science. Lafleur not only considered the probability of life in the solar system but also on what would later be called exoplanets. 'We may conclude, with a fair degree of assurance, that life in the universe is not confined to our planet.'[57]

The term astrobiology first appeared in a book title (*Astrobiologiya*) in 1953, a monograph written by the Russian scientist Gavriil Tikhov, who is today recognized as one of the founding fathers of modern astrobiology. There are other, more or less equivalent, names for the new science, one of them 'bioastronomy' and another 'exobiology'. The latter name was coined or at least promoted in a 1960 paper by the American molecular biologist Joshua Lederberg, a Nobel laureate in medicine of 1958.[58]

There is one more word for this science with no subject that deserves brief mention. In contrast to astro- and exobiology, *xenobiology* is currently the study of how new synthetic life forms can be created in the laboratory, that is, life forms based on a different biochemistry and biology than that known from terrestrial organisms. It is thus a study considerably more speculative than what concerns the majority of astrobiologists. The Greek prefix *xenos* (ξένος) means 'stranger' or 'foreigner' and is known from, for example, xenophobia and the inert gas xenon. Perhaps not surprisingly, the name xenobiology originated in the context of science fiction and not science. It was coined in 1954 by the American science fiction author Robert Heinlein, who later, when astrobiology and exobiology entered the scene, commented on the terminological aspects of the words. With regard to astrobiology he reasonably objected that '"astron" is a star—and stars are about the least likely places to find life, culture, etc.' Moreover:

'Exobiology' does not suffer from the innate self-contradiction found in 'astrobiology' but the prefix 'ex-' or 'exo-' has its own great shortcomings; it is tired and means too many things... . I submit that it is more sensible to use this almost-virgin prefix [xeno] in designating non-terrestrial things, concepts, and fields of study as it will minimize conflicts in meaning, since a neologism constructed with the prefix 'xeno-' is extremely unlikely to resemble or duplicate any other word already in existence... . But the situation is quite different with 'ex-' and 'exo-'; there are hundreds of probable conflicts with common words... . In my opinion, 'xeno-' is the best choice from the standpoint of derivation.[59]

The word *ecosphere* was effectively introduced in 1958 by the American ecologist and geographer LaMont Cole, who, referring to life on Earth, intended it to be a

[56] Lingam and Loeb (2020).
[57] Lafleur (1941).
[58] Lederberg (1960).
[59] Heinlein and Wooster (1961).

combination of 'ecosystem' and 'biosphere'.[60] However, other scientists had already used the term in a more comprehensive, astronomical sense. It seems to have been coined by the German-American pioneer of aviation and space medicine Hubertus Strughold in a book of 1953 in which he discussed the possibility of life on Mars. Other of Strughold's neologisms were *bioastronautics* and *spatiography* (for geography of space).

From the 1960s onwards there were two different meanings of ecosphere, one related to the Earth and its near environment and another to life on an astronomical scale.[61] Scientists still use the now popular term ecosphere in different meanings. According to *OED*, it refers to 'The global ecosystem of the earth or of another planet; the earth as a physical environment together with the organisms inhabiting it.' But it also refers to 'The region of space within a solar system that includes planets whose conditions are not incompatible with the existence of living things.' Although the word ecosphere was familiar to many American scientists by the late 1960s, apparently it was not well known in Europe where it was announced as a new term in *Nature* sixteen years after Strughold had coined it.[62] The dissemination of the term was slow and only accelerated well into the twenty-first century. In the thirty-year period 1958–1987 'ecosphere' only appeared in 29 scientific papers. By contrast, in the decade between 2011 and 2022 the number was 4,733 (Web of Science).

One of the roots of astrobiology, and one that has more than a little in common with the pluralism of the past, is speculations from the 1960s concerning extraterrestrial intelligences. SETI, an acronym for Search for Extra Terrestrial Intelligence, grew out of discussions among American and Russian scientists regarding technologically advanced extraterrestrial beings and, should they exist, how to communicate with them. Without using the acronym, in 1971 NASA (National Aeronautics and Space Administration) funded a SETI research program to look for civilizations on distant planets and since then SETI and related initiatives have developed into a major if somewhat controversial branch of research. As mentioned, most astrobiologists occupy themselves with possible primitive life forms in our planetary system, or for that matter with primitive life on Earth, and not with the hypothetical advanced civilizations beyond it.

Like other new fields of science, SETI researchers are in need of a unique and consistent terminology which they cannot simply adopt from other fields. Key terms have to be clarified and agreed upon, which is particularly important for a field that is not always recognized to have passed the line from fringe to proper science. For example, what does it mean to call a planet or something else 'habitable'? Can 'alien life' be defined in a suitable, unambiguous, and operational way? Are common terms such as 'intelligence', 'civilization', and 'consciousness' appropriate when speaking of extraterrestrials?[63]

In 1979 the term Extraterrestrial Life appeared as a separate category in *Astronomy and Astrophysics Abstracts*, and three years later a commission on 'bioastronomy' was formed under the International Astronomical Union. Not all scientists welcomed

---

[60] Cole (1958).
[61] Huggett (1999). Dick (1996), pp. 136–7.
[62] Gillard (1969).
[63] Almár (2011). On the origin and early development of SETI research, see Dick (1996), pp. 399–453.

the new field, though. Some critics objected that it was too speculative and science fiction-like, while others argued from scientific premises that we are the only intelligent species in the Milky Way. 'Bioastronomy resembles nothing so much as parapsychology... . Is it appropriate for this "science" ... to remain in the IAU?' asked the American physicist Frank Tipler.[64] His question was rhetorical.

Despite the critical voices astrobiology is today a flourishing research area with its own institutes, societies, and communication structures. Results from astrobiological research are published either in the traditional journals devoted to physics, astronomy, biology, and the earth sciences, or in new and more specialized journals such as *Astrobiology, International Journal of Astrobiology*, and *Astrobiology Magazine*. The field is highly interdisciplinary, covering not only aspects of the physical, chemical, and biological sciences, but also parts of psychology, neurology, sociology, and computer science.

Even linguistics has been seriously considered in relation to astrobiology and SETI research. *Astrolinguistics*, a term coined in 2013, is the study of artificial language systems designed to make us communicate with aliens. Universal languages intended for all earthlings are known in the versions of, for example, Esperanto and Ido, but the ambitions of the astrolinguists are higher. They want to construct a language comprehensible to all civilizations in the entire universe.[65] The field was pioneered in 1960 by Hans Freudenthal, a prominent German-Dutch mathematician, who called his invented language *Lincos* from an abbreviation of *lingua cosmica*. 'It is not unthinkable', Freudenthal wrote, 'that inhabitants of other planets have anticipated this project.' He went on:

> A language for cosmic intercourse might already exist. Messages in that language might unceasingly travel through the universe, maybe on wavelengths that are intercepted by the atmosphere of the earth and the ionosphere, but which could be received on a station outside. On such an outpost we could try to switch into the cosmic conversation.[66]

On the basis of formal logic and computer languages other scientists have constructed improved versions of Freudenthal's original system. 'It seems an unavoidable conclusion', writes one author, 'that some kind of cosmic language, a *Lingua Cosmica* (LINCOS) will have to be used for the purpose of nontrivial interstellar communication between intelligent species.'[67]

The *anthropic principle* is yet another term that sometimes comes up in relation to astrobiology or, what in this case might be a better term, 'cosmobiology', had it not been used in medical astrology.[68] This controversial principle or assumption exists in several versions—weak, strong, final, participatory, and more—but the general idea

---

[64] *Physics Today* **40** (December 1987): 92.

[65] DeVito (1992). For the creation and use of traditional universal languages, see Moon and Spencer (1948b) and Gordin (2017), pp. 105–18, 148–56.

[66] Freudenthal (1960), pp. 14–15.

[67] Ollongren (2013), p. 75.

[68] The term 'cosmobiology' typically refers to a school of so-called scientific astrology founded by the German astrologer Reinhold Ebertin in the 1940s.

is that the present epoch is privileged in the sense that it is the epoch in which carbon-based life originated. Life could not have originated in any other epoch, and this constrains the numerical values of the natural constants and the fundamental parameters of physical theory. In a nutshell, the world is as it is because we exist! In the more cautious formulation of the British astrophysicist Brandon Carter, who coined the term in 1973, 'what may be called the *anthropic principle* ... [is] that what we can expect to observe must be restricted by the conditions necessary for our presence as observers'.[69] Without going into the intricacies of this much-discussed principle it is worth noting that it is sometimes taken to be antithetical to the unwritten assumption of SETI research, namely that there are and have been numerous other 'observers' throughout the universe and at epochs different from ours.

The respected Dutch-American astronomer Peter van de Kamp announced in 1963 that he had identified a Jupiter-sized planet orbiting around Barnard's Star, one of the few eponymously named stars (after American astronomer Edward Barnard, who discovered it in 1916). According to an article of 19 April 1963 in the *New York Times*, the discovery 'adds support to the conviction of astronomers that a great many solar systems exist, some of them possibly supporting life'.[70] Alas, the sensational discovery evaporated a decade later when it turned out that the observational data, on which the claim was based, rested on instrument errors. Nonetheless, van de Kamp maintained throughout his life that he had discovered a planet moving around the star. He had not, but when he died in 1995 such planets had become real.

About a decade before the identification of the first exoplanets some astronomers began talking about the *Goldilocks zone* as a synonym of 'habitable zone'. The phrases refer to the range of distances that a planet can be from its star and have a surface temperature at which water is a liquid. The conditions are 'just right' for life in this zone, just as the temperature of only Baby Bear's bowl of porridge was just right for the taste of Goldilocks. Although the whimsical use of the popular fairy tale was originally just a colloquial metaphor meant for popular consumption, later on references to 'Goldilocks zone' and 'Goldilocks principle' also entered some research publications even in their titles. Surprisingly, the Goldilocks metaphor can be found as early as in a 1935 article in the *Los Angeles Times*, where the author wrote about the possibility of habitable planets around other stars: 'To find out, science today has turned Goldilocks. You remember Goldilocks, the original blonde terror who annoyed the Three Bears... . With identical curiosity and startlingly similar conclusions, the astronomer now is sampling the stars for conditions to support life.'[71]

The discoveries since the early 1990s of a large number of *exoplanets* revolving around stars other than the Sun is obviously of great interest to both astrobiology and SETI research as they have generally increased the probability that there is actually life on other planets in the Milky Way. Astronomers have not yet found a twin Earth, but a few of the exoplanets are believed to be habitable to at least primitive organisms. The term exoplanet seems to have been coined as a neologism in 1992, perhaps as a contraction of the word 'extrasolar planet' (or occasionally 'exosolar

---

[69] Kragh (2011), p. 224. Dick (1996), pp. 528–32.
[70] https://oklo.org/2012/02/. Dick (1996), pp. 201–12.
[71] Quoted in Zimmer (2013). Goldilocks zone and principle are accepted words according to *OED*.

planet') that astronomers had used for a couple of years. The American astronomer Bernard Burke may have introduced exoplanet and also the adjectival form 'exoplanetary' in an unpublished report of December 1992, stating that 'a serious search for planets orbiting stars other than our Sun (Exoplanets) is feasible'.[72] The prefix *exo-*, meaning external or outside, is known from other scientific fields, for instance the exothermic chemical reactions producing heat. The antonym is *endo-* as in endoscope or the endothermic reactions requiring heat, but of course there are no 'endoplanets' except the ordinary planets of the solar system.

In 2019, the Swiss astronomers Michel Mayor and Didier Queloz were awarded the Nobel Prize for the first definite detection of an exoplanet, but in their 1995 article in *Nature* announcing the discovery they wrote of an extrasolar planet and not of an exoplanet. The latter name only appeared in a regular scientific journal two years later and it took some time until the contracted form, a middle clipping, replaced extrasolar planet. With this term given in parentheses, the number of papers with exoplanet in the title increased as follows (Web of Science):

1995–9: 13 (11); 2000–4: 106 (62); 2005–9: 428 (98); 2010–14: 1,085 (61); 2015–19: 1,373 (6).

For the four years 2020–3 the numbers were 1,320 (1).

The planet discovered by Mayor and Queloz is called 51 Pegasi b. Today more than 5,500 exoplanets have been confirmed in about 4,000 planetary systems, and they all have a name. According to the naming standard of the IAU, the name of an exoplanet is formed by adding a lower case letter to the name of the host star starting with 'b' (and not 'a', which is reserved for the star).[73] The next discovered planet orbiting the same star is 'c', the next again 'd', and so on. Sometimes exoplanets are named not by the host star but by the instrument used for their discovery. Thus Kepler-62f, one of the many exoplanets identified by the Kepler Space Telescope, is named after NASA's instrument and not directly after the famous German astronomer.

In addition to the systematic or scientific catalogue names, since 2015 many of the exoplanets have been given proper names based on a public nomination procedure, not unlike to what happened with the naming of Pluto back in 1930. The proposed names have to agree with certain rules decided by the IAU. Among those rules are that a name proposal will *not* be accepted if it is an acronym or name of a pet animal (!), or if it refers to 'names of individuals, places or events known for political, military or religious activities'.[74] Thus, assumedly an exoplanet cannot be named Waterloo or Muhammad. Also excluded are 'names of living individuals'. That is, eponyms are acceptable but only with certain restrictions and if they refer to dead persons. Among the approved public names many are eponyms or toponyms. Others are mythological and others again neologisms. Examples are Galilei, Hypatia, Ganja, Poltergeist, and Babylonia.

[72] *Observatory* **113** (1993): 114–21.
[73] Hessman et al. (2010).
[74] https://www.nameexoworlds.iau.org/2022naming-rules

Speculations concerning extraterrestrial civilizations were revived in the wake of the discovery of 1I/Oumuamua referred to in Section 6.2. The nature and origin of the enigmatic interstellar object was unknown, so why not suggest that it was a remnant of an alien spacecraft? This was indeed what Abraham Loeb, a distinguished Harvard professor of astrophysics, suggested, not only in popular science magazines but also in the highly respected *Astrophysical Journal*. Together with his co-author Shmuel Bialy he considered the possibility 'that 'Oumuamua is a lightsail, floating in interstellar space as debris from advanced technological equipment... . [This] more exotic scenario is that 'Oumuamua may be a fully operational probe sent *intentionally* to Earth vicinity by an alien civilization.'[75] Loeb took the scenario seriously and elaborated on it in several later articles and in a popular book entitled *Extraterrestrial: The First Sign of Intelligent Life Beyond Earth*. Suffice to say that his advocacy of extraterrestrial life has not persuaded either astronomers or astrobiologists.

6.4 The universe at large

The Greek word *kosmos* (cosmos, world; κόσμος) had originally connotations such as 'order', 'regular behaviour', 'harmony', and 'beauty'.[76] It might also refer to the ornaments worn by women. Thus, it is not by accident that the word is found in 'cosmetics' or that 'cosmetology'—the art and practice of beauty treatment—is much like 'cosmology'. Another derived word is 'cosmopolitan', which literally means a citizen of the world and in this sense is a pleonasm (after all, who isn't a citizen of the world?). However, in common usage the word typically refers to a person who feels at home in several different countries and cultures. The adjectival form 'cosmological' is in most respects the same as 'cosmic', although the latter term is broader. The similarity in names reflects the belief commonly held in ancient Greek natural philosophy that the *universe,* the Latin equivalent of cosmos, is ordered and can be comprehended by the human mind. Cosmos is antithetical to *chaos*, the two words being antonyms. The Latin *universum* literally means 'turned into one' but was used early on in the sense 'all things' or 'the whole world'.

Cosmology in the sense of the science of the universe was not a term employed by the Greeks, but they thought it was a possibility. When the Austrian physicist Lise Meitner was appointed lecturer at the University of Berlin in 1922 she chose for her inaugural address 'The Significance of Radioactivity for Cosmic [*kosmische*] Processes'. When an academic press requested permission to publish the lecture, it referred to her address as being on 'the significance of radioactivity for cosmetic [*kosmetische*] processes'.[77] The latter subject was perhaps thought to be more appropriate for a female lecturer.

Although cosmology has always been part of natural philosophy, until the twentieth century the term was rarely used in a scientific context. The first academic book

---

[75] Bialy and Loeb (2018). In 2021 Loeb founded the non-profit Galileo Project with the aim of searching for extraterrestrial intelligence. See Loeb and Laukien (2022).

[76] But see Finkelberg (1998) who argues in erudite details that this traditional understanding of the Greek κόσμος is speculative and without philological and doxographical justification.

[77] Sime (1996), p. 110.

carrying the title may have been the prominent German philosopher Christian Wolff's *Cosmologia Generalis* from 1734, a work which today would be classified as philosophical rather than scientific. The same was the case with the French mathematician Pierre Louis Maupertuis' *Essai de Cosmologie* from 1751. A related term, namely the *cosmography* referring principally to the mapping of the universe or parts thereof, was older and more common. Blount's *Glossographia* said about cosmography that it was 'a Description of several Parts of the visible World, delineating them according to their Number, Positions, Motions, Magnitude, Figures, & c. the two parts of which are *Astronomy* and *Geography*'. When Ptolemy's famous *Geographica* was first translated into Latin in 1406, it was entitled *Cosmographia*. A work in the same tradition was published a century later, namely Martin Waldseemüller's *Cosmographiae Introductio* from 1507. This work is known in particular for having introduced the eponymous term 'America', a choice which the author justified as follows:

> Also another fourth part [of the world] has been discovered by Americus Vesputius ... and I do not see why anyone should justifiably forbid it to be called Amerige, as in 'Americus' Land', or America, from its discoverer Americus, a man of perceptive character; since both Europe and Asia have received their names from women.[78]

Whereas cosmology and cosmography were sciences or philosophical ideas dealing with an essentially static world, *cosmogony* means literally the study of how the universe came into being and thus includes a temporal dimension. The suffix *-gony* is from Greek meaning birth, genesis, or generation, connotations which are also known from the related suffix *-gen* (as in oxygen etc., cp. Section 5.3). However, the term cosmogony is not widely used any longer and today the beginning and evolution of the universe are included under the broader label cosmology.

This label or term was not included in either Harris' *Lexicon Technicum* from 1704, Blount's *Glossographia* from 1707, Phillips' *New World of Words* from 1720, or Johnson's *English Dictionary* from 1755. All four works referred to cosmography and Johnson's also to cosmogony. The *English Dictionary* defined cosmogony as 'the rise or birth of the world; the creation', while cosmography meant 'the science of the general system or affections of the world, distinct from geography, which delivers the situation and boundaries of particular countries'. As far as 'universe' is concerned, Johnson defined the term in a very general and vague sense—'the general system of things'—that did not suggest any connection at all to the science of astronomy.[79]

In contrast to the dictionaries of Harris, Blount, Phillips, and Johnson, Benjamin Martin's *Philosophical Grammar* from 1735 made frequent use of cosmology, which he classified as one of the four branches of natural philosophy, defining it as the general view of the universe including its spatial and temporal extension. Concerning the etymology and meaning of the term, he said: 'It is composed of the two Greek Words, Κοσμό, the World, and Λογό, a discourse; and therefore by *Cosmology*, is implied a

---

[78] Quoted in Beretta (2019), p. 228. In Greek mythology, Asia and Europa were both daughters of two Titans. The Italian explorer Amerigo Vespucci (1451–1512) called the continent the 'New World'.

[79] Johnson (1755).

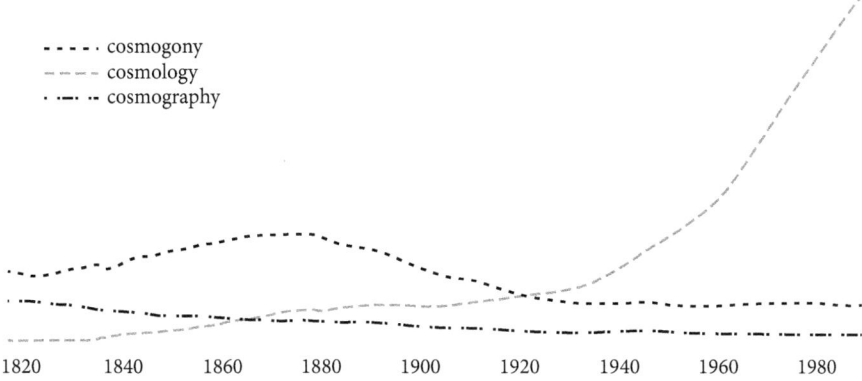

Figure 6.1 Ngram curves showing the relative frequency of the words 'cosmology', 'cosmography', and 'cosmogony' in written English ca. 1820–1990.
Data from Google Ngram.

philosophical and physiological *Discourse* of the *World*, or *Universe* in general.'[80] As to the question of the shape of the universe, his answer was this: 'It hath no determinate Form or Figure at all; forasmuch as it is every way infinite and unlimited.' In Kant's previously mentioned *Universal Natural History and Theory of the Heavens* from 1755 (Section 1.1), a most important treatise in the history of cosmology, there are four references to cosmogony and just a single one to cosmology: 'The physical part of cosmology may hopefully achieve in the future the perfection to which Newton has raised its mathematical half.'[81]

Until about 1860 the terms cosmogony and cosmography appeared more frequently in books than cosmology (Figure 6.1). Still in 1920, cosmogony was as common as cosmology, both words appearing with the low frequency of 0.5 per million words. As a further illustration, consider the astronomy professor John Nichol's *Cyclopaedia of the Physical Sciences*, which first appeared in 1857. This 900-page work included extensive articles on 'nebulae' and 'nebular hypothesis', but the reader would look in vain for entries on 'cosmology' and 'universe'.

Book and article titles referring to cosmogony in the early twentieth century were not about the universe as a whole but about the origin of the planetary system as in Henri Poincaré's *Hypothèses Cosmogoniques* from 1913, or they were about the nebulae as in James Jeans' *Astronomy and Cosmogony* from 1928. Still in the early part of the twentieth century there were but a few monographs with cosmology or derived words in the titles. The subject of Willem de Sitter's popular *Kosmos* from 1932 was scientific cosmology, and the same was the case with Richard Tolman's technical textbook *Relativity, Thermodynamics and Cosmology* from 1924. On the other hand, James A. Williams' *Cosmology* from 1939 was as much about philosophy and theology as about astronomy and physics. The author was a Jesuit 'professor of cosmology' at the Catholic St. Louis University.

[80] Martin (1735), p. 101. Notice Martin's use of 'physiological' (Section 1.5).
[81] Kant (1981), p. 88.

As the term cosmology traditionally belonged to the realm of philosophical and religious thought and only slowly migrated to science, so was it the case with 'cosmologist' for a person studying the cosmos. Traditionally, a cosmologist speculated about the universe and its meaning; he (or on a few occasions she) did not investigate it scientifically. In the latter sense the word only began to be used from about 1980 and then mostly by journalists and popular authors. The Austrian-British physicist Hermann Bondi was a leading figure in cosmological research in the 1950s and the author of the first modern textbook with cosmology as its title (Section 6.5). And yet he considered himself a physicist studying the universe, not a cosmologist. 'I always detest being referred to as a cosmologist', he said in his autobiography published in 1990.[82] According to a British astronomer and critic of modern cosmology, 'the word "cosmologist" should be expunged from the scientific dictionary and returned to the priesthood where it properly belongs'.[83]

Astronomers in the early twentieth century inherited one of the classical problems of cosmology later known as *Olbers' paradox*, which is the question of why the sky at night is dark rather than shining bright. The question, which goes all the way back to Kepler, was analysed in depth by the Swiss astronomer Jean-Phillippe de Chéseaux in 1744 and in 1826 by the better known Wilhelm Olbers in Germany, known also for his important observations of asteroids.[84] What matters here is neither the nature of the problem nor its solution, but its name.

For one thing, from a historical point of view it is a misnomer since it was not originally considered a paradox. Chéseaux and Olbers both thought that they could explain the darkness of the night sky which hence was not paradoxical. For another thing, the problem or paradox is named after Olbers, which may seem to be unfair to Chéseaux given that his analysis was largely the same as that later offered by the German astronomer. The double eponym 'Chéseaux–Olbers' or 'Olbers–Chéseaux' is used from time to time, but none of the designations are common. The standard term Olbers' paradox may well be an example not only of Stigler's law of eponymy but also of the Matthew effect in science (Section 2.4)—Olbers is a famous astronomer whereas Chéseaux is a more obscure figure in the annals of astronomy. As to the name itself, it is of surprisingly late origin as it was probably first introduced in Bondi's textbook *Cosmology* in 1952.

To the modern public the term *galaxy* (plural: galaxies) is much better known than *nebula* (plural: nebulae or nebulas). Through most of astronomical history the first term has been used as synonymous with our Milky Way such as reflected in its etymology. Galaxy is derived from Greek mythology with *gála* (γάλα), literally meaning 'milk' or 'milky'. The word is related to the Latin *lac* or *lact-* known not only from *via lactea* but also from organic compounds such as galactose (milk sugar) and the lactic acid ($CH_3CHOHCOOH$) first isolated by Scheele in 1780 from sour milk. Given that galaxy = Milky Way, well into the twentieth century the word was in singular and therefore usually written with a capital first letter. 'The stars stand out like jewels, and among them stretches, like the rim of a gigantic wheel, the Milky Way, or

---

[82] Bondi (1990), p. 63.
[83] Disney (2000).
[84] See Harrison (1987) for a careful examination of Olbers' paradox and its history.

"Galaxy", encircling the whole earth', wrote an American astronomer in 1926.[85] The phrase 'Milky Way galaxy' would be redundant—which it is not today—since the word galaxy literally meant the Milky Way. In English literature galaxy appears in a poem by Geoffrey Chaucer, the author of *Canterbury Tales*, from about 1380:

> See yonder, lo, the Galaxyë,
> Which men clepeth [called] the Milky Wey,
> For hit is whyt [white]: and some, parfey,
> Callen hit Watlinge Strete.[86]

The standard identification of the term galaxy with the Milky Way also appears from the entry in Blount's *Glossographia*:

Galaxy, in the Heavens, is that parcel of Stars, called the milky Way; it only calls a whiteness in the Sky to the naked Eye, but by the help of a Telescope, an innumerable number of little Stars are discovered, which appearing close together cause that whiteness which we see, and from thence call the Milky Way.

In contrast to galaxy, the term nebula could be used in plural as it was originally a designation for fuzzy cloud-like objects of 'nebulous stars', a term employed by Ptolemy in his star catalogue from about AD 150 where he registered five stars with the label *nepheloeidés*. Nebulous stars, we read in *New World of Words* from 1720, are 'certain fixed Stars, of a dull, pale, and dim Light; so call'd because they look cloudy or bring Clouds'.

Robert Hooke spoke of 'nebulae' (spelled *nebulæ*) in 1676, but in the sense of the indistinct rim around a sunspot for which he also used the term 'umbrae' (*umbræ*). The modern sense of nebula was introduced in a short paper of 1716 written by Edmond Halley of comet fame. As Halley noted, although the apparent size of the nebulae was small, they were so far away that 'they cannot fail to occupy Spaces immensely great, and perhaps not less than our whole Solar System'.[87] The adjective 'nebular' as in 'nebular hypothesis' was coined by Whewell in 1833 as a designation for Laplace's cosmogonical theory and its extension to nebulae by William Herschel. The nebular hypothesis was controversial because of its naturalistic explanation of cosmic evolution, but Whewell, an Anglican clergyman, argued that it was nonetheless in agreement with a divinely created universe.[88]

The British gentleman astronomer William Huggins and his wife Margaret were pioneers in spectroscopic studies relating to the nebular hypothesis. In 1864 William Huggins discovered two bright green lines in a planetary nebula that for a long time caused confusion because they could not be reproduced in the laboratory. Could the green lines be due to a gaseous element that only existed in the nebulae? By the

[85] Abbot (1926), p. 171.
[86] Skeat (1937), p. 335. The old English verb *clepe* meant to call or name. Chaucer used the obscure word *parfey* in the sense of 'verily' or 'truly'. Watling Street (Watlinge Strete) is a long historic route in England that passes a large part of the country.
[87] Halley (1716).
[88] Whewell (1833), p. 184. Brush (1996), pp. 8–9, 67–9.

late 1890s the Hugginses had convinced themselves that *nebulium* was a new heavenly element. In a brief note in *Astrophysical Journal* Margaret Huggins discussed the name for the 'as yet undiscovered gas', mentioning that her husband had occasionally used the term 'nebulum':

> Independently, Miss Agnes Clerke has made the suggestion to me of *nebulium* as an appropriate term, which 'though not unobjectionable from an etymological point of view', is on all fours with *coronium*'. If, however, the Greek nomenclature adopted for *helium* and *argon* is to be followed, the term *nephelium* or *nephium* may be suggested as suitable... . It is most desirably that the name chosen should be one universally acceptable to astrophysicists, and so exclusively adopted.[89]

Agnes Mary Clerke, to whom Huggins ascribed the name nebulium, was a respected astronomer and prolific writer of astronomical texts. Although she believed in the existence of the nebular element, she never referred to her coining of the term. The nebulium hypothesis was taken seriously until about 1915 and entered as a crucial component of a nuclear atomic theory proposed in 1911 by the British physicist John Nicholson. However, nebulium belongs to the class of spurious elements (Section 5.5). It took until 1927 before the green lines, upon which the hypothesis rested, were fully explained on the basis of quantum mechanics and then ascribed to O^{2+}, an unusual form of the oxygen ion.[90]

By the early decades of the twentieth century astronomers recognized that many of the enigmatic nebulae were clusters of stars. The burning question now concerned their location relative to the Milky Way system, which was only resolved by Edwin Hubble in the mid-1920s when he introduced a new term in the astronomers' dictionary. His landmark paper of 1926 published in the *Astrophysical Journal* carried the title 'Extra-Galactic Nebulae'.

Hubble had at first used the phrase 'non-galactic nebulae', but the influential American astronomer Harlow Shapley, who thought it was too neutral and clumsy, urged Hubble to call the objects just 'galaxies'. Although Hubble agreed to drop the term 'non-galactic', he insisted instead on calling them 'extra-galactic' nebulae. He consistently avoided the term 'galaxies' and always spoke of 'nebulae'.[91] His widely read semi-popular book published in 1936 carried the title *The Realm of the Nebulae*. Shapley, on the other hand, campaigned for 'galaxies' but without his campaign having much immediate success. In 1943 he published the popular work *Galaxies*, probably the first monograph ever with the word in its title. Only fifteen English-language scientific articles were published from 1926 to 1940 with galaxies in their title, and of these nine were written by Shapley (Web of Science).

---

[89] Huggins (1898). The Greek word *nephos* (νέφος) means cloud. *Nephology*, a name for the study of clouds, was introduced in the mid-nineteenth century but is no longer commonly used. Coronium was another hypothetical element so named because its presence was based on spectral lines found in the Sun's corona. At around 1900, Mendeleev placed coronium in his periodic table with an atomic weight less than that of hydrogen.

[90] Hirsh (1979). Coronium was likewise found to be due to a highly ionized form of iron, namely the ion Fe^{13+}.

[91] Dick (2013), p. 133.

In *The Universe Around Us* and other popular works from about 1930, James Jeans wrote of galaxy in singular and as equivalent to the Milky Way. For the plural form galaxies he, like Hubble and most other astronomers, kept to nebulae. It was only with Hubble's vindication of Kant's island-universe conjecture (Section 1.1) that the adjective *intergalactic*, as an analogue of interstellar, made sense and slowly began to appear in the scientific literature. Jeans ignored the term and spoke instead of 'internebular space'. When we look at the universe as a whole, 'from the giant nebulae and the vast interstellar and internebular spaces down to the tiny structure of the atom, little but vacant space passes before our mental gaze'.[92] The modern term intergalactic first appeared in 1928, in a *Nature* review of Millikan's theory of the cosmic rays.[93] Today the term has entered more than 10,000 scientific papers.

The term galaxies won broad acceptance only after Hubble's death in 1953 and within a decade or two it had largely replaced extra-galactic nebulae. Data from Web of Science show that in the first half of the twentieth century the ratio of papers with galaxies in the title to those with nebulae was 0.16, whereas the ratio in the second half was 5.41. During the decade 2013–22 the ratio was about 15. Google Ngram provides another way of illustrating the marked shift from nebula/nebulae to galaxy/galaxies through the twentieth century: In 1910 the frequency in ppm of nebula was 3.3 and that of galaxy 0.99, whereas in 2010 the frequencies of the two words were 1.3 and 6.6, respectively. With the popularity of the term galaxy followed figurative uses of the same kind as noted in Section 6.1. Instead of speaking of 'an astronomical sum of money' one can speak of 'a galactic sum of money'. The term may also refer to something illustrious and glamorous, as when we are told about 'a festival with a galaxy of famous artists'.

It needs to be pointed out that galaxy is not just another and more modern term for nebula. The two are not synonyms. Nebulae of different kinds, including planetary nebulae (Section 6.2), are well and alive, and they continue to be investigated by astronomers who differentiate between galaxies and nebulae. In a nutshell, what is referred to as nebulae are clouds of dust and ionized gases which generally are much smaller than the huge collections of stars and other celestial objects called galaxies. While a galaxy can and often does contain nebulae, a nebula cannot contain galaxies.

The paradigm shift that cosmology experienced in the 1920s was in part due to Einstein's general theory of relativity and in part due to advances in observational astronomy. Continuing the spectroscopic studies of extra-galactic nebulae pioneered by the American astronomer Melvin Slipher, in 1929 his compatriot Hubble famously established a linear relationship between the redshifts and distances of a number of spiral nebulae. With the redshifts interpreted as Doppler shifts the 'Hubble law' states that $v = Hr$, where v is the radial velocity, r the distance, and H a constant factor soon called the 'Hubble constant'. However, neither in 1929 nor later did Hubble unambiguously interpret his empirical law as an expression for the expanding universe. In other words, it is nothing but a myth, and unfortunately a persistent one, that Hubble discovered the expansion of the universe in the generally understood sense of discovery.

---

[92] Jeans (1930), p. 109.
[93] *Nature* **122** (1928): 555–6.

The recognition that the Hubble law is, in fact, an expansion law only came a few years later when it became known that the Belgian priest and physicist Georges Lemaître in 1927 had derived a similar but differently interpreted law on the basis of Einstein's cosmological field equations. Moreover, in this paper Lemaître also estimated a numerical value for the H constant in approximate agreement with the one later found observationally by Hubble. Incidentally, the still commonly used term 'Hubble constant' is unfortunate because H is a measure of the age of the universe and therefore, by its very meaning, a quantity that slowly decreases. (The inverse of H, $T = 1/H$, has the dimension of time and is known as the 'Hubble time'). For this reason, the alternative term 'Hubble parameter' is preferable and indeed, there has been a shift towards the latter term in the more recent scientific literature. While in the period 1970–89 the ratio between Hubble constant and Hubble parameter was 5.75, in the period 2000 to 2019 it decreased drastically to 0.95 (Google Scholar).

The nomenclature Hubble law, or the possessive Hubble's law, was originally used only in a few cases and became widely adopted only in the 1950s after Hubble's death. A decade later it was routinely used by all astronomers, physicists, and cosmologists, and then not for the empirical redshift-distance law but for the expansion of the universe. Before Hubble monopolized the name it appeared in several books and articles under the name 'Hubble–Humason law' as a reminder that the observational data collected 1929–31 were to a large extent due to Hubble's assistant and close collaborator Milton Humason.

In the autumn of 2018 members of the General Assembly of the IAU decided by simple majority (78%) to support a resolution recommending that instead of referring to Hubble's law, 'from now on the expansion of the universe [should] be referred to as the "Hubble–Lemaître law"'.[94] Given that Lemaître suggested the law two years before Hubble, one should perhaps have reversed the order of names to Lemaître–Hubble, but this version was not considered. Hubble–Lemaître law is now officially sanctioned by IAU, but it is a recommendation only and the large majority of astronomers and physicists just don't care. As shown by data in Google Scholar and Web of Science they happily continue to use the traditional term Hubble('s) law.

As far as eponymies are concerned, Hubble has been richly awarded, examples being not only the law and the constant (or parameter), but also the Hubble time and the phenomenal Hubble Space Telescope launched into space in 1990. The space telescope is often personified as just Hubble. Such personification, where one speaks of a non-human thing or concept as if it were a human, is common in daily language ('the Sun smiled') and also appears, if not frequently, in scientific communication. The James Webb Space Telescope mentioned in Section 1.2 is routinely referred to as Webb or sometimes James Webb. Similarly, when modern astronomers refer to Kepler, in most cases they have the Kepler Space Telescope in mind, not the person behind the name.

The revelation of the expanding universe was not accepted by all astronomers. During the 1930s there were several proposals of alternatives which maintained the traditional static universe and explained the observed redshifts accordingly. The most popular of the alternatives assumed that photons on their journey through space

---

[94] O'Raifeartaigh and O'Keeffe (2020). Trimble (2012).

slowly lost their energy and hence were observed with a wavelength λ greater than that at emission ($E = h\nu = hc/\lambda$). The light gets 'tired'. The theory was consequently known as the 'tired light' theory, an apt anthropomorphic metaphor. The founder of tired-light theories was Fritz Zwicky, who proposed the idea in a paper of 1929. However, Zwicky did not use the catchy term which may have been coined by Howard P. Robertson, a theoretical physicist in favour of the expanding universe. In an address of 1932, Robertson referred to Zwicky's hypothesis that 'the observed red shift would be due to the properties of "tired" light rather than the nebulae themselves'.[95] Tired-light theories continued to be discussed until the 1980s after which they vanished from peer-reviewed scientific journals.

With the acceptance of the expanding universe most theoretical cosmologists agreed that space (or space-time) was curved and could be described by a so-called metric fundamental to relativistic cosmology. The full and somewhat cumbersome name is the Friedmann–Lemaître–Robertson–Walker metric or model, a multi-eponym referring in both alphabetical and chronological order to the four scientists responsible for it. In many cases the names are not spelled out but instead appear acronymically as FLRW or often as just RW (Robertson–Walker) with Friedmann and Lemaître left out. Robertson–Walker is about four times as frequent in scientific papers as Friedmann–Lemaître–Robertson–Walker.

Science eponyms with four or even more names are rare and understandably so. Another example from astrophysics is what is called the B2FH (or B^{2}FH) theory and almost never the Burbidge–Burbidge–Fowler–Hoyle theory. This very important theory of stellar nuclear synthesis was introduced in a 1957 landmark paper by the four physicists entering the name in alphabetic order (Geoffrey Burbidge, Margaret Burbidge, William Fowler, Fred Hoyle). I am only aware of a single science eponym or abbreviation referring to more than four scientists (the 'Brout–Englert–Guralnik–Hagen–Higgs–Kibble particle' mentioned in Section 3.5 was just a joke). In the literature on theoretical particle physics in the 1970s there are several references to the 'ABFST model', where the letters stand for five Italian physicists, Daniele Amati, Luciano Bertocchi, Sergio Fubini, Antonio Stanghellini, and Mario Tonin. The paper in which the five-letter acronym was proposed referred in its title to 'the Amati–Bertocchi–Fubini–Stanghellini–Tonin multiperipheral model'.[96]

The notion of a closed expanding universe was difficult to comprehend, for what did the universe expand into? And how could the universe be of finite size and yet without limits? To make the notion comprehensible to lay people, the scientists made use of metaphors and analogies to ordinary experiences. The most successful and still standard analogy was due to Eddington, who in 1931 suggested 'to imagine the nebulae to be embedded in the surface of a rubber balloon which is being inflated'.[97] It is a powerful metaphor which works for anyone who has witnessed a balloon being inflated. A much later analogy used pedagocically belongs to the same culinary type as the plum-pudding model of the Thomson model mentioned in Section 1.3. A recent paper on representations in cosmology says about this metaphor or analogy:

---

[95] Kragh (2017).
[96] Tow (1970).
[97] Eddington (1931). He repeated the analogy in the semi-popular and widely read Eddington (1933).

One of the most used analogies is that of a bread loaf or cake dough with embedded chocolate bits or raisins. As the dough cooks it expands and so the space between each raisin increases. However, the raisins themselves do not expand. In this analogy, the dough is an analogical representation of the Universe, and the raisins are the clusters of galaxies.[98]

The raisin bread analogy is in some respects poorer than Eddington's, but it has the advantage that it avoids the impression that the galaxies are themselves expanding along with the expansion of space. The same year that Eddington came up with the balloon analogy, Lemaître proposed the first expanding model of the universe with a beginning in time, a big-bang theory *avant la lettre*. He was entering a totally new territory which he could only explain or rather illustrate by using metaphoric language. His theory of the exploding universe is often misdated to 1927, an error unfortunately repeated in *OED*'s entry on 'Big Bang'.

Without using mathematics Lemaître announced in his letter to *Nature* of 9 May 1931 how he conceived 'the beginning of the universe in the form of a unique quantum, the atomic weight of which is the total mass of the universe'. The original quantum or 'highly unstable atom would divide in smaller and smaller atoms by a kind of super-radioactive process'.[99] He later spoke of the 'primeval atom'—his favoured name for the origin of everything—as an 'isotope of the neutron'. Lemaître tried to explain in words what could not, even in principle, be expressed in exact mathematical terms. The earliest development of the originally undifferentiated quantum or primeval atom was subject to the laws of quantum mechanics and therefore not deterministic. He again expressed his point in a carefully chosen metaphorical language:

Clearly the initial quantum could not conceal in itself the whole cause of evolution; but, according to the general principle of indeterminacy, that is not necessary. Our world is now understood to be a world where something really happens; the whole story of the world need to have been written down in the first quantum *like the song on the disc of a phonograph*. The whole matter of the world must have been present at the beginning, but the story it has to tell may be written step by step.[100]

On other occasions Lemaître spoke of his explosion theory as a 'fireworks theory' of evolution. As a result of the initial 'super-radioactive process' smaller radioactive particles would be formed and some of these particles, or their decay products, would still remain in the form of cosmic rays. The then-enigmatic cosmic rays, he said, were fossils from the cosmic past, 'ashes and smoke of bright but very rapid fireworks'. He realized of course that the apt fireworks metaphor was flawed,

[98] Salimpour et al. (2021).
[99] Lemaître (1931).
[100] Lemaître (1931). Emphasis added. Recall that William Thomson as early as 1884 had made use of the phonograph (Greek 'sound writing') to illustrate his suggestion of the word 'mho' as a unit of conductivity (Section 2.3). Lemaître's phonograph was what most people at the time would call a gramophone.

as fireworks explode into the surrounding space (just as Eddington's inflating balloon does) whereas there was no space into which the primeval universe could expand.

Referring to Lemaître, Eddington, and other innovative scientists, in 1989 the physicist and science writer Alan Lightman emphasized that 'Metaphor in science serves not just as a pedagogical device, but also as an aid to scientific discovery.'[101] But he also pointed out, as others have done, that metaphors are double-edged swords that need to be handled with care because they are limited by our sensual experiences. A metaphor is a way of speaking about an object or phenomenon about which most ordinary people have some previous and shared knowledge. But things are different in at least some branches of science, where such knowledge is absent. As Lightman explains:

> When we hear that 'the chairman plowed through the discussion', we already know a good deal about chairmen, committees, and tiring discussion. But when we say that the photon scattered off an electron, what concrete experience do we have with electrons? When we say that the universe is shaped like the surface of a balloon, what do we really know about how space curves in three dimensions?

The metaphorical conceptualization of atoms as miniature planetary systems popular in the early twentieth century served a purpose but if taken too seriously it tended to restrict the physicists' imagination. For Jean Perrin, the first scientist to suggest a planetary atomic model in 1901, the analogy was more than just a manner of speech. The atom, he suggested, consists of 'one or more masses strongly charged with positive electricity, sort of *positive suns* whose charge is much higher than that of a corpuscle [electron] and ... by a multitude of corpuscles, sort of *little negative planets*'.[102] However, planets revolve around the Sun in definite orbits and cannot pass abruptly from one orbit to another. A different version of the planetary atom was suggested a few years later by the Japanese physicist Hantaro Nagaoka, who metaphorically likened the atom to the planet Saturn surrounded by the numerous particles in the ring system. Nagaoka's picture of the atom came to be known as the 'Saturnian model', a term he did not use and which appeared in the literature only in the 1950s.

Had Niels Bohr stayed with a planetary or Saturnian model he would never had come up with the fundamental concept of quantum jumps (the word is itself a metaphor, see Section 4.3) which has no analogue in macroscopic physics and cannot be easily translated into a metaphor. And yet, in a popular book of 1927, Bertrand Russell came up with such a metaphorical translation to explain this 'quite unintelligible' feature of the Bohr atom. 'An electron', Russell said, 'is like a man who, when he is insulted, listens at first apparently unmoved, and then suddenly hits out.'[103]

---

[101] Lightman (1989). See also Kramar and Ilchenko (2021) and Section 1.2.
[102] Cited in Schirrmacher (2012).
[103] Russell (1927), p. 63.

6.5 Big Bang

Despite Lemaître's early theory of the primeval atom, Big Bang cosmology in its modern meaning only took off with a series of papers in the late 1940s written by the Russian-American nuclear physicist George Gamow and his associates Ralph Alpher and Robert Herman. Rather than focusing on the cosmic rays as fossils of the cosmic past, as Lemaître did, Gamow's fossils were the chemical elements whose distribution and abundance he thought could be explained by thermonuclear processes in the very early and very hot universe. His research programme of 'nuclear archaeology', as it has been called, was to reconstruct the history of the universe by means of hypothetical nuclear processes, and to test these by the resulting pattern of element abundances. Metaphorical references to archaeology and fossils were frequent in the works by Gamow and his research group.

For the primordial soup of nuclear particles, perhaps consisting purely of neutrons, Gamow adopted the term *ylem* which Alpher had come across in 1948. 'According to Webster's New International Dictionary, 2nd Ed., the word "ylem" is an obsolete noun from which the elements were formed', Alpher wrote. 'It seems highly desirable that a word of so appropriate a meaning be resurrected.'[104] Gamow quickly appropriated the term and used it frequently as if he had coined it himself, but it was rarely used by other physicists. 'Ylem' is the accusative form of the word 'hyle' used in the Renaissance to denote the first matter in the universe (*OED*). It was from this word that William Prout in 1815 constructed the term 'proto-hyle' contracted to 'protyle'. As mentioned in Section 3.3, Prout's protyle inspired Rutherford to suggest the name 'proton' for the hydrogen nucleus.

Late in life Gamow suggested a unified terminology for the large numbers appearing in cosmology and nuclear physics which he wanted to base on the factor 10^9, one billion in American English and one thousand million in British English. Apart from reusing the old word *eon* ('aeon' in British English) in the precise sense of 1 eon = 10^9 years, Gamow wrote,

> I suggest the introduction of a unit of distance equal to 10^9 light years to be called hubble in honour of the American astronomer, Edwin Hubble, who was the first to use it for measuring the distances between the galaxies of stars. Conveniently, in these units one hubble per eon becomes the speed of light in vacuum. I also propose tentatively that we use the term one inferno (°I) for the temperature of 10^9 Kelvin.[105]

For the energy 10^9 electron volts the term GeV (giga electron volts) had been officially recommended in 1966 to replace the BeV used by American physicists. The 'B' in the latter unit did not stand for a prefix but for the American 'billion' as in the 'Bevatron' accelerator constructed at the Lawrence Berkeley National Laboratory in the early 1950s. The name was formed from 'Billions of *eV* Synchro*tron*' with the 'a' added. Thus, BeV did not give rise to Bevatron, but the other way around. However, whereas billion meant 10^9 in the United States, in most other countries it meant 10^{12}.

---

[104] Alpher (1948).
[105] Gamow (1968).

Disliking both BeV and GeV, Gamow suggested to 'name 10^9 electron volts after a great nuclear physicist, one Rutherford'. Strangely, he seems to have been unaware that there already was a unit named after Rutherford, namely for the rate of radioactive decay (Section 4.2). In any case, his proposal of a new nomenclature for units fell on deaf ears as Gamow was largely the only one to use it. One physicist responded that the new units were utterly unnecessary: 'Giving new names to different magnitudes of already familiar units of time, distance, temperature and electrical energy will add to vocabulary without adding to knowledge.'[106]

Gamow's theory of the early universe was an early version of the Big Bang, a name which was coined by the young British astrophysicist Fred Hoyle, who ironically was an arch antagonist to this kind of cosmology.[107] In collaboration with Hermann Bondi and Thomas Gold, in 1948 Hoyle announced a radically different theory of the universe, the so-called steady-state theory. The Hoyle–Bondi–Gold theory rested on what Gold rather presumptuously called the 'perfect cosmological principle', namely that on a large scale the universe was not only homogeneous in space but also in time. It had always looked the same and would remain to do so eternally. With the expansion of space and a constant average density of matter it followed that matter must be created continually, which was a most controversial claim because it seemed to violate the fundamental principle of energy conservation. While according to Gamow and other evolutionary cosmologies one could speak of the 'early universe' and the 'old universe', these phrases were meaningless within the framework of steady-state cosmology which posited an eternal universe with no beginning and no end.

Before proceeding with the origin and history of the term Big Bang it will be relevant to consider another word that played a crucial role in the cosmological controversy. Both of the rival theories were *creation* theories although of very different kinds. The two main antagonists in the controversy, Gamow and Hoyle, were keenly aware that ex nihilo creation in a cosmological context was an explosive concept because of its historical association with Genesis. They stressed, each in his own way, that when they spoke of creation it was in an innocent technical sense quite different from the usage of theologians and philosophers. Gamow's semi-popular *The Creation of the Universe* caused raised eyebrows by its title alone. In the preface to the second edition of 1952 he explained that 'the author understands this term [creation], not in the sense of "making something out of nothing", but rather as "making something shapely out of shapelessness", as, for example, in the phrase "the latest creation of Parisian fashion"'.[108]

Hoyle recalled that when he was attending a conference in Moscow in 1958, he 'was told in all seriousness by Russian scientists that my ideas would have been more acceptable in Russia if a different form of words had been used. The words "origin" or "matter forming" would be O.K., but creation in Soviet Union was definitely out'.[109] Creation (or the verb create) is a polysemic word which has been and still is used extensively in a variety of meanings, some related to science but most not. A chef may

[106] *Nature* **220** (1968): 311–12.
[107] This section relies on Kragh (2014b).
[108] Gamow (1952). Kragh (1996), pp. 227–33.
[109] Quoted in Kragh (1996), p. 263.

create an exquisite meal; a statue made by a sculptor is a *creation*; the person who suggests a new word, a neologism, *creates* the word; as mentioned by Gamow, fashion houses regularly present their newest *creations*. And, not to forget, God *created* the universe in six days. So-called *creationism* is not about the origin of the universe in the far past but a religiously based, non-scientific theory of the short history of the divinely created cosmos.

Chemists in the late nineteenth century synthesizing non-natural chemical compounds were in the business of creation, and they were later followed by nuclear chemists creating transuranic elements (Section 5.5). Quantum physicists in the 1930s began speaking of 'creation operator' and 'pair creation'. In 1948, Hoyle introduced a 'creation tensor' in the equations of his cosmological theory. It is widely accepted that the term creation is unfortunate as a denomination for the ultimate birth of the universe (so is the anthropomorphic 'birth') and that a more neutral term such as 'origin' or 'beginning' is preferable. Lemaître often spoke of 'creation', but only in a theological context and never in relation to scientific cosmology. According to Web of Science, since 1968 there have appeared 224 research papers in physics and astronomy which include in their title both of the words 'universe' and 'creation'.

On 28 March 1949, Hoyle gave a BBC radio talk in which he contrasted the two views of the universe and referred to 'the hypothesis that all the matter in the universe was created in one big bang at a particular time in the remote past'. This was the origin of the term *big bang*. In his widely read *The Nature of the Universe* published in 1950 he repeated the big bang phrase, leaving no doubt that he disliked the concept and much preferred the steady-state universe with its slow and eternal continual creation of matter. Hoyle certainly dismissed the idea of a sudden origin of the universe, but he did not describe it in ridiculing or mocking terms as it is often stated. The supposedly derogatory part of the name big bang must be 'bang', a term that Eddington had used for finite-age cosmological models as early as 1928. 'As a scientist I simply do not believe that the universe began with a bang', he said, inventing half of the later term.[110] No one felt Eddington's designation to be pejorative.

There is no solid evidence that Gamow, Lemaître, or other protagonists of explosion cosmologies at the time felt offended by the big bang term or that they paid attention to it at all. In fact, for a considerable period of time it was largely forgotten. In an interview of 1989, Alan Lightman asked Hoyle if he was really the source of the name, to which Hoyle replied:

> Well, I don't know whether that's correct, but nobody has challenged it ... I was constantly striving over the radio—where I had no visual aids, nothing except the spoken word—for visual images. And that seemed to be one way of distinguishing the steady-state and the explosive big bang. And so that was the language I used.[111]

Was Hoyle's big bang really a neologism? Both yes and no, I guess. In non-cosmological but scientific contexts one can find the term a few times prior to Hoyle's radio talk, first perhaps in a note of 1924 referring to studies of sound waves

---

[110] Eddington (1928), p. 85.
[111] Lightman and Brawer (1990), p. 60.

produced by large explosions. A paper received by the *Journal of Meteorology* in early 1949—shortly before Hoyle's BBC lecture—carried the title 'Upper-Atmosphere Temperatures from Helgoland Big Bang'. When it comes to cosmology, Hoyle's term was definitely a neologism if not one that caught on immediately or was conceived as catchy. Far from, for it only appeared insignificantly until the 1970s and Hoyle himself only returned to it in 1965. Lemaître never used the term, and Gamow only once. The number of references to the cosmological Big Bang before that year seems to have been restricted to just a few dozen, of which most appeared in the popular literature, especially in the United States. Altogether I have counted thirty-four sources that mention the name before 1965 and of these twenty-three are of a popular or general nature, seven are scientific contributions, and four are philosophical studies. Remarkably, none of the seven scientific contributions appeared in astronomical journals.

There are some misconceptions in the earlier literature concerning the origin of the term Big Bang, which more than once has been ascribed to Gamow rather than Hoyle. Given their different views of the universe the mistake is understandable, but it is nevertheless a mistake. The entry on Big Bang in Isaac Asimov's *Words of Science* from 1974 says, 'Thinking of that first tremendous explosion, Gamow called it the big bang theory of the universe's origin, and the name has stuck.'[112] In fact, Gamow much disliked the name, which he only referred to in one of his many publications on cosmology.

The first time Big Bang appeared in a research publication was in a 1957 paper by William Fowler, the nuclear astrophysicist who was a close collaborator of Hoyle. Five years later the term made its entry in *Physical Review* in a paper written by Steven Weinberg. While most of the few references were neutral, the eminent British-American cosmologist George McVittie disliked the term and what it stood for. In 1961 he wrote of 'an initial "nuclear explosion" or a "big bang" which initiates the start of the expansion' but only to distance himself from such 'notions [which] have been woven round the predictions of general relativity by imaginative writers'. In a later paper he stated that the term was 'loaded with inappropriate connotations', which made him conclude that 'it is unfortunate that the term "big bang", so casually introduced by Hoyle, has acquired the vogue which it has achieved.'[113]

Among the inappropriate connotations mentioned by McVittie is that the term big bang invites images in the form of an enormous and very noisy explosion, which is a vivid but also grossly misleading image. As a metaphor it shares the problems of Lemaître's fireworks metaphor referred to in Section 6.4. Nonetheless, it is routinely used in popular presentations and perhaps, for want of a more appropriate metaphoric name, unavoidable.

In several cases big bang appeared in nonscientific contexts and mostly independent of the term's meaning in cosmology. American newspapers in the late 1940s and through the 1950s often used the phrase for nuclear weapons and not for the origin of the universe. The threat of a big bang nuclear war real during the Cold War

---

[112] Asimov (1974), p. 54, where Lemaître's primeval-atom hypothesis is also misdated to 1927 instead of 1931 (as it is in the *OED*). For other misconceptions, see Kragh (2014b).

[113] McVittie (1961). McVittie (1974).

period, such as illustrated by John Osborne's play *Look Back in Anger*, where one of the characters says: 'If the big bang does come, and we all get killed off, it ... will just be ... as pointless and inglorious as stepping in front of a bus.' The public mind may have perceived some kind of link between big bang cosmology and nuclear warfare. At least, this is what the *Observer* suggested in an article from 1950 stating that 'The present popular interest in cosmology seems to have a connection with the atomic bomb.'[114]

The American philosopher Norwood Russell Hanson used repeatedly the term big bang in an analysis of 1963 dealing with the concept of creation in the two competing world systems. He coined his own word for the advocates of what he called the 'Disneyoid' picture of the exploding early universe, namely 'big bangers.'[115] Hanson used Disneyoid, possibly a neologism, again eight years later, but then in a different context.

With the discovery of the cosmic microwave background (CMB) in the spring of 1965 the steady-state theory was effectively abandoned and within a couple of years the hot big bang model emerged as something like a paradigm in cosmology. Nonetheless, still in the decade after 1965 big bang was not widely used. The first research paper referring to Hoyle's name in its title dates from 1966. As an alternative but not quite synonymous name, the 'primordial fireball' suggested by John Wheeler appeared in several American publications. Textbooks in cosmology published in the early 1970s showed no unity with regard to the nomenclature. Some of them included the term big bang, others mentioned it only *en passant*, and others again avoided it altogether.

When Big Bang cosmology including the name became popular also outside physics and astronomy, the name was adopted in other fields of scholarship. Until 2012 some 500 papers devoted to studies in the humanities and social sciences referred to Big Bang and only rarely in the cosmological meaning of the word (Figure 6.2). Hoyle's term proliferated and began to appear also in a variety of commercial, cultural, and artistic contexts that had only the word in common with what physicists and astronomers associated with it. Numerous music albums, television series, films, comics, and commercial products of any kind carry the name that Hoyle casually coined in 1949. In an interview of 1995 he said, referring to the name, 'Words are like harpoons, once they get in, they are very hard to pull out.'[116] He was right, but even harpoon-like words sometimes have a limited lifetime.

Many people felt and still feel that Big Bang is an unfortunate name, not only because of its association with a primordial explosion, but also because it is such an undignified label for what is unquestionably the most momentous event ever in the history of the universe. When the popular astronomy magazine *Sky and Telescope* ran a competition in 1993 to find a more suitable name, the judges received no less than 13,099 responses. None of them were found worthy of supplanting Hoyle's misleading and 'inappropriately bellicose' name.[117] The English word has been directly

---

[114] For both quotations, see Kragh (2014b)
[115] Hanson (1963).
[116] Horgan (1995).
[117] Ferris (1993).

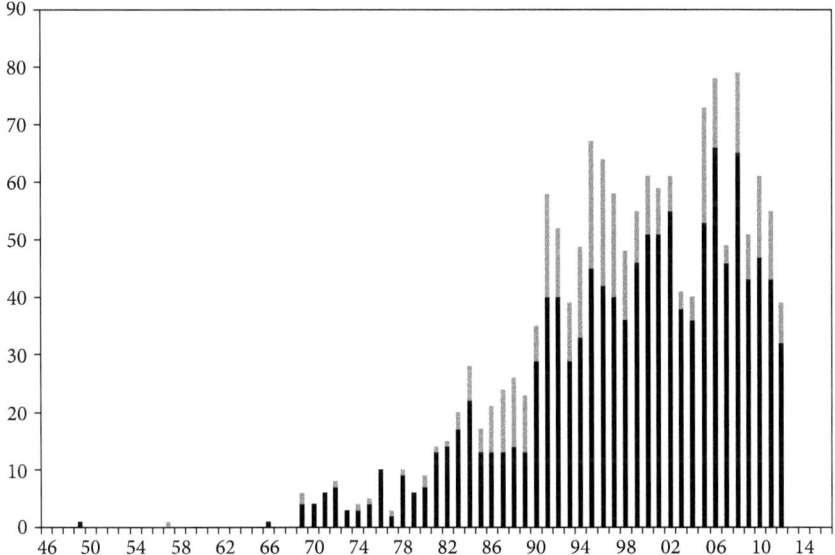

Figure 6.2 Web of Science data showing the number of science papers (black) with 'big bang' in their title. Total number = 1,205. The greyish extensions of the upward lines refer to papers in the humanities, arts, and social sciences.

adopted in many languages other than English, including French (théorie du big bang) and Italian (teoria del big bang). Germans have constructed their own version, namely *Urknall* meaning the original bang, a word which is close to the Dutch *oerknal*.

As black hole is used figuratively and colloquially, so is it the case with Big Bang, which is often used to express a forceful beginning or radical change of any kind. A sentence such as 'the big bang of country music was 1927' is readily understood. The jumps of Poland and Hungary to market economies in the early 1990s were described as economic big bangs. In Great Britain the term was used for a major transformation of the London Stock Exchange that took place in 1986. 'After the Big Bang tomorrow, the City will never be the same again', wrote *Sunday Express Magazine* on 26 October that year (*OED*). In its series 'Among the New Words', the linguistic journal *American Speech* included in the autumn of 1987 big bang, but without relating the term directly to cosmology. According to this source, the new word referred to 'the immediate consequences of the October 1986 deregulation of the London Stock Exchange; any such forceful beginning'. It did note, however, that 'The term alludes to the *Big Bang* theory of the origin of the universe, itself a specialization of the general sense "great explosion", perhaps with allusion to the explosion of atomic bombs'. *American Speech* defined the neologism 'big banger' not in the sense used by Hanson, but as 'one involved with the Big Bang on the London Stock Exchange'.[118]

---

[118] Algeo (1991), pp. 26 and 225. Green (1991), p. 26. For neologisms of the twentieth century, see also Aygo (1999).

The balloon metaphor introduced by Eddington back in 1931 is still much alive but now mostly applied to describe a very different kind of cosmic expansion limited to the first split second of the life of the universe. A majority of modern cosmologists believe that the universe began in an extremely brief *inflation* era during which space increased by an incomprehensibly large factor. Inflation theory, as it is called, dates from about 1980 when it was pioneered by the American physicist Alan Guth, who was also responsible for the term 'inflation': 'I will now describe a scenario, which I call the inflationary universe ...'.[119] In a critical review of inflation cosmology, two philosophers of science paid attention to the name:

> Perhaps inspired by the double-digit rises in the cost of living at the time, Guth soon came up with a catchy name for the period of exponential expansion—he called it inflation. The importance of a good name in the social diffusion of a new scientific idea cannot be overestimated. Catchy names like *big bang* (from Hoyle) or *black hole* (from Wheeler) stuck and conferred a high recognition value to their respective theories. Whatever its scientific merits, *inflation* sounds just great.[120]

Of course, most people will associate inflation with the rate of increase in prices, but to physicists and astronomers it has an additional and completely different meaning. It is in some sense an unfortunate name since it brings to mind either the well-known economic inflation or the equally well-known inflation of a balloon or a car tyre.

The cosmological inflation theory of Guth and others was very successful and soon developed into a confusing variety of versions. In fact, today this class of theories is so popular that currently more research papers are written on cosmic inflation than on economic inflation. During the forty years between 1981 and 2020, the total number of academic papers in economy and cosmology (or 'particles and fields') with inflation in the title amounted to 13,240. In the first half of the period the economy-to-cosmology ratio was 1.34 and in the second half 0.85 (Web of Science). How can one possibly make the abstruse concept of cosmic inflation understandable to the general public? Popularizers inevitably invoke Eddington's old balloon metaphor, which is reused with 'such consistency that it has become the definitive metaphor for detailing the inflationary epoch'.[121]

According to some physicists and philosophers, the Big Bang with or without inflation was not the beginning of everything. Perhaps it was preceded by a 'big crunch' (or 'big squeeze' as Gamow called it), belonging to an earlier cosmic cycle starting with its own Big Bang. And perhaps there is an endless series of cycles in time stretching from minus infinity to plus infinity—no absolute creation, no absolute end. The old philosophical idea of an eternally cyclic, periodic, oscillatory, or 'phoenix' universe—a term referring to the old legend of the immortal bird phoenix—has recently been revived as possible cosmic scenarios and given rise to several new names.

For example, the American theoretical physicist Paul Steinhardt and his British collaborator Neil Turok developed in the early years of the twenty-first century a

---

[119] Guth (1981). Guth (1997), p. 176.
[120] Earman and Mosterin (1999), p. 9.
[121] Phipps (2020).

cyclic 'ekpyrotic' model as an alternative to the inflation theory. They derived the name from the ancient Greek word *ekpyrosis* (ἐκπύρωσις; *ek*, out of; *pyro*, fire or heat). Steinhardt and Turok explained how they came to the name: 'We considered the question of what to call the new scenario. Friends frivolously suggested the Big Splat, the Brane Smash, and other humorous names. But, as many physicists do, we had a predilection for homage to the ancients.' The two authors then approached classical scholars, and according to them

> the new cosmological scenario sounded like the ancient Greek notion of *ekpyrosis*, in which the universe is born from out of fire. The word doesn't exactly roll off the tongue. [Burt] Ovrut thought it sounded like some sort of skin disease. But we eventually settled for the *ekpyrotic universe*, and the name has stuck.[122]

Steinhardt and his colleagues in theoretical cosmology have also introduced the so-called *quintessence* as a repulsive cosmic force over very long distances, an alternative to Einstein's cosmological constant known since 1917. This name too alludes to ancient natural philosophy, in this case to Aristotle's ether (*aither*) or fifth element or what medieval philosophers called the *quinta essentia*. Once a term belonging to natural philosophy, it has crept into common language in the form of the adjective *quintessential* with connotations such as 'archetypical', 'representative', and 'fundamental'. If we speak about a 'quintessential English country pub', we are not referring to Aristotle.

At about the same time that the new cyclic cosmologies were discussed, a heated debate about universes other than ours started. The term 'universe' may seem to be in the singular by necessity and the plural form to be meaningless. Nevertheless, inspired by string theory and the anthropic principle, some high-ranking physicists have suggested the existence of a huge number of hidden universes with natural laws, space-times, and matter components different from those we know. In one universe there may be no protons and in another no electrons. Several names for this (so far purely hypothetical) class of universes have been proposed—pluriverse, megaverse, parallel words are some of them—but the generally adopted name is the now popular *multiverse*.

The word may have been coined by the famous American philosopher William James, who as early as 1895 used it for a world without order and a guiding power. According to James: 'Visible nature is all plasticity and indifference, a multiverse, as one might call it, and not a universe. To such a harlot we owe no moral allegiance; with her as a whole we can establish no sentimental communion.'[123] In an earlier essay James introduced another term, the *nulliverse*, which has not, to my knowledge, been adopted by either modern cosmologists or other scientists. He referred to 'the tendency to enthrone mere juxtaposition as lord of all and to make the Universe what has been well styled a Nulliverse'.[124] To the mind of James, the multiverse was essentially a nulliverse.

---

[122] Steinhardt and Turok (2007), p. 149. The ekpyrosis of the ancient Greeks was a very slow combustion, not an explosive or catastrophic event.

[123] James (1895).

[124] James (1879): 317–46. According to *OED*, the first use of 'nulliverse' dates from 1847.

For most of a century multiverse was rarely used and then only in a philosophical and moral sense. It took until 1998 before the term appeared in a scientific paper and since then it has proliferated to more than 1,400 research papers written by physicists and cosmologists. The term may have migrated from philosophy to cosmology or, more likely, have been independently reinvented by modern cosmologists. At this stage it is impossible to say whether the word and the concept it refers to will survive for long. Like other catchy names it has spread to other fields, some scientific and others not. The formally similar term 'multiversity' was coined in the 1960s, but as a critical denomination of the large American university with its multifarious institutions and lack of focus on universal learning.[125]

At about the same time that multiverse made its entry in the scientific literature, astronomers realized that not only is the universe expanding, it is expanding at an increasing speed. The 'accelerating universe' apparently required a new force of nature to blow it up, which physicists found in a new concept called *dark energy*, a repulsive cosmic force that is due to the energy of empty space as expressed by Einstein's cosmological constant or perhaps to some other mechanism. Because the cosmological constant has always been designated by the Greek letter lambda (Λ, sometimes λ), the so-called concordance cosmological model that emerged in the early years of the twenty-first century is known as the ΛCDM model with CDM standing for cold dark matter. The widely used acronym, pronounced lambda-CDM, is unusual by combining a Greek letter with three Latin letters. Many of the key observational data leading to the ΛCDM model were provided by a satellite instrument called by another acronym, the eponymous WMAP or the Wilkinson Microwave Anisotropy Probe, named in honour of the American cosmologist David Wilkinson. In agreement with the generally accepted rule not to name scientific instruments after living scientists, the microwave probe was originally called MAP and only renamed WMAP after Wilkinson's death in 2002.

Modern cosmologists believe that the major part of the matter–energy content of the universe is made up of the still somewhat mysterious dark energy. The term is credited to the American astrophysicist Michael Turner, who introduced it, possibly in analogy with Zwicky's dark matter, in a paper of 1999. He did not comment on the neologism but merely used it as if it were well known to his colleagues in astrophysics and cosmology.[126] As usual, one can find a few earlier cases in which the same term appeared in different scientific contexts. Today, more than 6,000 papers have been published with dark energy in the title, practically all of them in astrophysics, cosmology, and particle physics.

---

[125] Day (1973).
[126] Huterer and Turner (1999).

Bibliography

Aad, G. et al. (2015). 'Combined measurement of the Higgs boson mass in pp collisions at s = 7 and 8 TeV with the ATLAS and CMS experiments'. *Physical Review Letters* **114**: 191803.

Abbot, Charles G. (1926). *The Earth and the Stars*. London: Chapman & Hall.

Abbri, Ferdinando (2019). 'Discovering elements in a Scandinavian context: Berzelius' *Lärbok i Kemien* and the order of chemical substances'. *Substantia* **3**, Suppl. 5: 49–58.

Abel, Ernest L. (2014). 'Medical eponym Angst'. *Names: A Journal of Onomastics* **62**: 76–85.

Adams, Frank D. (1954). *The Birth and Development of the Geological Sciences*. New York: Dover Publications.

Adler, F. and Hans von Halban (1939). 'Control of the chain reaction involved in fission of the uranium nucleus'. *Nature* **143**: 793.

Aldersey-Williams, Hugh (1995). *The Most Beautiful Molecule: The Discovery of the Buckyball*. New York: John Wiley & Sons.

Alexander, Jerome (1946). 'Prior use of the Rutherford Unit'. *Science* **104**: 276.

Alfvén, Hannes (1966). *Worlds-Antiworlds: Antimatter in Cosmology*. San Francisco: Freman and Company.

Algeo, John (1991). *Fifty Years among the New Words: A Dictionary of Neologisms 1941–1991*. Cambridge, UK: Cambridge University Press.

Algom, Daniel (2021). 'The Weber–Fechner law: A misnomer that persists but that should go away'. *Psychological Review* **128**: 757–65.

Allbutt, T. Clifford (1905). *Notes on the Composition of Scientific Papers*. London: Macmillan.

Almár, Iván (2011). 'SETI terminology: Do we interpret SETI terms correctly?' *Acta Astronautica* **68**: 351–7.

Alpher, Ralph A. (1948). 'A neutron-capture theory of the formation and relative abundance of the elements'. *Physical Review* **74**: 1577–89.

Altschuler, Daniel R., and Fernando J. Ballasteros (2019). *The Women of the Moon: Tales of Science, Love, Sorrow, and Courage*. Oxford: Oxford University Press.

Ammon, Ulrich (2012). 'Linguistic inequality and its effects on participation in scientific discourse and on global knowledge accumulation—with a closer look at the problems of the second-rank language communities'. *Applied Linguistics Review* **3**: 333–55.

Anderson, Carl D. and Herbert L. Anderson (1983). 'Unraveling the particle content of cosmic rays', in *The Birth of Particle Physics*, eds. Laurie M. Brown and Lillian Hoddeson, pp. 131–154. Cambridge, UK: Cambridge University Press.

Anderson, Carl D. and Seth H. Neddermeyer (1938). 'Mesotron (intermediate particle) as name for the new particle of intermediate mass'. *Nature* **142**: 878.

Anderson, J. B. (1996). 'The language of eponyms'. *Journal of the Royal College of Physicians* **30**: 174–7.

Anderson, Wilda (1985). 'Rhetoric and nomenclature in Lavoisier's chemical language'. *Topoi* **4**: 165–9.

Angel, J. R. P. and H. S. Stockman (1980). 'Optical and infrared polarization of active extragalactic objects'. *Annual Review of Astronomy and Astrophysics* **18**: 321–61.

[Anon.] (1884). 'The problematical satellite of Venus'. *Observatory* **7**: 222–6.

[Anon.] (1896). 'Wanted, a name'. *British Medical Journal* **1**: 677–8.

[Anon.] (1937). 'The first international acoustical conference'. *Nature* **140**: 370.

Armstrong, Henry (1896). 'Osmotic pressure and ionic dissociation'. *Nature* **55**: 78–9.

Arnold, William A. (1991). 'Experiments'. *Photosynthesis Research* **27**: 73–82.

Asimov, Isaac (1974). *The Words of Science*. London: Book Club Associates.

Askar'yan, G. A. (1984). 'Investigation of the Earth by means of neutrinos. Neutrino geology'. *Soviet Physics Uspekhi* **27**: 896–900.

Assis, Andre K. T. (2010). *The Experimental and Historical Foundations of Electricity*. Montreal: Apeiron.

Atkinson, Paul (2007). 'The best laid plans of mice and men: The computer mouse in the history of computing'. *Design Issues* **23**: 46–61.

Authier, André (2013). *Early Days of X-Ray Crystallography*. Oxford: Oxford University Press.

Aygo, John (1999). *Movers and Shakers: A Chronology of Words that Shaped Our Age*. New York: Oxford University Press.

Baade, Walter and Fritz Zwicky (1934). 'On super-novae'. *Proceedings of the National Academy of Sciences* **20**: 259–63.

Bagioli, Mario (2012). 'From ciphers to confidentiality: Secrecy, openness and priority in science'. *British Journal for the History of Science* **45**: 213–33.

Baity, Elizabeth C. (1973). 'Archaeoastronomy and ethnoastronomy so far'. *Current Anthropology* **14**: 389–431.

Baldwin, Melinda (2015). *Making Nature: The History of a Scientific Journal*. Chicago: University of Chicago Press.

Barany, Michael J. (2020). 'Impersonation and personification in mid-twentieth century mathematics'. *History of Science* **58**: 1–20.

Barnett, Adrian and Zoe Doubleday (2020). 'Meta-research: The growth of acronyms in the scientific literature'. *eLife* **9**: e60080.

Barney, Stephen A. et al., eds. (2006). *The Etymologies of Isidore of Seville*. Cambridge, UK: Cambridge University Press.

Barry, R. G., J. Jania, and K. Birkenmejer (2011). 'A. B. Dobrowolski—the first cryospheric scientist and the subsequent development of cryospheric science'. *History of Geo- and Space Sciences* **2**: 75–9.

Bartusiak, Marcia (2015). *Black Hole*. New Haven, CT: Yale University Press.

Bassett, H. (1960). 'Nomenclature of inorganic chemistry'. *Journal of the American Chemical Society* **82**: 5523–44.

Batchen, Geoffrey (1993). 'The naming of photography'. *History of Photography* **17**: 22–32.

Baum, Richard and William Sheehan (1997). *In Search of Planet Vulcan: The Ghost in Newton's Clockwork Universe*. New York: Basic Books.

Bazerman, Charles (1988). *Shaping Written Knowledge: The Genre and Activity of the Experimental Article in Science*. Madison: University of Wisconsin Press.

Beaver, Donald (1976). 'Reflections on the natural history of eponymy and scientific law'. *Social Studies of Science* **6**: 89–98.

Beckman, Jenny (2016). 'The publication strategies of Jöns Jacob Berzelius (1779–1848): Negotiating national and linguistic boundaries in chemistry'. *Annals of Science* **73**: 195–207.

Beinert, Helmut (2002). 'Bioinorganic chemistry: A new field or discipline? Words, meanings, and reality'. *Journal of Biological Chemistry* **277**: 37967–72.

Belyi, Vilen V. (1997). 'Rafinesque's linguistic activity'. *Anthropological Linguistics* **39**: 60–73.

Ben-Naim, Arieh (2008). *Statistical Thermodynamics Based on Information: A Farewell to Entropy*. Singapore: World Scientific.

Bensaude-Vincente, Bernadette (2009). 'The chemists' style of thinking'. *Berichte zur Wissenschaftsgeschichte* **32**: 365–78.

Bentley, I. Madison (1906). 'The psychology of organic movements'. *American Journal of Psychology* **17**: 293–305.

Beretta, Marco (2019). 'Names as rewards: The ambiguous role of eponyms in the history of science'. *Nuncius* **34**: 219–35.

Bergson, Henri (1908). *L'Évolution Créatrice*. Paris: Félix Alcan.

Bernstein, Jeremy (1984). *Three Degrees above Zero: Bell Laboratories in the Information Age*. Cambridge, UK: Cambridge University Press.

Bertoni, Gianfranco (2018). 'History of dark matter'. *Reviews of Modern Physics* **90**: 045002.

Bialy, Shmuel and Abraham Loeb (2018). 'Could solar radiation pressure explain 'Oumuamua's peculiar acceleration?' *Astrophysical Journal Letters* **868**: L1.

Bickel, Lennard (1979). *The Deadly Element: The Story of Uranium*. New York: Stein and Day.

Billings, John S. (1886). 'Scientific men and their duties'. *Science* **8**: 541–51.

Birtwistle, George (1928). *The New Quantum Mechanics*. Cambridge, UK: Cambridge University Press.

Blair, Ann (2000). 'Mosaic physics and the search for a pious natural philosophy in the late renaissance'. *Isis* **91**: 32–58.

Blake, George G. (1926). *History of Radio Telegraphy and Telephony*. London: Radio Press.

Bloomfield, Leonard (1935). 'Linguistic aspects of science'. *Philosophy of Science* **2**: 499–517.

Blount, Thomas (1707). *Glossographia Anglicana Nova*. London.

Bockheim, James G. (2015). *Cryopedology*. Cham, Switzerland: Springer.

Bodenstein, Max and Carl Wagner (1929). 'Ein Vorschlag für die Bezeichnung der Lichtmenge in der Photochemie'. *Zeitschrift für physikalische Chemie* **3B**: 456–8.

Bohm, David (1980). *Wholeness and the Implicate Order*. London: Routledge.

Bohr, Niels (1928). 'The quantum postulate and the recent development of atomic theory'. *Nature* (Supplement) **121**: 580–90.

Bondi, Hermann (1990*). Science, Churchill and Me. The Autobiography of Hermann Bondi*. Oxford: Pergamon Press.

Born, Max (1924). 'Über Quantenmechanik'. *Zeitschrift für Physik* **26**: 379–95.

Born, Max, Werner Heisenberg, and Pascual Jordan (1925). 'Zur Quantenmechanik, II'. *Zeitschrift für Physik* **35**: 557–615.

Born, Max and Emil Wolf (1970). *Principles of Optics: Electromagnetic Theory of Propagation, Interference and Diffraction of Light*. Oxford: Pergamon Press.

Borrelli, Arianna (2015). 'The story of the Higgs boson: The origin of mass in early particle physics'. *European Physical Journal H* **40**: 1–52.

Bourdillon, A. J. (2015). 'The science of science and quasi-science'. *Journal of Physical Chemistry & Biophysics* **5**: 182–3.

Bowler, Peter J. (1975). 'The changing meaning of "evolution" '. *Journal for the History of Ideas* **36**: 95–114.

Breit, Gregory (1928). 'The principle of uncertainty in Weyl's system'. *Physical Review* **32**: 570–9.

Brink, Daniel (1989). 'The linguistic theories of Simon Stevin'. *Journal of Germanic Linguistics* **1**: 133–52.

Brock, William H. (1985). *From Protyle to Proton: William Prout and the Nature of Matter, 1785–1985*. Bristol: Adam Hilger.

Brock, Wiliam H. (1993). *The Norton History of Chemistry*. New York: Norton & Company.

Brown, Laurie M. (1978). 'The idea of the neutrino'. *Physics Today* **31** (9): 23–8.

Browne, Charles A. (1928). 'Emerson and chemistry, Part I'. *Journal of Chemical Education* **5**: 269–79.

Brush, Stephen G. (1996). *A History of Modern Planetary Physics: Nebulous Earth*. Cambridge, UK: Cambridge University Press.

Bryan, Kirk (1946). 'Cryopedology—the study of frozen ground and intensive frost-action with some suggestions on nomenclature'. *American Journal of Science* **244**: 622–642.

Bud, Robert (1993). *The Uses of Life: A History of Biotechnology*. Cambridge, UK: Cambridge University Press.

Buffon [Georges-Louis Leclerc] (2018). *The Epochs of Nature*. Chicago: University of Chicago Press.

Bunge, Mario (1966). 'On null individuals'. *Journal of Philosophy* **63**: 776–8.

Burke, John G. (1986). *Cosmic Debris: Meteorites in History*. Berkeley: University of California Press.

Burke-Gaffney, Michael W. (1963). 'Pogson's scale and Fechner's law'. *Journal of the Royal Astronomical Society of Canada* **57**: 3–8.

Cabanac, Guillaume (2014). 'Extracting and quantifying eponyms in full-text articles'. *Scientometrics* **98**: 1631–45.

Calder, Nigel (1977). *They Key to the Universe: A Report on the New Physics*. London: British Broadcasting Corporation.

Campbell, Norman (1924). 'The word "scientist" or its substitute'. *Nature* **114**: 788.

Canales, Jimena (2020). *Bedeviled: A Shadow History of Demons in Science*. Princeton: Princeton University Press.

Caneva, Kenneth L. (2021). *Helmholtz and the Conservation of Energy: Contexts of Creation and Reception*. Cambridge, MA: MIT Press.

Cantor, Geoffrey N. (1987). 'Weighing light: The role of metaphor in eighteenth-century optical discourse', in *The Figural and the Literal: Problems of Language in the History of Science and Philosophy*, eds. Andrew E. Benjamin, Geoffrey N. Cantor, and John R. R. Christie, pp. 124–46. Manchester: Manchester University Press.

Caso, Arthur (1980). 'The production of new scientific terms'. *American Speech* 55: 101–11.

Cassidy, Davis (1988). 'When did the indeterminacy principle become the uncertainty principle?' *American Journal of Physics* 66: 278–9.

Cat, Jordi (2001). 'On understanding: Maxwell on the methods of illustration and scientific metaphor'. *Studies in History and Philosophy of Modern Physics* 32: 395–441.

Chamberlin, Thomas C. (1889). 'Lord Kelvin's address on the age of the Earth as an abode fitted for life'. *Science* 10: 11–18.

Chanowitz, Michael and Stephen Sharpe (1983). 'Glueballs and meiktons which decay to multi-kaon final states'. *Physics Letters B* 132: 413–18.

Charlton, Anne (2004). 'Medicinal uses of tobacco in history'. *Journal of the Royal Society of Medicine* 97: 292–6.

Chateau-Smith, Carmela (2022). 'Language, thought, and the history of science'. *Topoi* 41: 573–86.

Chatt, Joseph (1979). 'Recommendations for the naming of elements of atomic number greater than 100'. *Pure and Applied Chemistry* 51: 381–4.

Chenevix, Richard (1811). 'Reflexions on some mineralogical systems'. *Philosophical Magazine* 37: 39–51.

Chiu, Hong-Yee (1964). 'Gravitational collapse'. *Physics Today* 17 (3): 21–4.

Christensen-Dalsgaard, Jørgen and Douglas Gough (1976). 'Towards a heliological inverse problem'. *Nature* 259: 89–92.

Clark, J. Latimer and Charles T. Bright (1861). 'Measurement of electrical quantities and resistance'. *The Electrician* 1: 3–4.

Clausius, Rudolf (1867). *The Mechanical Theory of Heat*, ed. T. A. Hirst. London: T. Van Voorst.

Coetzee, John Maxwell (1982). 'Newton and the ideal of a transparent scientific language'. *Journal of Literary Semantics* 11: 3–13.

Cohen, Fred (1987). 'Computer viruses—theory and experiments'. *Computers & Security* 6: 22–35.

Cohen, I. Bernard (1985). *Revolution in Science*. Cambridge, MA: Harvard University Press.

Cohen, Julius B. and James Miller (1904). 'The influence of substitution in the nucleus on the rate of oxidation of the side chain. II: Oxidation of the halogen derivatives of toluene'. *Journal of the Chemical Society* 85: 1622–30.

Cohen, Seymour S. (1942). 'The electrophoretic mobilities of desoxyribose and ribose nucleic acids'. *Journal of Biological Chemistry* 146: 471–3.

Cole, LaMont C. (1958). 'The ecosphere'. *Scientific American* 198: 83–96.

Compton, Arthur H. (1921). 'The magnetic electron'. *Journal of the Franklin Institute* 192: 145–55.

Compton, Arthur H. (1939). 'Foreword to symposium on cosmic rays'. *Reviews on Modern Physics* 11: 122.

Comte, Auguste (1835). *Cours de Philosophie Positive*, vol. 2. Paris: Bachelier.

Condon, Edward and Leon F. Curtiss (1946). 'New units for the measurement of radioactivity'. *Journal of Chemical Physics* 14: 399.

Constable, Edwin C. and Catherine E. Housecroft (2020). 'Before radicals were free—the radical particulier of de Morveau'. *Chemistry* 2: 1–11.

Cooper, James K. (1986). 'Electrocardiography 100 years ago'. *New England Journal of Medicine* 315: 461–4.

Cook, O. F. (1912). ' "Genes" not made in Germany'. *Science* 36: 115–16.

Cooper, John M., ed. (1997). *Plato: Complete Works*. Indianapolis: Hackett Publishing Company.

Copernicus, Nicolaus (1995). *On the Revolution of Heavenly Spheres*, trans. C. G. Wallis. Amherst, NY: Prometheus Books.

Coster, Dirk and George Hevesy (1923). 'On the new element hafnium'. *Nature* 111: 218.

Craik, Alex (1999). 'Calculus and analysis in early 19th-century Britain: The work of William Wallace'. *Historia Mathematica* 26: 239–67.

Crawford, Elisabeth (1996). *Arrhenius: From Ionic Theory to the Greenhouse Effect*. Canton, MA: Science History Publications.

Crease, Robert P. (2000). 'Physics, metaphorically speaking'. *Physics World* **13**:1: 17.

Crease, Robert P. and Charles C. Mann (1996). *The Second Creation: Makers of the Revolution in 20th-Century Physics*. New York: Collier Books.

Crepeau, John (2009). 'Loschmidt, Stefan and Stigler's law of eponymy'. *Physics in Perspective* **11**: 357–78.

Crick, Francis H. C. (1966). 'The genetic code—yesterday, today and tomorrow', in *The Genetic Code, Proceedings of the XXXI Cold Spring Harbor Symposium on Quantitative Biology*, ed. Leonora Frisch, pp. 3–9. Cold Spring Harbor.

Cronin, Blaise (2001). 'Hyperauthorship: A postmodern perversion or evidence of a structural shift in scholarly communication practices?' *Journal of the American Society for Information Science and Technology* **52**: 558–69.

Crosland, Maurice P. (1978). *Historical Studies in the Language of Chemistry*. New York: Dover Publications.

Crowe, Michael J. (1999). *The Extraterrestrial Life Debate 1750–1900*. Mineola, NY: Dover Publications.

Crowe, Michael J. (2011). 'The surprising history of claims for life on the Sun'. *Journal of Astronomical History and Heritage* **14**: 169–79.

Crystal, David (1987). *The Cambridge Encyclopedia of Language*. Cambridge, UK: Cambridge University Press.

Cunningham, Clifford J. and Wayne Orchiston (2011). 'Who invented the asteroid: William Herschel or Stephen Weston?' *Journal of Astronomical History and Heritage* **14**: 230–4.

Dahl, Per F. (1997). *Flash of the Cathode Rays: A History of J. J. Thomson's Electron*. Bristol: Institute of Physics Publishing.

Dalitz, Richard H. and Rudolf Peierls (1986). 'Paul Adrien Maurice Dirac'. *Biographical Memoirs of Fellows of the Royal Society* **32**: 139–85.

Dalton, John (1808). *A New System of Chemical Philosophy*. Manchester: R. Bickerstaff.

Danielson, Dennis (2001). 'Scientist's birthright'. *Nature* **410**: 1031.

Darrow, Karl K. (1926). 'Contemporary advances in physics. XIV. Introduction to wave-mechanics'. *Bell System Technical Journal* **6**: 653–701.

Darwin, Charles (1964). *On the Origin of Species by Means of Natural Selection*. Cambridge, MA: Harvard University Press.

Darwin, Charles G. (1922). 'Discussion on the quantum theory'. *British Association for the Advancement of Science, Report*, pp. 473–5. London: John Murray.

Darwin, Charles G. (1936). 'Terminology in physics'. *Nature* **138**: 908–11.

Darwin, Charles G. (1939). 'Use of the termination—tron in physics'. *Nature* **143**: 602.

Davis, Watson (1933). 'Free positive electrons'. *Science* **77**, Supplement: 5.

Davis, Watson (1936). 'Discovery of the positron'. *Science* **84**, Supplement: 8–9.

Davy, Humphry (1808). 'Electro-chemical researches, on the decomposition of the earths'. *Philosophical Transactions of the Royal Society* **98**: 333–70.

Dawkins, Richard (1989). *The Selfish Gene*. Oxford: Oxford University Press.

Day, Dwight H. (1973). 'The origin of multiversity'. *American Speech* **48**: 299–300.

Dean, Dennis R. (1979). 'The word "geology"'. *Annals of Science* **36**: 35–43.

Dear, Peter (1985). 'Totius in verba'. *Isis* **76**: 145–61.

Deltete, Robert J. (2003). 'Energetics', in *The Oxford Companion to the History of Modern Science*, ed. John L. Heilbron, 82–84. New York: Oxford University Press.

Deming, Alison H. (1998). 'Science and poetry: A view from the divide'. *Creative Nonfiction* **11**: 11–29.

Dennis, John G. and Tanya M. Atwater (1974). 'Terminology of geodynamics'. *AAPG Bulletin* **58**: 1030–6.

DeVito, Carl L. (1992). 'Languages, science and the search for extraterrestrial intelligence'. *Leonardo* **25**: 13–16.

Dewey, John (1883). 'Knowledge and the relativity of feeling'. *Journal of Speculative Philosophy* **17**: 56–70.

Diament, Henri (1991). 'Politics and nationalism in the naming of chemical elements'. *Names: A Journal of Onomastics* **39**: 203–16.

Dick, Steven J. (1996). *The Biological Universe: The Twentieth-Century Extraterrestrial Life Debate and the Limits of Science*. Cambridge, UK: Cambridge University Press.

Dick, Steven J. (2013). *Discovery and Classification in Astronomy*. Cambridge, UK: Cambridge University Press.

Dietz, Robert S. (1961). 'Continent and ocean basin evolution by spreading of the sea floor'. *Nature* **190**: 854–7.

Dingle, Herbert (1934). 'Designation of the positive electron'. *Nature* **133**: 330.

Dirac, Paul A. M. (1926a). 'On the theory of quantum mechanics'. *Proceedings of the Royal Society A* **112**: 661–77.

Dirac, Paul A. M. (1926b). 'Quantum mechanics and a preliminary investigation of the hydrogen atom'. *Proceedings of the Royal Society A* **110**: 561–78.

Dirac, Paul A. M. (1930). *The Principles of Quantum Mechanics*. Oxford: Clarendon Press.

Dirac, Paul A. M. (1935). *The Principles of Quantum Mechanics*, 2nd edn. Oxford: Clarendon Press.

Dirac, Paul A. M. (1939). 'A new notation for quantum mechanics'. *Proceedings of the Cambridge Philosophical Society* **35**: 416–18.

Dirac, Paul A. M. (1977). 'Recollections of an exciting era', in *History of Twentieth Century Physics*, ed. Charles Weiner, pp. 109–46. New York: Academic Press.

Dirac, Paul A. M. (1995). *The Collected Works of P.A.M. Dirac 1924–1948*, ed. Richard H. Dalitz. Cambridge, UK: Cambridge University Press.

Disney, Michael J. (2000). 'The case against cosmology'. *General Relativity and Gravitation* **32**: 1125–34.

Dixon, James (1859). *A Guide to the Practical Study of Diseases of the Eye*. London: John Churchill.

Dolby, R. G. A. (1976). 'Debates over the theory of solutions: A study of dissent in physical chemistry in the English-speaking world in the late nineteenth and early twentieth centuries'. *Historical Studies in the Physical Sciences* **7**: 297–404.

Drake, Stillman (1984). 'Galileo, Kepler, and phases of Venus'. *Journal for the History of Astronomy* **15**: 198–208.

Draper, John W. (1880). 'On the phosphorograph of a solar spectrum and on the lines in its infra-red region'. *Proceedings of the American Academy of Arts and Sciences* **16**: 223–34.

Drummond, Jack C. (1920). 'The nomenclature of the so-called accessory food factors (vitamins)'. *Biochemical Journal* **14**: 660.

DuBay, Shane, Daniela Palmer, and Natalia Piland (2020). 'Global inequality in scientific names and who they honor'. https://www.biorxiv.org/content/10.1101/2020.08.09.243238v5.abstract

Dufay, Charles F. (1734). 'A discourse concerning electricity'. *Philosophical Transactions of the Royal Society* **38**: 258–66.

Duhem, Pierre (1974). *The Aim and Structure of Physical Theory*. New York: Atheneum.

Duncan, Robert and Christopher Thompson (1992). 'Formation of very strongly magnetized neutron stars: Implications for gamma-ray bursts'. *Astrophysical Journal* **392**: L9–L13.

Duveen, Denis I. and Herbert S. Klickstein (1954). 'The introduction of Lavoisier's chemical nomenclature into America'. *Isis* **45**: 278–92.

Eamon, William (1994). *Science and the Secrets of Nature: Books of Secrets in Medieval and Early Modern Culture*. Princeton: Princeton University Press.

Earman, John and Mosterin, Jesus (1999). 'A critical look at inflationary cosmology'. *Philosophy of Science* **66**: 1–49.

Eckert, Michael (2010). *Heinrich Hertz*. Hamburg: Ellert & Richter Verlag.

Eddington, Arthur S. (1928). *The Nature of the Physical World*. Cambridge, UK: Cambridge University Press.

Eddington, Arthur S. (1931). 'The expansion of the universe'. *Monthly Notices of the Royal Astronomical Society* **91**: 412–16.

Eddington, Arthur S. (1933). *The Expanding Universe*. Cambridge, UK: Cambridge University Press.

Ehl, Rosemary G. and Aaron J. Ihde (1954). 'Faraday's electrochemical laws and the determination of equivalent weights'. *Journal of Chemical Education* **31**: 226–32.

Einstein, Albert (1916). 'Die Grundlage der allgemeinen Relativitätstheorie'. *Annalen der Physik* **49**: 769–822.

Einstein, Albert (1918). 'Prinzipielles zur allgemeinen Relativitätstheorie'. *Annalen der Physik* **55**: 241–4.

Einstein, Albert (1924). 'Quantentheorie des einatomigen idealen Gases'. *Sitzungsberichte der Preussischen Akademie der Wissenschaften*: 261–7.

Einstein, Albert (1982). *Ideas and Opinions*. New York: Three Rivers Press.

Einstein, Albert (1989). *The Collected Papers of Albert Einstein*, vol. 2, ed. John Stachel. Princeton: Princeton University Press.

Ellis, Fred (1953). 'Naming of chemical elements'. *Names: A Journal of Onomastics* **1**: 163–76.

Ewing, Ann (1964). ' "Black holes" in space'. *Science News Letter* **86** (18 January): 39.

Fahie, John J. (1899). *A History of Wireless Telegraphy 1838–1899*. Edinburgh: Blackwood and Sons.

Fairbridge, Rhodes W. (1999). 'History of geochemistry', in *Encyclopedia of Geochemistry*, eds. Clare P. Marshall and Rhodes W. Fairbridge, pp. 315–22. Dordrecht: Springer.

Fara, Patricia (2002) *An Entertainment for Angels: Electricity in the Enlightenment*. Cambridge, UK: Icon Books.

Faraday, Michael (1839). *Experimental Researches in Electricity*. London: Taylor.

Faraday, Michael (2011). *The Correspondence of Michael Faraday*, vol. 6, ed. Frank James. Bristol: IOP Publishing.

Favrholdt, David (1993). 'Niels Bohr's views concerning language'. *Semiotica* **94**: 5–34.

Feinberg, Gerald (1970). 'Particles that go faster than light'. *Scientific American* **222** (2): 68–77.

Feld, Bernard T. and Gertrud W. Szilard, eds. (1972). *The Collected Works of Leo Szilard: Scientific Papers*. Cambridge, MA: MIT Press.

Fermi, Enrico (1951). *Elementary Particles*. New Haven: Yale University Press.

Ferris, Timothy (1993). 'Needed: A better name for the big bang'. *Sky and Telescope* **86** (no.2): 4–5.

Findlen, Laura, ed. (2005). *Athanasius Kircher: The Last Man Who Knew Everything*. New York: Routledge.

Finkelberg, Aryeh (1998). 'On the history of the Greek κόσμος'. *Harvard Studies in Classical Philology* **98**: 103–36.

Flood, Walter E. (1960). *Scientific Words: Their Structure and Meaning*. London: Oldbourne.

Fölsing, Albrecht (1997). *Albert Einstein: A Biography*. New York: Penguin Books.

Fontani, Marco, Mariagrazia Costa, and Mary V. Orna (2015). *The Lost Elements: The Periodic Table's Shadow Side*. Oxford: Oxford University Press.

Ford, Alan and F. David Peat (1988). 'The role of language in science'. *Foundations of Physics* **18**: 1233–42.

Forgacs, R. L. and A. Warnick (1966). 'Lock-on magnetometer utilizing a superconducting sensor'. *IEEE Transactions on Instrumentation and Measurement* **15** (3): 113–20.

Fowler, Ralph H. (1926). 'General forms of statistical mechanics with special reference to the requirements of the new quantum mechanics'. *Proceedings of the Royal Society A* **113**: 432–49.

Fraps, G. S. (1931). 'Hybrid words'. *Science* **74**: 438.

Fred, Herbert L. and Tsung O. Cheng (2003). 'Acronymesis: The expanding misuse of acronyms'. *Texas Heart Institute Journal* **30**: 255–7.

Frenkel, Yakov (1931). 'What does Einstein mean?' *Science* **74**: 609–18.

Frenkel, Yakov (1932). *Wave Mechanics: Elementary Theory*. Oxford: Oxford University Press, 1932.

Frenkel, Yakov (1936). 'On the absorption of light and the trapping of electrons and positive holes in crystalline dielectrics'. *Physikalische Zeitschrift der Sowjetunion* **9**: 158–86.

Freudenthal, Hans (1960). *Lincos: Design of a Language for Cosmic Intercourse, Part I*. Amsterdam: North-Holland Publishing Company.

Friend, J. Newton (1927). 'Helium or helion?' *Nature* **119**: 199.

Funk, Casimir (1912). 'The etiology of the deficiency diseases'. *Journal of State Medicine* **20**: 341–68.

Futscher, Klaus (1949). 'Elementarteilchen'. *Physikalische Blätter* **5**: 258–67.

Gal, Joseph (2019). 'Louis Pasteur, chemical linguist: Founding the language of stereochemistry'. *Helvetica Chimica Acta* **102**: e1900098.

Galilei, Galileo (1989). *Sidereus Nuncius or The Sidereal Messenger*, ed. Albert van Helden. Chicago: University of Chicago Press.

Gamow, George (1952). *The Creation of the Universe*. New York: Viking Press.

Gamow, George (1968). 'Naming the units'. *Nature* **219**: 765.

Garber, Elizabeth, Stephen Brush, and C. Everitt, eds. (1995). *Maxwell on Heat and Statistical Mechanics*. London: Associated University Presses.

Gell-Mann, Murray (1995). *The Quark and the Jaguar: Adventures in the Simple and the Complex*. New York: Henry Holt and Co.

Gest, Howard (2002). 'History of the word photosynthesis and evolution of its definition'. *Photosynthesis Research* **73**: 7–10.

Ghosh, Abhik and Paul Kiparsky (2019). 'The grammar of the elements'. *American Scientist* **107**: 350–2.

Gillard, A. (1969). 'On the terminology of biosphere and ecosphere'. *Nature* **223**: 500–1.

Glaser, Otto (1993). *Wilhelm Conrad Röntgen and the Early History of the Roentgen Rays*. San Francisco: Norman Publishing.

Glasstone, Samuel (1958). *Sourcebook on Atomic Energy*. Princeton: Van Nostrand Company.

Goldhaber, Maurice (1956). 'Speculations of cosmogony'. *Science* **124**: 218–19.

Goldsmith, Robert H. (1982). 'Metalloids'. *Journal of Chemical Education* **59**: 526–27.

Goldstein, Bernard and Giora Hon (2005). 'Kepler's move from *orbs* to *orbits*: Documenting a revolutionary scientific concept'. *Perspectives on Science* **13**: 74–111.

Golinski, Jan V. (1987). 'Robert Boyle: Scepticism and authority in seventeenth-century chemical discourse', in Benjamin, Cantor, and Christie, pp. 58–82.

Golinski, Jan V. (1992). 'The chemical revolution and the politics of language'. *The Eighteenth Century* **33**: 238–51.

Good, Gregory A. (2000). 'The assembly of geophysics: Scientific disciplines as frameworks of consensus'. *Studies in History and Philosophy of Modern Physics* **31**: 259–92.

Gordin, Michael D. (2004). *A Well-Ordered Thing: Dmitrii Mendeleev and the Shadow of the Periodic Table*. New York: Basic Books.

Gordin, Michael D. (2017). *Scientific Babel: The Language of Science from the Fall of Latin to the Rise of English*. London: Profile Books.

Gordin, Michael D. (2022). 'Einsteinian language: Max Talmay, Benjamin Lee Whorf and linguistic theory'. *British Journal for the History of Science* **55**: 145–65.

Gorelik, Gennady E. and Victor Ya. Frenkel (1994). *Matvei Petrovich Bronstein and Soviet Theoretical Physics in the Thirties*. Basel: Birkhäuser.

Gough, Douglas (1983). 'Oscillations as a probe of the Sun's interior'. *Nature* **304**: 689–90.

Gough, Douglas (1996). 'Astereoasteroseismology'. *Observatory* **116**: 313–15.

Gough, Douglas (2022). 'The privileged life of a theoretical observer'. *Solar Physics* **297**: 95.

Gozzi, Raymond (2017). 'The computer "virus" as metaphor'. *ETC: A Review of General Semantics* **74**: 502–5.

Grant, Edward (2007). *A History of Natural Philosophy: From the Ancient World to the Nineteenth Century*. Cambridge, UK: Cambridge University Press.

Green, Jonathan (1991). *Neologisms: New Words Since 1960*. London: Bloomsbury Publishing.

Greenaway, Frank (1966). *John Dalton and the Atom*. Ithaca, NY: Cornell University Press.

Greene, Mott T. (2015). *Alfred Wegener: Science, Exploration, and the Theory of Continental Drift*. Baltimore: Johns Hopkins University Press.

Grieder, Peter (2001). *Cosmic Rays at Earth*. Amsterdam: Elsevier.

Grier, David A. (2005). *When Computers Were Human*. Princeton: Princeton University Press.

Grimes, James S. (1858). *Outlines of Geonomy: A Treatise on the Physical Laws of the Earth and the Creation of the Continents*. Boston: Phillips, Sampson & Company.

Gross, Alan G. (1990). *The Rhetoric of Science*. Cambridge, MA: Harvard University Press.

Gross, Alan G., Joseph E. Harmon, and Michael S. Reidy (2002). *Communicating Science: The Scientific Article from the 17th Century to the Present*. Oxford: Oxford University Press.

Guedes, Patricia et al. (2023). 'Eponyms have no place in 21st-century biological nomenclature'. *Nature Ecology & Evolution* **7**: 1157–60.

Gunnarsson, Britt-Louise, ed. (2011). *Languages of Science in the Eighteenth Century*. Berlin: De Gruyter.

Guth, Alan H. (1981). 'Inflationary universe: A possible solution to the horizon and flatness problems'. *Physical Review D* **23**: 347–56.

Guth, Alan H. (1997). *The Inflationary Universe*. Reading, MA: Addison-Wesley.

Gutman, Ivan (1985). 'On the origin of the name of the element indium'. *Journal of Chemical Education* **62**: 674.

Haack, Susan (2019). 'The art of scientific metaphors'. *Revista Portuguesa de Filosofia* **75**: 2049–66.

Hadamard, Jacques (1954). *The Psychology of Invention in the Mathematical Field*. New York: Dover Publications.

Haeckel, Ernst (1866). *Generelle Morphologie der Organismen*. Berlin: Georg Reimers.

Hall, A. Rupert and Marie Boas Hall, eds. (1965–86). *The Correspondence of Henry Oldenburg*. Madison: University of Wisconsin Press.

Hallam, Anthony (1989). *Great Geological Controversies*. Oxford: Oxford University Press.

Halley, Edmond (1716). 'An account of several nebulæ or lucid spots like clouds, lately discovered among the fixt stars by help of the telescope'. *Philosophical Transactions of the Royal Society* **29**: 390–2.

Halliday, Michael A. K. (2003). *On Language and Linguistics*. London: Bloomsbury Publishing.

Halmos, Paul R. (1967). *A Hilbert Space Problem Book*. New York: Springer-Verlag.

Hampson, G. F. (1919). 'Systematic papers published in the German language'. *Science* **49**: 193.

Hannaway, Owen (1975). *The Chemists and the Word: The Didactic Origins of Chemistry*. Baltimore: John Hopkins University Press.

Hanson, Norwood Russell (1963). 'Some philosophical aspects of contemporary cosmologies', in *Philosophy of Science: The Delaware Seminar*, vol. 2, ed. Bernard H. Baumrin, pp. 465–82. New York: Interscience.

Hargrove, James L. (2006). 'History of the calorie in nutrition'. *Journal of Nutrition* **136**: 2958–61.

Harris, John (1704). *Lexicon Technicum, or, an Universal English Dictionary of Arts and Sciences*. London.

Harrison, Edward (1987). *Darkness at Night: A Riddle of the Universe*. Cambridge, UK: Cambridge University Press.

Haste, Helen (1993). 'Dinosaur as metaphor'. *Modern Geology* **18**: 349–70.

Hawthorne, Robert M. (1970). 'Avogadro's number: Early values by Loschmidt and others'. *Journal of Chemical Education* **47**: 751–64.

Haymet, Anthony (1986). 'Footballene: A theoretical prediction for the static, truncated icosahedral molecule C60'. *Journal of the American Chemical Society* **108**: 319–21.

Heathcote, Niels H. (1967). 'The early meaning of *electricity*: Some *Pseudodoxia Epidemica*'. *Annals of Science* **23**: 261–75.

Heaviside, Oliver (1951). *Electromagnetic Theory*. London: E. & F. N. Spon.

Hecht, Jeff (2005). *Beam: The Race to Make the Laser*. Oxford: Oxford University Press.

Heilbron, John (2002). 'Coming to terms'. *Nature* **415**: 585.

Heinlein, Robert and Harold Wooster (1961). '"Xenobiology"'. *Science* **134**: 223–5.

Heisenberg, Werner (1971). *Physics and Beyond: Memoires of a Life in Science*. London: George Allen & Unwin.

Hendry, John (1983). 'Monopoles before Dirac'. *Studies in History and Philosophy of Science* **14**: 81–7.

Henry, John (2000). 'Magic and the origin of modern science'. *The Lancet* **354**: 23.

Hentschel, Klaus (2002). 'Why not one more imponderable? John William Draper's tithonic rays'. *Foundations of Chemistry* **4**: 5–59.

Hentschel, Klaus (2018). *Photons: The History and Mental Models of Light Quanta*. Cham: Springer.

Herbert, John et al. (2022). 'Words matter: On the debate over free speech, inclusivity, and academic excellence'. *Journal of Physical Chemistry Letters* **13**: 7100–4.

Herdeiro, Carlos and José Lemos (2019). 'The black hole fifty years after: Genesis of the name'. ArXiv:1811.06587v2 [physics.hist-ph].

Hermes, Matthew E. (1996). *Enough for One Lifetime: Wallace Carothers, Inventor of Nylon*. Washington, DC: Chemical Heritage Foundation.

Herschel, John (1840). 'On the chemical action of the rays of the solar spectrum on preparations of silver and other substances, both metallic and non-metallic, and on some photographic processes'. *Philosophical Transactions of the Royal Society* **130**: 1–59.

Herschel, John (1851). *Preliminary Discourse on the Study of Natural Philosophy*. London: Longman etc.

Herschel, William (1789). 'Remarks on the construction of the heavens'. *Philosophical Transactions of the Royal Society* **79**: 212–26.

Herschel, William (1795). 'On the nature and construction of the Sun and fixed stars'. *Philosophical Transactions of the Royal Society* **85**: 46–72.

Herschel, William (1802). 'Observations on the two lately discovered celestial bodies'. *Philosophical Transactions of the Royal Society* **92**: 213–32.

Hessman, Frederic et al. (2010). 'On the naming convention used for multiple star systems and extrasolar planets'. ArXiv:1012.0707 [astro-ph. SR].

Heynick, Frank (1983). 'From Einstein to Whorf: Space, time, matter, and reference frames in physical and linguistic relativity'. *Semiotica* **45**: 35–64.

Hill, Archibald V. (1956). 'Why biophysics?' *Science* **124**: 1233–7.

Hind, John R. (1861). 'On the nomenclature of the minor planets'. *Monthly Notices of the Royal Astronomical Society* **21**: 233–5.

Hirsh, Richard F. (1979). 'The riddle of the gaseous nebulae'. *Isis* **70**: 197–212.

Hjertholm, Peter (2023). *A History of the Cultural Travels of Energy: From Aristotle to the OED*. New York: Routledge.

Hoddeson, Lillian H. et al., eds. (1992). *Out of the Crystal Maze: Chapters from the History of Solid-State Physics*. New York: Oxford University Press.

Hoffleit, Dorrit (1939). 'Observations of supernovae'. *Proceedings of the American Philosophical Society* **81**: 265–76.

Hofstadter, Robert (1956). 'Electron structure and nuclear structure'. *Reviews of Modern Physics* **28**: 214–54.

Hogben, Lancelot (1970). *The Vocabulary of Science*. New York: Stein and Day.

Holberg, Jay B. (2007). *Sirius: Brightest Diamond in the Night Sky*. Chichester, UK: Praxis Publishing.

Holloway, Marshall G. and Charles P. Baker (1972). 'How the barn was born'. *Physics Today* **25** (July): 9.

Holton, Gerald (1988). *Thematic Origins of Scientific Thought: Kepler to Einstein*. Cambridge, MA: Harvard University Press.

Hon, Giora and Bernard R. Goldstein (2013). 'J. J. Thomson's plum-pudding atomic model: The making of a myth'. *Annalen der Physik* **525**: A129–A133.

Horgan, John (1995). 'The return of the maverick'. *Scientific American* **272** (March): 46–8.

Hoskin, Michael (2014). 'William Herschel and the planetary nebulae'. *Journal for the History of Astronomy* **45**: 209–25.

Howard, Don (2004). 'Who invented the Copenhagen interpretation? A study in mythology'. *Philosophy of Science* **71**: 669–82.

Howard, Irmgaard K. (2002). 'H is for enthalpy, thanks to Heike Kamerlingh Onnes and Alfred W. Porter'. *Journal of Chemical Education* **79**: 697.

Howarth, Richard J. (2020). 'Etymology in the earth sciences'. *Earth Sciences History* **39**: 1–27.

Hudson, John A. (2019). 'Dr Margaret Todd and the introduction of the term "isotope" ', in *Women in their Element: Selected Women's Contributions to the Periodic System*, eds. Annette Lykknes and Brigitte Van Tiggelen, pp.280–9. New Jersey: World Scientific.

Huggett, Nick (2003). 'Quarticles and the identity of indiscernibles', in *Symmetries in Physics: Philosophical Reflections*, eds. Katherine Brading and Elena Castellani, pp. 239–49. Cambridge, UK: Cambridge University Press.

Huggett, R. J. (1999). 'Ecosphere, biosphere, or Gaia? What to call the global ecosystem'. *Global Ecology and Biogeography* **8**: 425–31.

Huggins, Margaret L. (1898). '... Teach me how to name the ... light'. *Astrophysical Journal* **8**: 54.

Hughes, David W. and Brian G. Marsden (2007). 'Planet, asteroid, minor planet: A case study in astronomical nomenclature'. *Journal of Astronomical History and Heritage* **10**: 21–30.

Hufbauer, Karl (1991). *Exploring the Sun: Solar Science since Galileo*. Baltimore: Johns Hopkins University Press.

Hunt, Bruce J. (1991). *The Maxwellians*. Ithaca, NY: Cornell University Press.

Hurwic, Józef (1987). 'Reception of Kasimir Fajans' quanticule theory of the chemical bond'. *Journal of Chemical Education* **64**: 122–3.

Huterer, Dragan and Michael Turner (1999). 'Prospects for probing the dark energy via supernova distance measurements'. *Physical Review D* **60**: 081301.

Huxley, Thomas (1877). 'On the study of biology'. *American Naturalist* **11**: 210–21.

Igea, J. M. (2013). 'The history of the idea of allergy'. *Allergy* **68**: 966–73.

Ihde, Aaron J. (1964). *The Development of Modern Chemistry*. New York: Dover Publications.

IUPAC (1997). 'Names and symbols of transfermium elements'. *Pure and Applied Chemistry* **69**: 2471–3.

Jackson, J. David (2008). 'Examples of the zeroth theorem of the history of science'. *American Journal of Physics* **76** (2008): 704–19.

Jackson, J. David and Lev B. Okun (2001). 'Historical roots of gauge invariance'. *Reviews of Modern Physics* **73**: 704–19.

Jacobs, J. A. (1975). *A Textbook on Geonomy*. New York: Halsted Press.

Jakobson, Roman (1997). 'Einstein and the science of language', in *Albert Einstein: Historical and Cultural Perspectives*, eds. Gerald Holton and Yehuda Elkana, pp. 139–50. Mineola, NY: Dover Publications.

James, William (1879). 'The sentiment of rationality'. *Mind* **4**: 317–46.

James, William (1895). 'Is life worth living?' *International Journal of Ethics* **6**: 1–24.

Jammer, Max (1966). *The Conceptual Development of Quantum Mechanics*. New York: McGraw-Hill.

Janković, Vladimir (2000). *Reading the Skies: A Cultural History of English Weather, 1650–1820*. Manchester: Manchester University Press.

Jeans, James (1901). 'The mechanism of radiation'. *Philosophical Magazine* **2**: 421–55.

Jeans, James (1930). *The Universe around Us*. Cambridge, UK: Cambridge University Press.

Jensen, Kai A., ed. (1971). *Nomenclature of Inorganic Chemistry*. London: Butterworths.

Jensen, William B. (2004). 'Why helium ends in "–ium" '. *Journal of Chemical Education* **81**: 944.

Jensen, William B. (2005). 'The origin of the name "nylon" '. *Journal of Chemical Education* **82**: 676.

Jensen, William B. (2007a). 'How and when did Avogadro's name become associated with Avogadro's number?' *Journal of Chemical Education* **84**: 223.

Jensen, William B. (2007b). 'The origin of the oxidation-state concept'. *Journal of Chemical Education* **84**: 1418–19.

Jensen, William B. (2008). 'Why tungsten instead of wolfram?' *Journal of Chemical Education* **85**: 488–9.

Jerrard, H. G. and D. B. McNeill (1992). *Dictionary of Scientific Units*. London: Chapman & Hall.

Jespersen, Otto (1929). 'Nature and art in language'. *American Speech* **5**: 89–103.

Jørgensen, Sophus M. (1908). *The Fundamental Conceptions of Chemistry*. London: Society for the Promotion of Christian Knowledge.

Johannsen, Wilhelm (1909). *Elemente der exakten Erblichkeitslehre*. Jena: Verlag Gustav Fischer.

Johnson, George (1999). *Strange Beauty: Murray Gell-Mann and the Revolution in Twentieth-Century Physics*. New York: Alfred A. Knopf.

Johnson, Samuel (1755). *A Dictionary of the English Language*, 2 vols. London.

Johnson, Sean F. (2012). *The Neutron's Children: Nuclear Engineers and the Shaping of Identity*. Oxford: Oxford University Press.

Jones, Richard F. (1932). 'Science and language in England in the mid-seventeenth century'. *Journal of English and Germanic Philology* **30**: 315–31.

Joravsky, David (1986). *The Lysenko Affair*. Chicago: University of Chicago Press.

Kamminga, Harmke (1982). 'Life from space—a history of panspermia'. *Vistas in Astronomy* **26**: 67–86.

Kant, Immanuel (1981). *Universal Natural History and Theory of the Heavens*, trans. Stanley Jaki. Edinburgh: Scottish Academic Press.

Kapitsa, Peter (1938). 'Viscosity of liquid helium below the λ–point'. *Nature* **141**: 74.

Kauffman, George B. (1985). 'The role of gold in alchemy, part III'. *Gold Bulletin* **18**: 109–19.

Kennelly, Arthur E. (1931). 'The oersted considered as a new magnetic unit'. *Scientific Monthly* **32**: 378–80.

Ketner, Kenneth L. (1981). 'Peirce's ethics of terminology'. *Transactions of the Charles S. Peirce Society* **17**: 327–47.

Keyes, Ralph (2021). *The Hidden Story of Coined Words*. Oxford: Oxford University Press.

Klemun, Marianne (2015). 'Geognosie versus Geologie: Nationale Denkstile und kulturelle Praktiken bezüglich Raum und Zeit im Widerstreit'. *Berichte zu Wissenschaftsgeschichte* **38**: 227–42.

Knott, Cargill G. (1911). *Life and Scientific Work of Peter Guthrie Tait*. Cambridge, UK: Cambridge University Press.

Kojevnikov, Alexei (1999). 'Freedom, collectivism, and quasiparticles: Social metaphors in quantum physics'. *Historical Studies in the Physical and Biological Sciences* **29**: 295–331.

Kojevnikov, Alexei (2004). *Stalin's Great Science: The Times and Adventures of Soviet Physicists*. London: Imperial College Press.

Kolb, Leigh (2019). 'Etymological fallacy', in *Bad Arguments: 100 of the Most Important Fallacies in Western Philosophy*, eds. Robert Arp, Steven Barbone, and Michael Bruce, pp. 266–9. Oxford: John Wiley & Sons.

Kollerstrom, Nicholas (2009). 'The naming of Neptune'. *Journal of Astronomical History and Heritage* **12**: 66–71.

Koopman, Bernard O. (1931). 'Dirac on quantum mechanics'. *Bulletin of the American Mathematical Society* **37**: 495–6.

Kragh, Helge (1980). 'Anatomy of a priority conflict: The case of element 72'. *Centaurus* **23**: 275–301.

Kragh, Helge (1989). 'Concept and controversy: Jean Becquerel and the positive electron'. *Centaurus* **32**: 203–40.

Kragh, Helge (1990). 'From "electrum" to positronium'. *Journal of Chemical Education* **67**: 196–7.

Kragh, Helge (1992). 'Unifying quanta and relativity? Schrödinger's attitude to relativistic quantum mechanics', in *Erwin Schrödinger: Philosophy and the Birth of Quantum Mechanics*, eds. Michel Bitbol and Olivier Darrigol, pp. 315–38. Paris: Editions Frontieres.

Kragh, Helge (1995). 'From curiosity to industry: The early history of cryolite soda manufacture'. *Annals of Science* **52**: 285–301.

Kragh, Helge (1999). 'The decibel: Historical roots of a technical unit'. *Polhem* **16**: 157–66.

Kragh, Helge (2000). 'The chemistry of the universe: Historical roots of modern cosmochemistry'. *Annals of Science* **57**: 353–68.

Kragh, Helge (2001). 'From geochemistry to cosmochemistry: The origin of a scientific discipline 1915-1955', in *Chemical Sciences in the 20th Century*, ed. Carsten Reinholdt, pp. 160–92. Weinheim: Wiley-VCH.

Kragh, Helge (2008a). *Entropic Creation: Religious Contexts of Thermodynamics and Cosmology*. Aldershot: Ashgate.

Kragh, Helge (2008b). *The Moon that Wasn't: The Saga of Venus' Spurious Satellite*. Basel: Birkhäuser.

Kragh, Helge (2009a). 'The solar element: A reconsideration of helium's early history'. *Annals of Science* **66**: 157–82.

Kragh, Helge (2009b). 'The spectrum of the Aurora Borealis: From enigma to laboratory science'. *Historical Studies in the Natural Sciences* **39**: 377–17.

Kragh, Helge (2011). *Higher Speculations: Grand Theories and Failed Revolutions in Physics and Cosmology*. Oxford: Oxford University Press.

Kragh Helge (2012). *Niels Bohr and the Quantum Atom: The Bohr Model of Atomic Structure 1913-1925*. Oxford: Oxford University Press.

Kragh, Helge (2013). 'Nordic cosmogonies: Birkeland, Arrhenius and fin-de-siècle cosmical physics'. *European Physical Journal H* **38**: 549–72.

Kragh, Helge (2014a). 'The names of physics: plasma, fission, photon'. *European Physical Journal H* **39**: 262–82.

Kragh, Helge (2014b). 'Naming the big bang'. *Historical Studies in the Natural Sciences* **44**: 3–36.

Kragh, Helge (2016). *Julius Thomsen. A Life in Chemistry and Beyond*. Copenhagen: Royal Danish Academy of Sciences and Letters.

Kragh, Helge (2017). 'Is the universe expanding? Fritz Zwicky and early tired-light hypotheses'. *Journal of Astronomical History and Heritage* **20**: 2–12.

Kragh, Helge (2018a). 'The Lorenz-Lorentz formula: Origin and early history'. *Substantia* **2**(2): 7–18.

Kragh, Helge (2018b). *From Transuranic to Superheavy Elements: A Story of Dispute and Creation*. Berlin: Springer.

Kragh, Helge (2023). 'A terminological history of early particle physics'. *Archive for History of Exact Sciences* **77**: 73–120.

Kragh, Helge and Stephen J. Weininger (1996). 'Sooner silence than confusion: The tortuous entry of entropy into chemistry'. *Historical Studies in the Physical Sciences* **27**: 91–130.

Kragh, Helge et al. (2008). *Science in Denmark: A Thousand-Year History*. Aarhus: Aarhus University Press.

Kramar, Natalie and Olga Ilchenko (2021). 'From intriguing to misleading: The ambivalent role of metaphor in modern astrophysical and cosmological terminology'. *Amazonia Investiga* **10**: 92–100.

Krasnodebski, Marcin (2018). 'Throwing light on photonics: The genealogy of a technological paradigm'. *Centaurus* **60**: 3–24.

Kronick, David A. (1988). 'Anonymity and identity: Editorial policy in the early scientific journal'. *Library Quarterly* **58**: 221–37.

Kroto, Harry et al. (1985). 'C60: Buckyminsterfullerene'. *Nature* **318**: 162–3.

Kubbinga, Henk (2001). *L'Histoire du Concept de 'Molecule'*, 3 vols. Paris: Springer-Verlag France.

Kühne, Wilhelm (1877). 'Über das Verhalten verschiedener organisirter und sog. ungeformter Fermente'. *Verhandlungen des Naturhistorisch-Medicinischen Vereins zu Heidelberg* **1**: 190–3.

Kuhn, Thomas S. (1993). 'Metaphor in science', in *Metaphor and Thought*, ed. Andrew Ortony, pp. 533–42. Cambridge, UK: Cambridge University Press.

Kurti, Nicholas (1965). 'Notes about terminology and nomenclature'. *Journal of the Royal Aeronautical Society* **69**: 768.

Kurti, Nicholas (1970). 'Low temperature terminology'. *Cryogenics* **10**: 183–5.

Lafleur, Laurence J. (1941). 'Astrobiology'. *Astronomical Society of the Pacific Leaflets* **3**: 333–40.

Lagemann, Robert (1956). 'Pseudonyms of physicists'. *Scientific Monthly* **83**: 130–4.

Laidler, Keith J. (1993). *The World of Physical Chemistry*. Oxford: Oxford University Press.

Lake, Philip (1923). 'Wegener's hypothesis of continental drift'. *Geographical Journal* **61**: 179–87.

Lamb, Susan (2018). '(Not) a bromide story: Myth-busting bromide of potassium to create a case study of change and continuity in nineteenth-century medicine'. *Journal of Pharmacy and Pharmaceuticals* **60**: 108–23.

Langmuir, Irving (1928). 'Oscillations in ionized gases'. *Proceedings of the National Academy of Science* **14**: 627–37.

Langmuir, Irving (1989). 'Pathological science'. *Physics Today* **42** (10): 36–48.

Laplace, Pierre-Simon (1902). *A Philosophical Essay on Probability*. New York: John Wiley & Sons.

Lapointe, H. (1970). 'Origin and evolution of the term "psychology" '. *American Psychologist* **25**: 640–6.

Larmor, Joseph (1927). *Mathematical and Physical Papers*, vol. 1. Cambridge, UK: Cambridge University Press.

Lavoisier, Antoine-Laurent (1965). *Elements of Chemistry*, trans. Robert Kerr. New York: Dover Publications.

Lawrence, Christopher (2010). 'Historical keyword: Degeneration'. *The Lancet* **375**: 975.

Lawrence, Ernest O., Edwin McMillan, and Robert L. Thornton (1935). 'The transmutation functions for some cases of deuteron-induced radioactivity'. *Physical Review* **48**: 493–9.

Lay, George W. (1930). 'The language of scientists'. *Science* **72**: 567–9.

Lederberg, Joshua (1960). 'Exobiology: Approaches to life beyond the Earth'. *Science* **132**: 393–400.

Lederman, Leon (1993). *The God Particle: If the Universe is the Answer, What is the Question?* London: Bantam Press.

Lefèvre, Wolfgang (2018). 'The Méthode de nomenclature chimiques (1787): A document of transition'. *Ambix* **65**: 9–29.

Leicester, Henry M., ed. (1968). *Source Book in Chemistry 1900–1950*. Cambridge, MA: Harvard University Press.

Leicester, Henry M. and Herbert S. Klickstein, eds. (1952). *A Source Book in Chemistry 1400–1900*. Cambridge, MA: Harvard University Press.

Lemaître, Georges (1931). 'The beginning of the world from the point of view of quantum theory'. *Nature* **127**: 706.

Leone, Matteo, Alessandro Paoletti, and Nadia Robotti (2004). 'A simultaneous discovery: The case of Johannes Stark and Antonino Lo Surdo'. *Physics in Perspective* **6**: 271–94.

Letourneau, Charles (1878). *Biology*. London: Chapman and Hall.

Levere, Trevor L. (1971). *Affinity and Matter: Elements of Chemical Philosophy*. Oxford: Oxford University Press.

Leverington, David (2003). *Babylon to Voyager and Beyond: A History of Planetary Astronomy*. Cambrige, UK: Cambridge University Press.

Lévy-Leblond, Jean-Marc (1988), 'Quantum physics and language', *Physica B* **151**: 314–18.

Lévy-Leblond, Jean-Marc (1999). 'Quantum words for a quantum world', in *Epistemological and Experimental Perspectives on Quantum Physics*, eds. Daniel Greenberger, Wolfgang L. Reiter, and Anton Zeilinger, 75–87. Dordrecht: Kluwer Academic.

Lévy-Leblond, Jean-Marc (2003). 'On the nature of quantons'. *Science and Education* **12**: 495–502.

Lewis, Gilbert N. (1926). 'The conservation of photons'. *Nature* **117**: 874–5.

Lewis, Gilbert N. and Joseph Mayer (1929). 'The thermodynamics of gases which show degeneracy (Entartung)'. *Proceedings of the National Academy of Sciences* **15**: 208–18.

Lightman, Alan P. (1989). 'Magic on the mind: Physicists' use of metaphor'. *American Scholar* **97**: 97–101.

Lightman, Alan P. and Roberta Brawer (1990). *Origin: The Lives and Worlds of Modern Cosmologists*. Cambridge, MA: Harvard University Press.

Lindenmann, J. (1984). 'Origin of the terms "antibody" and "antigen" '. *Scandinavian Journal of Immunology* **19**: 281–5.

Lingam, Manasvi and Abraham Loeb (2020). 'What's in a name: The etymology of astrobiology'. *International Journal of Astrobiology* **19**: 379–85.

Linström, Bård and Lars Pettersson (2003). 'A brief history of catalysis'. *CATTECH* **7**: 130–8.

Little, Joseph (2008). 'The role of analogy in George Gamow's derivation of drop energy'. *Technical Communication Quarterly* **17**: 220–38.

Livingston, M. Stanley, ed. (1966). *The Development of High-Energy Accelerators*. New York: Dover Publications.

Lockard, E. N. (1950). 'Fertile virgins and fissile breeders: Nuclear neologisms'. *American Speech* **25**: 23–7.

Locke, John (1758). *Elements of Natural Philosophy*. Glasgow: R. Urie.

Locke, John (1836). *An Essay Concerning Human Understanding*. London: Tegg and Son.

Lockyer, J. Norman (1896). 'The story of helium'. *Nature* **53**: 319–22.

Lodge, Oliver (1893). 'Thoughts on the bifurcation of the sciences suggested by the Nottingham meeting of the British Association'. *Nature* **48**: 564–6.

Lodge, Oliver (1906). *Electrons or the Nature and Properties of Negative Electricity*. London: George Bell and Sons.

Lodge, Oliver (1924). *Atoms and Rays: An Introduction to Modern Views on Atomic Structure & Radiation*. New York: George Doran Company.

Lodge, Oliver (1931). *Advancing Science: Personal Reminiscences of the British Association in the Nineteenth Century*. London: Ernest Benn.

Loeb, Abraham and Frank H. Laukien (2022). 'Overview of the Galileo Project'. ArXiv:2209.02479 [physics.pop-ph].

Lorius, Claude et al. (1992). 'The icy core record: Past archive of the climate and signpost to the future'. *Philosophical Transactions: Biological Sciences* **338**: 227–34.

Lovelock, James (1972). 'Gaia as seen through the atmosphere'. *Atmospheric Environment* **6**: 579–80.

Lovelock, James (1989). 'Geophysiology, the science of Gaia'. *Reviews of Geophysics* **27**: 215–22.

Loyson, Peter (2009). 'Influences of ancient Greek on chemical terminology'. *Journal of Chemical Education* **86**: 1195–9.

Lyell, Charles (1830–1833). *Principles of Geology*, 3 vols. London: John Murray.

McIntosh, Robert P. (1975). '"Ecology": A clarification'. *Science* **188**: 1258.

McKenzie, Dan P. and Robert L. Parker (1967). 'The North Pacific: An example of tectonics on a sphere'. *Nature* **216**: 1276–80.

McKendrick, John G. and Walter Colquhoun (1904). 'The Blondlot or *n*-rays'. *Nature* **69**: 534.

McKie, Douglas (1952). *Antoine Lavoisier: Scientist, Economist, Social Reformer*. New York: Henry Schuman.

McLaughlin, Peter (2002). 'Naming biology'. *Journal of the History of Biology* **35**: 1–4.

MacLeod, Miles (2016). 'How language became a tool: The reconceptualisation of language and the empirical turn in seventeenth-century Britain', in *Language as a Scientific Tool: Shaping Scientific Language Across Time and National Tradition*, eds. Miles MacLeod et al., pp. 25–40. London: Routledge.

McVittie, George C. (1961). 'Rationalism versus empiricism in cosmology'. *Science* **133**: 1231–6.

McVittie, George C. (1974). 'Distance and large redshifts'. *Quarterly Journal of the Royal Astronomical Journal* **15**: 246–63.

Mahootian, Farzad (2015). 'Metaphor in chemistry: An examination of chemical metaphor', in *Philosophy of Chemistry: Growth of a New Discipline*, eds. Eric Scerri and Lee McIntyre, pp. 121–39. Dordrecht: Springer.

Malley, Marjorie C. (2011). *Radioactivity: A History of a Mysterious Science*. Oxford: Oxford University Press.

Malmkjaer, Kirsten, ed. (2006). *The Linguistic Encyclopedia*. London: Routledge.

Martin, Benjamin (1735). *The Philosophical Grammar*. London.

Martin, Craig (2015). 'The invention of the atmosphere'. *Studies in History and Philosophy of Science A* **52**: 44–54.

Mason, Paul H. (2015). 'Degeneracy: Demystifying and destigmatizing a core concept in systems biology'. *Complexity* **20**: 12–21.

Masson, Orme (1921). 'The constitution of atoms'. *Philosophical Magazine* **41**: 281–5.

Mattiello, Elisa (2017). *Analogy in Word-Formation: A Study in English Neologisms and Occasionalisms*. Berlin: De Gruyter.

Maurer, David W. and Ellesa C. High (1980). 'New words: Where do they come from and where do they go?' *American Speech* **55**: 184–94.

Maxwell, James C. (1965). *The Scientific Papers of James Clerk Maxwell*, ed. W. D. Niven. New York: Dover Publications.

Maxwell, James C. (1995). *The Scientific Letters and Papers of James Clerk Maxwell*, vol. II, ed. Peter M. Harman. Cambridge: Cambridge University Press.

Meckling, William (1961). 'Economic potential of communication satellites'. *Science* **133**: 1885–92.

Mendeleev, Dmitrii I. (1904). *An Attempt Towards a Chemical Conception of the Ether*. London: Longmans, Green & Co.

Menzies, Alan W. and C. A. Sloat (1926). 'Millikan rays and the acceleration of radioactive change'. *Science* **63**: 44–5.

Mermin, N. David (1981). 'E pluribus boojum: The physicist as neologist'. *Physics Today* **34** (4): 46–53.

Merton, Robert K. (1968). 'The Matthew effect in science'. *Science* **159**: 56–63.

Merton, Robert K. (1973). *The Sociology of Science: Theoretical and Empirical Investigations*. Chicago: University of Chicago Press.

Merton, Robert K. and Elinor Barber (2004). *The Travels and Adventures of Serendipity*. Princeton: Princeton University Press.

Metha, Arpan R. et al. (2020). 'Etymology and neuron(e)'. *Brain: A Journal of Neurology* **143**: 374–9.

Michaelis, Leonor and Louis B. Flexner (1928). 'Oxidation-reduction systems of biological significance'. *Journal of Biological Chemistry* **79**: 689–722.

Miller, Arthur I. (2005). *Empire of the Stars: Obsession, Friendship, and Betrayal in the Quest for Black Holes*. Boston: Houghton Mifflin Company.

Miller, David P. (2017). 'The story of "The Story of a Word" '. *Annals of Science* **74**: 255–61.

Millikan, Robert A. (1924). 'Atomism in modern physics'. *Journal of the Chemical Society, Transactions* **125**: 1405–17.

Millikan, Robert A. (1939). 'Mesotron as the name of the new particle'. *Physical Review* **55**: 105.

Mines, George R. (1911). 'The relation of the heart-beat to electrolytes and its bearing on comparative physiology'. *Journal of the Marine Biological Association of the United Kingdom* 9: 177–90.

Mirowski, Philip (1989). *More Heat than Light: Economics as Social Physics, Physics as Nature's Economics*. Cambridge, UK: Cambridge University Press.

Moggbridge, Bill (2006). *Designing Interactions*. Cambridge, MA: MIT Press.

Monk, Ray (2012). *Robert Oppenheimer: A Life Inside the Center*. New York: Doubleday.

Moon, Parry and Domina E. Spencer (1948a). 'Modern terminology for physics'. *American Journal of Physics* 16: 100–4.

Moon, Parry and Domina E. Spencer (1948b). 'Languages for science'. *Journal of the Franklin Institute* 246: 1–12.

Morris, Peter J. T., Anthony S. Travis, and Carsten Reinhardt (2001). 'Research fields and boundaries in twentieth-century organic chemistry', in *Chemical Sciences in the 20th Century: Bridging Boundaries*, ed. Carsten Reinhardt, pp. 14–42. Weinheim: Wiley-VCH.

Morris, Robert J. (1972). 'Lavoisier and the caloric theory'. *British Journal for the History of Science* 6: 1–38.

Morse, William R. (1927). 'Stanford expressions'. *American Speech* 2: 274–9.

Mosskop, D. (1995). '"Ich röntge, du röntgst...." Eine vergleichende Untersuchung eponymischer Verben anlässlich des 100. Jahrestages der Entdeckung der Röntgenstrahlen'. *Radiologe* 35: 367–72.

Mottana, Annibale (2017). 'Bernardo Cesi (*Cæsius*) and his *Mineralogia* (1636): Naming a new science from an indiscriminate piling of mineral accounts'. *Rendiconti Lincei* 28: 435–48.

Mott-Smith, Harold (1971). 'History of "plasmas" '. *Nature* 233: 219.

Müller, Max (1862). *Lectures on the Science of Language*. New York: Charles Scribner.

Myers, Rollie J. (2010). 'One-hundred years of pH'. *Journal of Chemical Education* 87 (1): 30–2.

Munns, David P. (2017). 'From the algatron to the zootron: The history of the twentieth century is the story of trons'. *Annalen der Physik* 529 (6): 1700135.

Needham, James G. (1930). 'Scientific names'. *Science* 71: 26–8.

Nernst, Walther (1914). 'Über die Anwendung des neuen Wärmesatzes auf Gase'. *Zeitschrift für Elektrochemie* 20: 357–60.

Newman, Robert P. (1995). 'American intransigence: The rejection of continental drift in the great debates of the 1920s'. *Earth Science History* 14: 62–83.

Newman, William R. (1994). *Gehennical Fire: The Lives of George Starkey, an American Alchemist in the Scientific Revolution*. Cambridge, MA: Harvard University Press.

Newman, William R. and Lawrence M. Principe (1998). 'Alchemy vs. chemistry: The etymological origin of a historiographic mistake'. *Early Science and Medicine* 3: 32–65.

Nickel, Douglas R. (2002). 'Talbot's natural magic'. *History of Photography* 26: 132–40.

Nordmann, Alfred (2006). 'From metaphysics to metachemistry', in *Philosophy of Chemistry: Synthesis of a New Discipline*, eds. Davis Baird, Eric Scerri, and Lee McIntyre, pp. 347–62. Dordrecht: Springer.

Nye, Mary Jo (1980). 'N-rays: An episode in the history and psychology of science'. *Historical Studies in the Physical Sciences* 11: 125–56.

Nye, Mary Jo (1993). *From Chemical Philosophy to Theoretical Chemistry: Dynamics of Matter and Dynamics of Disciplines 1800–1950*. Berkeley: University of California Press.

Nye, Mary Jo (2000). 'Physical and biological modes of thought in the chemistry of Linus Pauling'. *Studies in History and Philosophy of Science* 31: 475–91.

Öhrström, Lars and Jan Reedijk (2016). 'Names and symbols of the elements with atomic numbers 113, 115, 117 and 118'. *Pure and Applied Chemistry* 88: 1225–9.

Ørsted, Hans Christian (1998). *Selected Scientific Works of Hans Christian Ørsted*, eds. Karen Jelved, Andrew D. Jackson, and Ole Knudsen. Princeton: Princeton University Press.

O'Hara, J. G. (1975). 'George Johnstone Stoney, F.R.S., and the concept of the electron'. *Notes and Records of the Royal Society* 29: 265–76.

Oliveira, Mário J. de (2018). 'Elementary concepts and fundamental laws of the theory of heat'. *Brazilian Journal of Physics* 48: 299–313.

Ollongren, Alexander (2013). *Astrolinguistics: Design of a Linguistic System for Interstellar Communication Based on Logic*. New York: Springer.

O'Raifeartaigh, Cormac and Michael O'Keeffe (2020). 'Redshifts versus paradigm shifts: Against renaming Hubble's law'. *Physics in Perspective* **22**: 215–25.

Osterbrock, Donald E. (2001). 'Who really coined the word supernova? Who first predicted neutron stars?' *Bulletin of the American Astronomical Society* **33**: 1330.

Owen, Richard (1842). 'Report on British fossil reptiles'. *British Association for the Advancement of Science, Report*, pp. 60–204. London: John Murray.

Pais, Abraham (1953). 'On the baryon–meson–photon system'. *Progress of Theoretical Physics* **10**: 457–69.

Pancaldi, Giuliano (2003). *Volta: Science and Culture in the Age of Enlightenment*. Princeton: Princeton University Press.

Paneth, Friedrich A. (1950). 'Radioactive standards and units'. *Nature* **166**: 931–3.

Partington, James R. (1961). *A History of Chemistry*, vol. 2. London: Macmillan.

Patterson, Austin M. and H. V. Knorr (1933). 'A glossary of German-English equivalents relating to atomic structure'. *American Journal of Physics* **1**: 82–5.

Patterson, Andrew H. (1921). 'The elementary particle of positive electricity'. *Nature* **107**: 75.

Payne-Gaposchkin, Cecilia, and Sergei Gaposchkin (1938). *Variable Stars*. Cambridge, MA: The Observatory.

Pearson, Karl (1900). *The Grammar of Science*. London: Adam and Charles Black.

Pell, Morris B. (1872). 'On the constitution of matter'. *Philosophical Magazine* **43**: 161–85.

Penrose, Penrose (1968). 'Twistor quantisation and curved space-time'. *International Journal of Theoretical Physics* **1**: 61–99.

Perl, Martin L. (1978). 'The tau heavy lepton—a recently discovered elementary particle'. *Nature* **275**: 273–8.

Perrin, Jean (1910). *Brownian Movement and Molecular Reality*. London: Taylor & Francis.

Petersen, Aage (1963). 'The philosophy of Niels Bohr'. *Bulletin of the Atomic Scientists* **19** (7): 8–14.

Petruccioli, Sandro (1993). *Atoms, Metaphors and Paradoxes: Niels Bohr and the Construction of a New Physics*. Cambridge, UK: Cambridge University Press.

Phillips, Denise (2012). *Acolytes of Nature: Defining Natural Science in Germany, 1770–1850*. Chicago: University of Chicago Press.

Phillips, Edward (1720). The New World of Words, or, a General Dictionary, *compiled by John Kersey*. London.

Phipps, Gregory (2020). 'The narratives and metaphor of the balloonverse: A literary reading of the big bang theory'. *Journal of Literature and Science* **13**: 38–51.

Pines, David (1956). 'Collective energy losses in solids'. *Reviews of Modern Physics* **28**: 184–98.

Planck, Max (1900). 'Ueber irreversible Strahlungsvorgänge'. *Annalen der Physik* **1**: 69–122.

Popper, Karl R. (1935). *Logik der Forschung: Zur Erkenntnistheorie der modernen Naturwissenschaft*. Vienna: Springer.

Popper, Karl R. (2010). *Conjectures and Refutations: The Growth of Scientific Knowledge*. London: Routledge.

Powell, Alexander et al. (2007). 'Disciplinary baptism: A comparison of the naming stories of genetics, molecular biology, genomics, and systems biology'. *History and Philosophy of the Life Sciences* **29**: 5–32.

Poynting, John H. (1907). 'On prof. Lowell's method for evaluating the surface-temperatures of the planets'. *Philosophical Magazine* **14**: 749–60.

Pramling, Nikolas (2008). 'The role of metaphor in Darwin and the implications for teaching evolution'. *Science Education* **93**: 535–47.

Preece, William (1898). 'Aetheric telegraphy'. *Journal of the Institution of Electrical Engineers* **27**: 869–86.

Priestley, Joseph (1794). *History and Present State of Electricity*, 5th edn. London: Johnson and Rivington.

Pulaczewa, Hanna (1999). *Aspects of Metaphor in Physics: Examples and Case Studies*. Tübingen: Max Niemeyer Verlag.

Pynchon, Thomas R. (1874). *Introduction to Chemical Physics*. New York: Van Nostrand.

Quiller-Couch, Arthur (1923). *On the Art of Writing*. Cambridge, UK: Cambridge University Press.

Raad, B. L. (1989). 'Modern trends in scientific terminology: Morphology and metaphor'. *American Speech* **64**: 128–36.

Rafinesque, Constantine S. (1840). *Amenities of Nature, or Annals of Historical and Natural Science*. Philadelphia.

Ramsay, William (1912). 'Experiments with kathode rays'. *Nature* **89**: 502.

Ramsay, William and Morris W. Travers (1898). 'On the companions of argon'. *Proceedings of the Royal Society* **63**: 389–400.

Rankine, William (1853). 'On the general law of the transformation of energy'. *Philosophical Magazine* **5**: 106–17.

Rankine, William (1864). 'On the history of energetics'. *Philosophical Magazine* **28**: 404.

Rao, B. V. Raghavendra (1937). 'Dispersion of sound waves in liquids'. *Nature* **139**: 885.

Rayner-Canham, Geoff and Zheng, Zheng (2008). 'Naming elements after scientists: An account of a controversy'. *Foundations of Chemistry* **10**: 13–18.

Reidy, Michael S. (1996). 'Masters of tidology: The cultivation of the physical sciences in early Victorian Liverpool'. *Perspectives on Science* **4**: 231–47.

Renn, Jürgen and Tilman Sauer (2003). 'Eclipses of the stars: Mandl, Einstein, and the early history of gravitational lensing', in *Revisiting the Foundations of Relativistic Physics*, eds. Abhay Ashtekar et al., pp. 69–93. Dordrecht: Kluwer Academic.

Richards, William C. (1858). *Electron; or, the Pranks of the Modern Puck: A Telegraphic Epic for the Times*. New York: Appleton and Company.

Richardson, Owen W. (1909). 'Thermionics'. *Philosophical Magazine* **17**: 813–33.

Richeson, A. W. (1946). 'On Faraday's terminology in electrolysis'. *Isis* **36**: 160–2.

Richter, Charles F. (1935). 'An instrumental earthquake magnitude scale'. *Bulletin of the Seismological Society of America* **25**: 1–32.

Riordan, Michael and Lillian Hoddeson (1997). *Crystal Fire: The Birth of the Information Age*. New York: Norton & Company.

Roberts, Lissa (1991). 'A word and the world: The significance of naming the calorimeter'. *Isis* **82**: 198–222.

Roberts, Lissa (1992). 'Condillac, Lavoisier, and the instrumentalization of science'. *The Eighteenth Century* **33**: 252–71.

Rocke, Alan J. (2010). *Image and Reality: Kekulé, Kopp, and the Scientific Imagination*. Chicago: University of Chicago Press.

Roelli, Phillip (2021). *Latin as the Language of Science and Learning*. Berlin: De Gruyter.

Roller, Duane (1947). 'An approach to the study of physical terminology'. *American Journal of Physics* **15**: 178–86.

Romer, Alfred, ed. (1970). *Radioactivity and the Discovery of Isotopes*. New York: Dover Publications.

Romer, Robert H. (2001). 'Heat is not a noun'. *American Journal of Physics* **69**: 107–9.

Röntgen, Wilhelm C. (1896). 'On a new kind of rays'. *Nature* **53**: 274–6.

Rosenfeld, Albert (1964). 'What is quasi-stellars? Heavens' new enigma'. *Life Magazine* 24 (January): 11.

Rosenfeld, Léon (1960). 'Heisenberg, physics and philosophy'. *Nature* **186**: 830–1.

Rosenfeld, Louis (1997). 'Vitamine—vitamin. The early years of discovery'. *Clinical Chemistry* **43**: 680–5.

Roseneau, Milton J. (1935). 'Serendipity'. *Journal of Bacteriology* **29**: 91–8.

Ross, Sydney (1961). 'Faraday consults the scholars: The origin of the terms of electrochemistry'. *Notes and Records of the Royal Society* **16**: 187–220.

Ross, Sydney (1962). '"Scientist": The story of a word'. *Annals of Science* **18**: 65–85.

Ross, Sydney (1991). *Nineteenth-Century Attitudes: Men of Science*. Dordrecht: Kluwer Academic.

Rott, N. (1985). 'Jakob Ackeret and the history of the Mach number'. *Annual Review of Fluid Dynamics* **70**: 1–9.

Rowlinson, John S. (2012). *Sir James Dewar, 1842–1923: A Ruthless Chemist*. Farnham: Ashgate.

Ruark, Arthur E. (1928). 'Heisenberg's indetermination principle'. *Physical Review* **31**: 311–12.

Ruark, Arthur E. (1945). 'Positronium'. *Physical Review* **68**: 278.

Ruark, Arthur E. and Harold C. Urey (1930). *Atoms, Molecules and Quanta*. New York: McGraw-Hill Book Company.

Rubin, Mordechai B. (2009). 'The history of ozone. VII. Mythical spawn of ozone: Antozone, oxozone, and ozohydrogen'. *Bulletin for the History of Chemistry* **34**: 39–49.

Rudwick, Martin J. S. (1976). *The Meaning of Fossils: Episodes in the History of Palaeontology*. Chicago: University of Chicago Press.

Russell, Bertrand (1927). *The ABC of Atoms*. London: Kegan Paul, Trench, Trubner & Co.

Russell, Bertrand (2010). *The Philosophy of Logical Atomism*. London: Routledge.

Russell, Colin A. (1999). 'Bunsen without his burner'. *Physics Education* **34**: 321–6.

Rutherford, Ernest (1910). 'Radium standards and nomenclature'. *Nature* **84**: 430–1.

Rutherford, Ernest (1920). 'Nuclear constitution of atoms'. *Proceedings of the Royal Society A* **97**: 374–400.

Rutherford, Ernest and Frederick Soddy (1903). 'Radioactive change'. *Philosophical Magazine* **5**: 576–91.

Rutherfurd, Lewis M. (1863). 'Astronomical observations with the spectroscope'. *American Journal of Science and Arts* **35**: 71–7.

Ryan, Janet N. (1985). 'The language gap: Common words with technical meanings'. *Journal of Chemical Education* **62**: 1098–9.

Sahulka, J. (1893). [No title]. *Transactions of the American Institute of Electrical Engineers* **10**: 430–2.

Salimpour, Saeed et al. (2021). 'Cosmos visualized: Development of a qualitative framework for analyzing representations in cosmology education'. *Physical Review Physics Education Research* **17**: 013104.

Savory, Theodore H. (1967). *The Language of Science*. London: Andre Deutsch.

Schirrmacher, Arne (2012). 'Bohr's genuine metaphor: On types, aims and uses of models in the history of quantum theory', in *One Hundred Years of the Bohr Atom*, eds. Finn Aaserud and Helge Kragh, pp. 111–40. Copenhagen: Royal Danish Society of Sciences and Letters.

Schleiden, Matthias J. (1849). *Scientific Botany; or, Botany as an Inductive Science*. London: Longmans, Brown, Green, and Longmans.

Schmadel, Lutz D. (2012). *Dictionary of Minor Planet Names*, vol. 2. Berlin: Springer.

Schrödinger, Erwin (1926). 'An undulatory theory of the mechanics of atoms and molecules'. *Physical Review* **28**: 1049–70.

Schubert, Andrá, Wolfgang Glänzel, and Gábor Schubert (2022). 'Eponyms in science: Famed or framed?' *Scientometrics* **127**: 1199–207.

Schuster, Arthur (1895). 'The kinetic theory of gases'. *Nature* **51**: 293.

Schuster, Arthur (1898). 'Potential matter—a holiday dream'. *Nature* **58**: 367.

Scott-Taggart, J. (1933). *The Manual of Modern Radio*. London: Amalgamated Press.

Scurlock, Ralph G., ed. (1992). *History and Origins of Cryogenics*. Oxford: Clarendon Press.

Seaborg, Glenn T. (1994). 'Terminology of the transuranium elements'. *Terminology* **1**: 229–52.

Seligman, G. (1947). '"Cryology" '. *Journal of Glaciology* **1**: 35.

Senning, Alexander (2019). *The Etymology of Chemical Names*. Berlin: De Gruyter.

Severnyi, A. B., V. A. Kotov, and T. T. Tsap (1979). 'Solar oscillations and the problem of the internal structure of the Sun'. *Soviet Astronomy* **23**: 641–7.

Shama, Gilbert (2019). 'The "Petri dish": A case of simultaneous invention in bacteriology'. *Endeavour* **43**: 11–16.

Shapin, Steven (1989). 'The invisible technician'. *American Scientist* **77**: 554–63.

Shapin, Steven and Simon Schaffer (1985). *Leviathan and the Air-Pump: Hobbes, Boyle, and the Experimental Life*. Princeton: Princeton University Press.

Shea, Elizabeth (2001). 'The gene as a rhetorical figure: "Nothing but a very applicable little word" '. *Science as Culture* **10**: 505–29.

Sheldon, Eric (1986). 'Relativity or invariance?' *American Journal of Physics* **54**: 775.

Shockley, William (1952). 'Transistor electronics: Imperfections, unipolar and analog transistors'. *Proceeding of the I. R. E.* **40**: 1289–313.

Siegfried, Robert (1988). 'The chemical revolution in the history of chemistry'. *Osiris* **4**: 34–50.

Silver, Arnold H. (2006). 'How the SQUID was born'. *Superconductor Science and Technology* **19**: S173–S178.

Silver, Daniel S. (2017). 'The new language of mathematics'. *American Scientist* **105**: 364–71.

Sime, Ruth L. (1996). *Lise Meitner: A Life in Physics*. Berkeley: University of California Press.

Simonton, Roy W. (1999). 'Origin and acceptance of the term pedology'. *Soil Science Society of America Journal* **63**: 4–10.

Skeat, Walter W., ed. (1937). *The Complete Works of Geoffrey Chaucer*. New York: Oxford University Press.

Slater, John C. (1925). 'Physically degenerate systems and quantum dynamics'. *Physical Review* **26**: 419–30.

Smith, Crosbie (1998). *The Science of Energy: A Cultural History of Energy Physics in Victorian Britain*. London: Athlone Press.

Smith, David R. (2022). 'A species of any other name would sound as sweet'. *EMBO Reports* **23**: e54643.

Smithells, Arthur (1907). 'Presidential address to Section B'. *Nature* **76**: 352–7.

Snelders, Harry A. M. (1977). 'Dissociation, Darwinism and entropy: A case-study from the history of physical chemistry'. *Janus* **64**: 51–75.

Soddy, Frederick (1904). *Radio-Activity*. London: The Electrician.

Soddy, Frederick (1913). 'Intra-atomic charge'. *Nature* **92**: 399–400.

Soddy, Frederick (1935). 'The story of isotopes'. *Science* **82**: 235–40.

Sommerfeld, Arnold (1922). *Atombau und Spektrallinien*. Braunschweig: Vieweg & Sohn.

Sprat, Thomas (2003). *History of the Royal Society*. Montana: Kessinger Publishing.

Srerer, Paul A. (1985). 'The metabolon'. *Trends in Biochemical Sciences* **10**: 109–10.

Stahl, Frieda A. (1987). 'Physics as metaphor and vice versa'. *Leonardo* **20**: 57–64.

Stallo, John B. (1960). *The Concepts and Theories of Modern Physics*. Cambridge, MA: Harvard University Press.

Standage, Tom (2000). *The Neptune File: A Story of Astronomical Rivalry and the Pioneers of Planet Hunting*. New York: Walker & Company.

Stanley, H. M. (1884). 'On the classification of the sciences'. *Mind* **9**: 265–74.

Steinhardt, Paul J. (2018). *The Second Kind of Impossible: The Extraordinary Quest for a New Form of Matter*. New York: Simon & Schuster.

Steinhardt, Paul J. and Neil Turok (2007). *Endless Universe: Beyond the Big Bang*. New York: Doubleday.

Stewart, Balfour and Peter G. Tait (1875). *The Unseen Universe. Or, Physical Speculations on a Future State*. London: Macmillan.

Stigler, Stephen M. (1980). 'Stigler's law of eponymy', in *Science and Social Structure: A Festschrift for Robert K. Merton*, ed. Thomas F. Gieryn, pp. 147–58. New York: New York Academy of Sciences.

Stuewer, Roger H. (1994). 'The origin of the liquid-drop model and the interpretation of nuclear fission'. *Perspectives on Science* **2**: 76–129.

Subramanyam, K. (1979). 'Acronymania'. *Technical Communication* **26**, no. 3: 13–15.

Surman, Jan (2016). 'Linguistic precision and scientific accuracy', in *Language as a Scientific Tool: Shaping Scientific Language Across Time and National Traditions*, eds. Miles MacLeod et al., pp. 131–48. New York: Routledge.

Sutherland, William (1899). 'Cathode, Lenard, and Röntgen rays'. *Philosophical Magazine* **47**: 269–84.

Sutton, Clive (1992). *Words, Science and Learning*. Buckingham: Open University Press.

Sutton, Clive (1994). ' "Nullius in verba" and "Nihil in verbis": Public understanding of the role of language in science'. *British Journal for the History of Science* **27**: 55–64.

Sylvester, James J. (1853). 'On a theory of the syzygetic relations of two rational integral functions'. *Philosophical Transactions of the Royal Society* **143**: 407–548.

Tait, Peter G. (1868). *Sketch of Thermodynamics*. Edinburgh: Edmonston and Douglas.

Talmey, Max (1932). 'Einstein's theory and rational language'. *Scientific Monthly* **35**: 254–7.

Tanford, Charles and Jaqueline Reynolds (2001). *Nature's Robots: A History of Proteins*. Oxford: Oxford University Press.

Tanner, H. G. (1924). 'Dalton as a name for the unit of atomic weight'. *Science* **59**: 460.

Terrall, Mary (2006). *The Man Who Flattened the Earth: Maupertuis and the Sciences in the Enlightenment*. Chicago: University of Chicago Press.

Terrall, Mary (2017). 'French in the siècle des lumières: A universal language?' *Isis* **108**: 636–42.

Thomas, Peter B. M. (2016). 'Are medical eponyms really dying out? A study of their usage in the historical biomedical literature'. *Journal of the Royal College of Physicians of Edinburgh* **46**: 295–9.

Thompson, Silvanus P. (1896). 'On hyperphosphorescence'. *Philosophical Magazine* **42**: 103–4.

Thompson, Silvanus P. (1898). *Michael Faraday: His Life and Work*. London: Cassell and Company.

Thomson, Joseph J. (1896). 'The Röntgen rays'. *Nature* **53**: 391–2.

Thomson, Joseph J. (1913). 'Cathode rays'. *Philosophical Magazine* **44**: 296–314.

Thomson, Joseph J. (1913). 'On the appearance of helium and neon in vacuum tubes'. *Nature* **90**: 645–7.

Thomson, Thomas (1817). *A System of Chemistry*. London: Baldwin, Cradock, and Joy.

Thomson, Thomas (1831). *A System of Chemistry of Inorganic Bodies*, vol. 1. London: Baldwin & Cradock.

Thomson, William (1879). 'The sorting demon of Maxwell'. *Nature* **20**: 126.

Thomson, William (1884). 'Electrical units of measurements', in *The Practical Applications of Electricity*, pp. 149–75. London: Institution of Civil Engineers.

Thomson, William (1897). 'Contact electricity and electrolysis according to Father Boscovich'. *Nature* **56**: 84–5.

Thomson, William (1902). 'Aepinus atomized'. *Philosophical Magazine* **3**: 257–83.

Thornthwaite, C. W. (1943). 'Meteorology and climatology'. *Science* **97**: 580–3.

Thornton, Brett F. and Shawn C. Burdette (2013). 'Recalling radon's recognition'. *Nature Chemistry* **5**: 804.

Todhunter, Isaac (1876). *William Whewell, D.D., Master of Trinity College, Cambridge: An Account of His Writings*, vol. 2. London: Macmillan.

Tolman, Richard C. (1924). 'Duration of molecules in upper quantum states'. *Proceedings of the National Academy of Sciences* **10**: 85–7.

Tow, Don M. (1970). 'Some predictions of the Amati-Bertocchi-Fubini-Stanghellini-Tonin multiperipheral model'. *Physical Review D* **2**: 154–63.

Trimble, Virginia (1995). 'A low-resolution view of high-resolution spectroscopy'. *Publications of the Astronomical Society of the Pacific* **107**: 1012–15.

Trimble, Virginia (2012). 'Eponyms, Hubble's law, and the three princes of parallax'. *Observatory* **132**: 34–40.

Trofast, Jan (2016*). Jac. Berzelius: The Discovery of Cerium, Selenium, Silicon, Zirconium and Thorium*. Lund, Sweden: Berzelius Society.

Truran, J. W. and A. G. W. Cameron (1970). 'The galactic halo'. *Nature* **225**: 710–11.

Tukey, John W. (1958). 'The teaching of concrete mathematics'. *American Mathematical Monthly* **65**: 1–9.

Tyndall, John (1868). *Faraday as a Discoverer*. London: Longmans, Green, and Co.

Uhlenbeck, George E. and Samuel Goudsmit (1925). 'Ersetzung der Hypothese vom unmechanischen Zwang durch eine Forderung bezüglich des inneren Verhaltens jedes einzelnen Elektrons'. *Naturwissenschaften* **13**: 953–4.

Vallery-Radot, René (1901). *The Life of Pasteur*. London: Constable and Company.

Van Berkel, Klaas, Albert van Helden, and Lodewijk Palm, eds. (1999). *A History of Science in The Netherlands: Survey, Themes and Reference*. Leiden: Brill.

Van der Waerden, Bartel, ed. (1968). *Sources of Quantum Mechanics*. New York: Dover Publications.

Van Dyke, Carolynn (1992). 'Old words for new worlds: Modern scientific and technological word-formation'. *American Speech* **67**: 383–405.

Van Helden, Albert (2006). *Huygens's Ring, Cassini's Division & Saturn's Children*. Washington, DC: Smithsonian Institution Libraries.

Van Laar, Johannes (1901). *Lehrbuch der mathematischen Chemie*. Leipzig: Barth.

Varney, Robert N. and Leon H. Fisher (1980). 'Electromotive force: Volta's forgotten concept'. *American Journal of Physics* **48**: 405–8.

Vickery, Hubert B. (1940). 'The origin of the word protein'. *Yale Journal of Biology and Medicine* **22**: 387–93.

Vlasov, N. A. (1965). 'Optical search for antimatter in the universe'. *Soviet Astronomy-AJ* **8**: 715–18.

Volta, Alessandro (1800). 'On the electricity excited by mere contact of conducting substances of different kinds'. *Philosophical Magazine* 7: 289–311.

Wade, Charles H. (1897). 'Wanted, a name—"electrography"'. *British Medical Journal* 2: 52.

Wagner, Kurt G. (1951). *Autoren-Namen als chemische Begriffe*. Weinheim: Verlag Chemie.

Walker, Charles T. and Glen A. Slack (1970). 'Who named the –ON's?' *American Journal of Physics* 38: 1380–9.

Walsingham, Lord (1918). 'German naturalists and nomenclature'. *Nature* 102: 4–6.

Watkins, Kenneth W. (1989). 'Chemical metaphors'. *Journal of Chemical Education* 66: 1020.

Weeks, Mary E. and Henry M. Leicester (1968). *Discovery of the Elements*. Easton, PA: Journal of Chemical Education.

Wegener, Alfred L. (1960). *Alfred Wegener: Tagebücher, Briefe, Erinnerungen*. Wiesbaden: Brockhaus.

Wegener, Alfred L. (1966). *The Origin of Continents and Oceans*. New York: Dover Publications.

Weiner, S. (1943). 'Chemical semantics'. *ETC: A Review of General Semantics* 1: 41–6.

Welch, G. Rickey (2009). 'Physiology, physiomics, and biophysics: A matter of words'. *Progress in Biophysics and Molecular Biology* 100: 4–17.

Wentrup, Curt (2022). 'Origins of organic chemistry and organic synthesis'. *European Journal of Organic Chemistry* 25: e202101492.

Wheeler, John A. (1968). 'Our universe: The known and the unknown'. *American Scientist* 56: 1–20.

Wheeler, John A. (1990). *A Journey into Gravity and Spacetime*. New York: Scientific American Library.

Whewell, William (1833). *Astronomy and General Physics Considered with Reference to Natural Theology*. London: Bohn.

Whewell, William (1837). *History of the Inductive Sciences*. London: John W. Parker.

Whewell, William (1840). *Philosophy of the Inductive Sciences, Founded Upon their History*. London: John W. Parker.

Whewell, William (1847). *Philosophy of the Inductive Sciences, Founded Upon their History*. London: John W. Parker.

Whitaker, Ewen A. (1999*). Mapping and Naming the Moon: A History of Lunar Cartography and Nomenclature*. Cambridge, UK: Cambridge University Press.

White, Paul (2002). *Thomas Huxley: Making the 'Man of Science'*. Cambridge, UK: Cambridge University Press.

Whitmore, Charles E. (1955). 'The language of science'. *Scientific Monthly* 80: 185–91.

Wilczek, Frank (2016). 'Time's (almost) reversible arrow'. Quanta Magazine, 7 January 2016.

Wilczek, Frank (2017). 'Inside the knotty world of "anyon" particles'. Quanta Magazine, 28 February 2017.

Wildman, E. A. (1933). 'Nomenclature of the electron'. *Science* 78: 191.

Wildt, Rupert (1940). 'Cosmochemistry'. *Scientia* 67: 85–90.

Wiley, H. W. (1928). 'Hybrid word'. *Science* 68: 15.

Williams, L. Pearce (1965). *Michael Faraday: A Biography*. New York: Basic Books.

Williams, Trevor I. (1993). 'Aluminium: Latecomer to the metal industry'. *Endeavour* 17: 89–93.

Wilson, J. Tuzo (1968). 'Static or mobile earth'. *Proceedings of the American Philosophical Society* 112: 309–20.

Winchell, Alexander (1883). *World-Life or Comparative Geology*. Chicago: Griggs & Company.

Wise, M. Norton (1981). 'German concepts of force, energy, and the electromagnetic ether, 1845–1880', in *Conceptions of Ether: Studies in the History of Ether Theories 1740–1900*, eds. G. Cantor and M. Hodge, pp. 269–308. Cambridge, UK: Cambridge University Press.

Witze, Alexandra (2022). 'Documents reveal NASA's internal struggles over renaming Webb telescope'. *Nature* 604: 15–16.

Wolff, Stefan (2012). 'Jüdische oder nichtjüdische Deutsche?' in *Heinrich Hertz vom Funkesprung zur Radiowelle*, eds. Ralph Burmeister and Andrea Niehaus, pp. 38–57. Munich: German Museum of Technology.

Wood, Robert M. (1985). *The Dark Side of the Earth*. London: Allen & Unwin.

Wothers, Peter (2019). *How the Elements Were Named: Antimony, Gold, and Jupiter's Wolf*. Oxford: Oxford University Press.

Woywodt, Alexander and Eric Matteson (2007). 'Should eponyms be abandoned?' *British Medical Journal* **335**: 335.

Yeo, Richard (1993). *Defining Science: William Whewell, Natural Knowledge, and Public Debate in Early Victorian Britain*. Cambridge, UK: Cambridge University Press.

Young, Thomas (1807). *A Course of Lectures on Natural Philosophy and the Mechanical Arts*, vol. 2. London: Joseph Johnson.

Zacharias, Helmut (2020). 'Key word: chromosome'. *Chromosome Research* **9**: 345–55.

Zangwill, Andrew (2021). *A Mind Over Matter: Philip Anderson and the Physics of the Very Many*. Oxford: Oxford University Press.

Zimmer, Ben (2013). 'A well-traveled metaphor: 'Goldilocks' visits many houses'. https://www.visualthesaurus.com/cm/wordroutes/a-well-traveled-metaphor-goldilocks-visits-many-houses/

Zwicky, Fritz (1933). 'Die rotverschiebung von extragalaktischen Nebeln'. *Helvetica Physica Acta* **6**: 110–27.

Zwicky, Fritz (1937). 'Nebulae as gravitational lenses'. *Physical Review* **51**: 290.

Index

Abbreviations used in the index:

ABFST	Amati-Bertocchi-Fubini-Stanghellini-Tonin
B2FH	Burbidge-Burbidge-Fowler-Hoyle
CDM	cold dark matter
DDT	dichlorodiphenyltrichloroethane
EPR	Einstein-Podolsky-Rosen
FLRW	Friedmann-Lemaître-Robertson-Walker
GUT	grand unified theory
IAU	International Astronomical Union
IEC	International Electrotechnical Commission
IUPAC	International Union of Pure and Applied Chemistry
IUPAP	International Union of Pure and Applied Physics
ΛCDM	lambda cold dark matter
MACHO	massive compact halo objects
MO	molecular orbital
NASA	National Aeronautics and Space Administration
OED	Oxford English Dictionary
PET	positron-electron tomography
QCD	quantum chromodynamics
QED	quantum electrodynamics
SETI	search for extra-terrestrial intelligence
SI	Système International (units)
SQUID	superconducting quantum interference device
WIMP	weakly interacting massive particles
WMAP	Wilkinson Microwave Anisotropy Probe